# American Nature Writing

This book is part of the Peter Lang Regional Studies list.
Every volume is peer reviewed and meets
the highest quality standards for content and production.

PETER LANG
New York • Berlin • Brussels • Lausanne • Oxford

Hong Cheng

# American Nature Writing: Thirty Lectures

Translated by Yan Qian and Jingwen Chen
Reviewed by Guohua Chen

PETER LANG
New York • Berlin • Brussels • Lausanne • Oxford

**Library of Congress Cataloging-in-Publication Control Number**: 2022050026

Bibliographic information published by **Die Deutsche Nationalbibliothek**.
**Die Deutsche Nationalbibliothek** lists this publication in the "Deutsche Nationalbibliografie"; detailed bibliographic data are available on the Internet at http://dnb.d-nb.de/.

ISBN 978-1-4331-7918-1 (hardcover)
ISBN 978-1-4331-7919-8 (ebook pdf)
ISBN 978-1-4331-7920-4 (epub)
DOI 10.3726/b16737

This edition is an authorized translation from the Chinese language edition
Published by arrangement with Foreign Language Teaching and Research Publishing Co. Ltd.

80 Broad Street, 5th floor, New York, NY 10004
www.peterlang.com

# Contents

# Introduction

## I

This book is based on a series of lectures I gave on American literature, but the subject of these lectures is not what you would normally expect to find in lectures about literature. Literature lectures usually revolve around themes such as love, war, and death, and explore the characterization of different individual human beings. The focus of my lectures, however, is on stories of places and the spiritual connection between people and these places. For a long time, nature has been a silent player in literature, treated only as a backdrop. Mark Christopher Allister once wrote: "Nature speaks. But in our discourse nature is silent, a symbolic presence only, a mute object" (Allister 2001, p. 30). Henry David Thoreau, in his essay *Walking* (1862), asked: "Where is the literature which gives expression to Nature?" He then declared: "I wish to speak a word for Nature" (Thoreau 2002, p. 59). What is pleasing is the fact that since the latter half of the twentieth century, especially since the 1980s, nature writing, a new force in American literature, a form of literature that tells the story of places and speaks on behalf of nature, has started to emerge. This literary form, however, is not simply a description of nature. From the earliest natural history essays to modern nature writing, American nature writing has developed a tradition, which seeks to tell the story of places and explore the connection between human spirituality and a particular place, blending natural history with the history of human development.

Thoreau's book *Walden* (1854) is a true portrayal of the natural history of Walden Pond and also that of the author's spiritual growth. Even today Thoreau's tradition still has a profound influence, with many American professors teaching nature writing courses at university. Let us look at two American literature professors, who are not only authors of nature writing but also teach the subject. They have even incorporated their own experiences of discovering nature into their teachings. John Tallmadge[1] based his book *Meeting the Tree of Life: A Teacher's Path* (1997) on his own experiences. His book was unique in that it chronicles his exploration of the wilderness as he followed in the footsteps of prominent nature writers. He linked these exploits to his early development as a young English teacher, and the book reveals his insights and his personal experience of exploring landscapes both in the real world and within his own mind. In nature, he saw a healthy community with its members dependent upon each other, a fair and sustainable model for human society. He wanted to learn from the tree of nature and to use the book of nature to teach others. To achieve this, he told stories of places. As he saw it, teachers and storytellers are very similar. "We too need teachers and storytellers to strengthen us. …for a place is a space with a story, and by sharing the story you make the place your own" (Tallmadge, p. 170). He combined his interests in literature, exploration of nature and language teaching, and built a career encompassing all three. Tallmadge's style of teaching and research is similar to what modern nature writers would describe as "narrative scholarship."[2] Ian Marshall[3] also based his book *Story Line: Exploring the Literature of the Appalachian Trail* (1998) on his own experiences. For more than 20 years, he often went hiking on the Appalachian Trail while reading literary works about the relationship between humans and nature in this region. He explored America's past, its landscape and its national experience, and he told

1 A contemporary nature writer and professor of American literature with more than 30 years of experience in higher education.

2 The term "narrative scholarship" was first coined by Professor Scott Slovic of the University of Nevada, and was used to describe the use of stories in enlivening literary criticism. In his book *Going Away to Think: Engagement, Retreat, and Ecocritical Responsibility* (2008), he writes that "Storytelling, combined with clear exposition, produces the most engaging and trenchant scholarly discourse. Nature writers themselves realize this…" (Slovic 2008: 34–35).

3 Former president of the Association for the Study of Literature and Environment.

vivid stories about American literature, ecological history and national identity. In the introduction to his book, he explained that he wanted to use his own experiences and writings as well as those of other people before him in order to write "a literary geo-history," and that his book was for him "practicing what has been called 'narrative scholarship'" (Marshall, pp. 3, 7). This genre of literature is precisely what my lectures on American nature writing tried to explore. However, this kind of literature tells not only stories about places. It seeks to answer a question at the forefront of our attention: "Does literature enable mankind to better adapt to the natural world or cause further alienation, and will reading such literature help to achieve a balanced ecology and harmonious coexistence between man and nature?" At the same time, nature writing is also a multi-disciplinary field, combining language, literature, philosophy, ecology, botany and ethics, with a broad scope for further study.

## II

Since 1995 when I first came across American nature writing, I have witnessed its flourishing again after languishing in obscurity for some time and witnessed its growth in popularity. It has now branched out into environmental literature and ecocriticism, and the "sense of place" has given way to the "sense of planet." The attention that this field is receiving is very exciting, and it is worth celebrating that more people are paying attention to and caring about nature. Of the many topics and fields of research relating to nature and literature, I have chosen American nature writing as the main topic of discussion for this book. This is because, despite the more recent emergence of ecocriticism, green studies, ecological literature and other new terms, almost all of them are intricately linked to nature writing. For example, when explaining "ecocriticism" in his monograph *Practical Ecocriticism: Literature, Biology, and the Environment* (2003), Glen A. Love specifically emphasized: "Nature writing, which has been the mainstay of literature-environment studies in the past, will continue to hold a central position" (p. 28).[4] Besides, nature writing has a tradition, a comparatively complete system, and a large number of representative authors and works, all of which provide the reader with a sense of beauty, a communication of the spirit and a connection of minds. Thoreau famously said, "Simplicity, simplicity, simplicity" (in *Walden,*

4 The topic of the relationship between nature writing and ecocriticism is discussed in further detail in Lecture 29—The Origin and Current Situation of Ecocriticism.

[1854] 2004, p. 78)! As modern society becomes ever more complex, as technical jargons increase and become ever more incomprehensible, as the number of adjectives describing the relationship between man and nature grows in complexity, we need to go back to basics. People today want to explore the wilderness and experience nature, not because they want to find a certain elusive object, but because they are looking for a constant. As Edward Abbey (1968) said, "Men come and go, cities rise and fall, whole civilizations appear and disappear—the earth remains, slightly modified. …man is a dream, thought an illusion, and only rock is real, Rock and Sun" (1971, p. 219). Sigurd F. Olson, another American nature writer, once wrote about his hobby of building stone walls. He described how, from the rocks that had not moved for millennia, he absorbed a sense of rooted stability that was unaffected by external factors. Thus, in order to discuss literary works that describe the relationship between man and nature, we must start with the basics; only after establishing the basics can we gain the confidence to expand and grow. This approach is not unique or extraordinary. For example, Robert Finch and John Elder, the editors of *The Norton Book of Nature Writing* (1990), decided to include in its first edition only non-fictional essays written in prose. Their reasoning was that prose essay was one of the most important literary forms in modern American literature, and a significant number of outstanding essayists were nature writers. When the book was republished in 2002, the editors changed the title of the book to *Nature Writing: The Tradition in English*, and wrote in the Introduction, "But our own goal remains to celebrate a distinct and limited, yet enormously rich, literary tradition in English—one whose principal expression has been the personal narrative essay and whose flowering, for various cultural, historical, and geographical reasons, has been especially remarkable in the United States" (Finch & Elder 2002, p. 17). This is also one of the main reasons why I have chosen *American Nature Writing: Thirty Lectures* as the title of my book.

## III

The term "writing" in "nature writing" is not used in the same way as in creative writing. It is not writing in a general sense, not the kind of general composition taught in a writing class, nor is it simply a form of writing. The connotations of the term "nature writing" are richer and more profound than what the two constituent words can convey. As mentioned before, it is steeped in a literary tradition with unique compositional characteristics, and it is itself a complete literary system, with its origins reaching back in history, and with many prominent representative authors and significant representative works of literature. From a literary historical standpoint, it is a distinct literary form. This view is briefly discussed in Chapter One, and its significance can be gleaned from the Contents page. In addition, the term "nature writing" is often interchangeable with "literature of nature" in English works. In *This Incomperable Lande: A Book of American Nature Writing* (Lyon 1989), its editor Thomas J. Lyon wrote a chapter especially explaining the meaning and history of nature writing. In it, he used the terms "American nature literature" and "our literature of nature" in place of "nature writing" (pp. 3, 23). In the Introduction to the 2002 edition of *Nature Writing: The Tradition in English*, Finch and Elder recognized that "Thoreauvian inheritance continues to leaven and transform literature of nature in English" and went on to point out that "…nature writing, like other forms of literature and the arts, has been opening its doors to an increasingly diverse group of writers" (p. 17). When Rachel Carson won the 1952 John Burroughs Medal,[5] she took the opportunity to define the concept of nature writing in her acceptance speech. She believed that the tradition of nature writing is "a long and honourable one," formed during the nineteenth century or even earlier, and if we follow this tradition or its spirit, we will create "a new type of literature" (Carson 1998, pp. 93–95). She emphasized the literary value and social function of nature writing. Barry Lopez in his *About This Life: Journey on the Threshold of Memory* (1998, p. 14) called nature writing "a literature of hope." Many others have referred to nature writing as literature. There is no need to list them all here.

In the Chinese classic *The Analects,* Confucius said, "People used to learn for the sake of improving themselves; they now learn for the sake of impressing others" (in Yang *The Analects with Annotations*, 1958, p. 161; "*gu zhi xue zhe wei ji, jin zhi xue zhe wei ren*"). When I first entered the field of nature writing, it was entirely to satisfy my own interest and I did not have any lofty ideas about my career. My experience is similar to what was described by Sigurd Olson in *The Singing Wilderness* (1956). He was particularly fond of red squirrels and devoted a chapter entitled "The Red Squirrel" to them. His love of red squirrels is unconnected with his academic interests and he never sought to publish any article about squirrels in an academic periodical. Therefore, zoologists would probably regard him as the least respected person in their field. Olson said that his interest

5 The John Burroughs Medal was first awarded in 1926 and is the highest prize awarded in America to the author of a distinguished book of natural history. In 1952, the medal was awarded to Rachel Carson for her book *The Sea Around Us* (1951). Carson was also the author of *Silent Spring* (1962).

in red squirrels was purely out of fondness and admiration. For him, these creatures symbolize the wilderness as he came to know it on the vast American continent. He wrote, "What better way of thinking of the ancient forests of the continent than to realize that a squirrel could have run through the tops of them from the Atlantic coast to the Mississippi without having to come to earth?" (p. 156) Dr Samuel Johnson once said: "The end of poetry is to instruct by pleasing." I think, at the beginning, it was the poetic feeling of nature writing or of nature that intrigued me and led to my falling completely in love with this field. This book is only one of the unintended results of that love. Therefore, I do not want this book to be too serious, but would like to share my enjoyment with the reader through a series of stories. If the reader can be persuaded to appreciate and love nature, and become unconsciously protective of our planet, then perhaps I can be a little proud of what I have chosen to do. After all, nearly two centuries ago, Henry David Thoreau wrote, "This curious world we inhabit is more wonderful than convenient; more beautiful than it is useful; it is more to be admired and enjoyed than used."

## References

Abbey, Edward. (1968) 1971. *Desert Solitaire: A Season in the Wilderness*. New York: Ballantine Book.

Allister, Mark. 2001. *Refiguring the Map of Sorrow: Nature Writing and Autobiography*. Charlottesville, VA: University Press of Virginia.

Carson, Rachel. 1998. *Lost Woods: The Discovered Writing of Rachel Carson*. Ed. Linda Lear. Boston: Beacon Press.

Finch, Robert, and John Elder, eds. 1990. *The Norton Book of Nature Writing*. New York: Norton.

Finch, Robert, and John Elder, eds. 2002. *Nature Writing: The Tradition in English*. New York: Norton.

Love, Glen A. 2003. *Practical Ecocriticism: Literature, Biology, and the Environment*. Charlottesville, VA: University of Virginia Press.

Lopez, Barry H. 1998. *About this Life: Journey on the Threshold of Memory*. New York: Knopf.

Lyon, Thomas J., ed. 1989. *This Incomperable Lande: A Book of American Nature Writing*. Boston: Houghton Mifflin Company.

Marshall, Ian. 1998. *Story Line: Exploring the Literature of the Appalachian Trail*. Charlottesville, VA: University Press of Virginia

Olson, Sigurd F. 1956. *The Singing Wilderness*. New York: Knopf.

Slovic, Scott. 2008. *Going Away to Think: Engagement, Retreat, and Ecocritical Responsibility*. Reno, NV: University of Nevada Press.

Tallmadge, John. 1997. *Meeting the Tree of Life: A Teacher's Path*. Salt Lake City, UT: University of Utah Press.

Thoreau, Henry David. (1862) 2002. *Wild Apples and Other Natural History Essays*. Ed. William Rossi. Athens, GA: University of Georgia Press.
Thoreau, Henry David. (1854) 2004. *Walden*, Boston: Shambhala.
Yang, Bojun. 1958. *The Analects with Annotations*. Beijing: Zhonghua Book Company.

1

# The Origin and History of American Nature Writing

## Lecture 1. The Concept and Origin of Nature Writing

Since the latter half of the twentieth century, and especially since the 1980s, there has appeared a new school of literature—American nature writing. It focuses on describing nature; examining the relationship between man and nature; and presenting beautiful scenes where the two merge, to tell stories of places that have begun to fade from the collective memory. Although American nature writing is influenced by the tradition of European Romanticism, as a literary genre it has its roots in America, the "Garden of Eden" and the "New World." Naturally, it has characteristics that are not to be found in any other country and differs from the naturalism once seen in Western literature. It is a contemporary literary genre with American characteristics, with its origins in the seventeenth century and its foundations in the nineteenth.

### *The Concept of American Nature Writing*

In terms of structure, nature writing belongs to the category of nonfiction prose literature, and most often uses essays, diaries, autobiographies, letters and other similar forms. In terms of content, it mainly seeks to explore the relationship between humanity and nature. In short, its most typical mode of expression is first person narrative, telling real and intimate stories of the authors' personal, physical and spiritual experiences when they leave the "civilized" world and enter a natural environment. American scholar Don Scheese, in his *Nature Writing: The Pastoral Impulse in America* (1996, p. 143), describes the genre as a "literature of the American wilderness that blends feeling with informed observation."

Nature writing has three main characteristics:

1. Emergence of a "land ethic." This philosophy encourages humanity to abandon its anthropocentric beliefs, emphasizes the equality between humanity and nature, and calls for a caring attitude towards nature and for seeking spiritual values in the wilderness.
2. Focus on a "sense of place." If race, class and gender were once hot topics in literature, then the "place" we rely on for survival should also deserve a position of equal importance.
3. Possession of its own unique literary form and language.

These characteristics are also reflected in nature writers themselves. First of all, they know and love nature, not just enjoying it, but feeling a spiritual connection. This connection, based on the "land ethic" and on a knowledge of the wilderness, has led to an ecological conscience, a moral value that nature writers abide by. Almost all nature writers live and write in a particular region, and each has their own unique style of discourse and language, which is why Marshall and Taylor in their article titled "A Catskills Dialogue: Looking for John Burroughs, from Wake Robin to Slabsides" (2006, p. 174) remark that "the nature-writer must be naturalist, moralist, and stylist" at the same time.

Nature writing is, literally, writing with nature as its theme. However, this concept has a very rich content, having grown from the unique American natural and cultural background. Well-known American historian Henry Nash Smith, in his book *Virgin Land: The American West as Symbol and Myth* ([1950], 1970, p. 187), states that "The character of American empire was defined not by streams of influence out of the past, not by a cultural tradition, not by its place in a world community, but by a relation between man and nature—or rather, even more narrowly, between American man and the American West."

In the seventeenth century, when the first group of European migrants arrived on the North American continent, they felt it was a second Garden of Eden gifted by God, and they therefore dreamed of building "a city set on a hill." However, the "Adams" of this new civilization soon discovered that they faced a silent scenery and an unsung land, very different from other places. This gave the initial tension in American literature: on the one hand, Americans felt the urge to sing the praises of their beautiful, rich and fertile land; on the other hand, this vast land did not have any suitable singers. Two centuries later writer Ralph Waldo Emerson, the philosopher of the American Renaissance, began to call for the land's own singers. Poet Walt Whitman, in the preface to the first edition of his *Leaves of Grass* ([1855] 1989 in Baym & Gottsman, p. 1960), proclaims that an American poet "incarnates its geography and natural life and rivers and lakes." American literature, growing from the country's unique natural environment, differed from other literatures in that it was a song of the land from the very beginning. In other words, it was deeply rooted in this "new" land, and reflected both the harmony and the conflicts that Americans experienced with this land.

After the American Renaissance, although there was no shortage of American writers who sang the praises of the American land and pursued spiritual sublimation in nature, there was no systematic school of nature writing for a long time. Nature writing as a concept was a product of modern times. Scheese, researching the origin of the term "nature writing," finds that it was first used in the twentieth century. In 1902, Francis H. Halsey published an essay titled "The Rise of the Nature Writers," commenting on the influential nature writers of the time (Scheese, p. 138). After that, more essays and books on the topic appeared. However, it was not until the 1980s that nature writing became a recognized term, referring to a nonfiction prose literary form relating to humanity's relationship with nature, and many academics and scholars began to teach and research the subject. In America alone, thousands of books on nature writing have been published, including anthologies of nature writing, and research monographs by scholars. Nature writing has entered the university classroom, with many institutions offering the subject, and most usually at postgraduate level, and many students in English departments choose nature writing as the focus of their master's or doctorate theses.

The significance of nature writing lies in the fact that it is innovative. In Western civilizations, people tend to separate the spiritual from the material, the self from the environment, and humanity from nature. In contrast, the driving force behind the development of nature writing is to blend the spiritual with the material, the self with the environment, and humanity with nature. From the early pure nature histories to literary nature histories, and from nature essays that mainly explored the relationship between nature and personal thoughts and behaviors to the nature writing which advocates coexistence between humanity and nature, we see the development of nature writing alongside the evolution of human civilization.

The significance of nature writing also lies in its being both inclusive, and also born of its time. Although nature writing emerged in America, its influence has not been limited to the country of its origin, for the damage that modern society has caused to nature has become an issue of international significance. Descriptions and praise of nature, and yearning for and seeking protection of the land under our feet, has crossed borders and gained universal significance. In addition, American nature writing has always encompassed multinational perspectives. For example, Michael Branch, a scholar of American nature writing—in his introduction to his anthology *Reading the Root: American Nature Writing before* Walden (2004, pp. xii–xxvii)—discusses the characteristics of the migrant country, and the descriptions of the land by settlers, explorers and travelers from Europe. Apart from writings by American authors, the anthology also contains writings by Italian, Spanish, French, Dutch, Belgian and British authors from the fifteenth to the mid-nineteenth century, demonstrating the multiculturality of nature writing from its beginning.

## *The Origin of American Nature Writing*

The concept of "nature writing" has emerged in modern times, but it has a long history of development, with ancient philosophical origins that can be traced back to Ancient Greece and the Roman Empire in the works of Aristotle and Virgil. American scholars Peter A. Fritzell, in his *Nature Writing and America: Essays upon a Cultural Type* (1990), and Don Scheese reflect upon the influences of, respectively, Aristotle's *History of Animals* and Virgil's *Eclogues* on nature writing in their monographs on the subject (Fritzell, p. 3; Scheese, pp. 13–14).

Several notable authors from England, including eighteenth-century natural historian Gilbert White, nineteenth-century Romantic poet William Wordsworth and naturalist Charles Darwin, and twentieth-century writer D. H. Lawrence, have also influenced nature writing. White's book *The Natural History and Antiquities of Selborne* (1788) influenced such renowned nature writers as Henry David Thoreau, John Burroughs and Rachel Carson, and an excerpt was included as the first model essay of nature writing in *The Norton Book of Nature Writing* (Finch & Elder 1990). Meanwhile Harvard English Professor Lawrence Buell, in his authoritative monograph *The Environmental Imagination: Thoreau, Nature Writing, and the Formation of American Culture* (1995), frequently mentions the influence Wordsworth and Darwin had on the development of nature writing. D. H. Lawrence's *Study in Classic American Literature* (1923) has been widely quoted in books on the subject.

However, it was in America, with its unique cultural background and its rapid modernization, that nature writing appeared and grew. In other words, nature writing is a form of literature that developed in a uniquely American natural and cultural background, which is why it is at its most representative in America.

When discussing the origins of American nature writing, American scholar Thomas Jefferson Lyon, editor of *This Incomperable Lande: A Book of American Nature Writing* (1989, p. 16), writes that "the first and greatest influence on nature writing, of course, is the land itself." American nature writing started with the first wave of European settlers arriving in the New World, when they encountered a unique landscape, almost unspoiled by human hands—both a luxuriant Garden of Eden and a desolate, devilish and wild land. Settlers endured much hardship and danger to cross the ocean and reach this wilderness. They left the old civilization for this remote corner of the Earth, and managed to build a new nation here—in itself an unprecedented success. This special relationship between humanity and nature determined the unique emotional connection between Americans and their new homeland. It was only when they had learned about the land under their feet that they could find themselves. Compared with other nations, they had a more urgent need to both understand and to immerse themselves in nature. They were not outsiders, but participants in nature. When faced with the wilderness of the New World, the settlers discarded the thinking of the old world and began to re-map their spiritual and cultural landscape in the wilderness. As Fritzell observes (p. 130), many early Americans, through writing diaries, travel journals, letters and essays, "turned to the business of identifying, describing and explaining those very 'outward objects' in their efforts to make sense of themselves, their preconceptions, and their American places"; and this gave rise to what is now called American nature writing.

In the seventeenth century, both John Smith—in his *A Description of New England* (1616)—and William Bradford—in *Of Plymouth Plantation* (1630–1650)—used contrasting images of a fertile Garden of Eden and a howling wilderness to present a picture of nature in the New World, making nature the focus of the attention of the inhabitants of this new civilization. These two works were both written in the form of descriptive prose, using simple and clear language, and their energy and vitality became the basis of later works in the unique genre of nature writing.

In the eighteenth century theologian Jonathan Edwards, in his *Personal Narrative* (1740) and *Images or Shadows of Divine Things* (ed. Perry Miller 1948), boldly connected inner spiritual experience with the natural landscape of the outside world. By analogy, he argued that God made the material world "shadows" of the spiritual world.

Another influential eighteenth-century author, William Bartram, described his "lonesome pilgrimage" through the wilderness in the American Southeast, showing a fresh aesthetic appreciation of the wilderness. As a result he became the first author to receive acclaim in both European and American literary circles, and has been widely acknowledged as the founder of American nature writing. Prose describing nature is, therefore, generally considered to have become a literary genre at the end of the eighteenth century, marked by the publication of Bartram's *Travels Through North and South Carolina* (1791). (It is worth mentioning here that as a naturalist, Bartram was also the first to use the word "sublime" to describe the beauty of nature in America, thereby distinguishing it from the picturesque beauty of Europe.)

If the seventeenth and eighteenth centuries saw the appearance of the founders of the themes, genre and style of nature writing, then the nineteenth witnessed that of the profound philosophical ideas of nature writing. This is evidenced by works such as painter Thomas Cole's "Essay on American Scenery" and writer Ralph Waldo Emerson's first work, *Nature*—both published in 1836—which saw American writers and artists starting to use the landscape of the New World as their literary and artistic inspiration. Before then, they had often looked towards Europe to find the source of their creativity. In his essay Cole concludes that to connect with itself America must not look backwards, but savour the present and look towards the future. If Europe represents culture, then America nature; and those born in nature's country should seek their cultural and artistic inspiration from nature. It is precisely because of this that Cole founded the Hudson River School art movement. His declaration that art should present natural scenery on the canvas (Cole, [1836] 2011, pp. 27–36) resonated with a group of landscape painters who Emerson described as having "new eyes" (Emerson [1836] [1903] 1979, p. 75), and works from this school complemented those of American nature writing.

Emerson's spirit and his thoughts have long been an important focus of study in the field of American culture and literature, although in China there is at present little knowledge of or interest in Emerson and his close connection with nature. In fact, Emerson's special connection with nature was evident in his books and diaries, his transcendentalist beliefs, and his life in general. In

*Nature* he improves on Edwards's view, stating: "Nature is the symbol of spirit" (Emerson, p. 25). In his published speech "The American Scholar" (1837) he states that "in fine, the ancient precept, 'Know thyself,' and the modern precept, 'Study nature,' become at last one maxim" (Emerson, p. 87). These views were the fundamental philosophical basis of nature writing, greatly influencing his contemporaries, later generations, and even the present generation of nature writers. Emerson's works are also known for their loose structure and poetic prose style. Emerson is regarded as the "father of American literature," which to this day retains his visions and attitudes. Of all the great American literary works, I believe Emerson's—with their aforementioned characteristics—are most clearly manifest in the genre of nature writing.

When Thoreau is mentioned, we usually think of Walden Pond and his most well-known work *Walden; or, Life in the Woods* (1854), which topped the list of "Ten Books That Shaped the American Character" published in *American Heritage* magazine in 1985. Thoreau is regarded as an idol in American culture and as the most influential American nature writer. His spirit is regarded as America's cultural heritage, and Walden Pond has become the destination of pilgrimage and the model of experiencing nature for many of his followers. People often think of Thoreau as Emerson's disciple; but from a modern perspective, Thoreau's contribution to nature writing is perhaps more significant than Emerson's. Thoreau not only put into practice Emerson's theories, but also went one step further. Foreseeing the conflict between an industrialized civilization and nature, he insists in his essay "Walking" ([1862] [1947] 1977, p. 609)) that "in Wildness is the preservation of the World." His argument for the value of the wilderness has profound significance in the development of nature writing.

The American nature writing described in this book has been influenced by various literary trends throughout its history, but its fundamentals hark from the era of Emerson. Although later nature writers would write about their experiences of their own times, they would always seek inspiration from Emerson, Thoreau and other writers of their generation. As Scheese (p. 22) explains, this is because Emerson and Thoreau were "the progenitors of modern nature writing and the authors most responsible for the transformation of natural history into nature writing."

In summary, in the nineteenth century, Emerson's *Nature* and Cole's "Essay on American Scenery" were the first works to define the philosophical ideas underpinning American nature writing. The vivid descriptions of expansive wilderness in the works of Thoreau and Whitman presented what the American literary critic F. O. Matthiessen, in his book *American Renaissance: Art and Expression in the Age of Emerson and Whitman* (1941, p. 76), refers to as "actual glory." At the same time, the Hudson River School brought the words and ideas of Emerson, Thoreau and Whitman to life in their paintings. Painters of nature on canvas and wilderness writers came together to present a mesmerizing vision that combined the human soul and nature, forming a unique trend and a modern atmosphere in which the culture of the New World was being created and incorporating the wilderness. This trend and atmosphere nurtured the development of American nature writing, fuelling a rise in its popularity which has led to its current status in American literature.

## Lecture 2. The Rise of Nature Writing and Its Features

### *The Rise of American Nature Writing*

Before the twentieth century, nature writing was full of idealism and its writers held an optimistic view of nature, hoping to attain freedom of the self, find the roots of their culture and the sublimation of their spirit in nature, but the focus of their thoughts and writings was limited to "nature and oneself" or "nature and personal thought and behavior." It was not until the twentieth century, with the appearance of John Muir, Mary Austin and other turn-of-the-century nature writers, that nature writing began to develop a much broader view, exploring the relationship between nature and humanity, and the harmony (or lack of) between humanity and the natural environment.

From the early to the mid-twentieth century works representative of the era, such as Mary Austin's *The Land of Little Rain* (1902), John Muir's *The Yosemite* (1912), Henry Beston's *The Outermost House: A Year of Life on the Great Beach of Cape Cod* (1928) and Aldo Leopold's *A Sand County Almanac* (1949) were published. These works drew on their respective authors' personal experiences of nature—in the desert, in mountainous and coastal areas, and on land undergoing desertification—to describe a magnificent wilderness that is gradually disappearing with the advance of modernization, thus seeking to remind their readers of the need to care for the land. The classic works of nature writing from this period also have another characteristic. Where Burroughs's first collection of essays *Wake-Robin* (1871), set in eastern America and published in 1871, demonstrates the genre's initial love of the green landscapes typical of the eastern states of America, Austin's *The Land of Little Rain* (1902) is set in the American Southwest and shows it beginning to appreciate the red rocks and desolate deserts of the Southwest, making this unique genre of literature even more rich and colorful.

Rachel Carson's *Silent Spring*, which was first published in 1962 and describes the threat posed to both environment and humankind by the reckless use of pesticides, contributed to the establishment of the first Earth Day. Edward Abbey's *Desert Solitaire: A Season in the Wilderness* (1968)—set in the deserts of Utah—conveys a yearning for quiet tranquility by a restless modern man, and puts forward the author's views on the wilderness and modern civilization. Annie Dillard's *Pilgrim at Tinker Creek* (1974) depicts the author's nature pilgrimages in the Blue Ridge Mountains of Virginia and sets out her reflections on tradition. Barry Lopez's *Arctic Dreams: Imagination and Desire in a Northern Landscape* (1986) describes his adventures in the Arctic and explores the intrinsic and extrinsic needs of humanity. Terry Tempest Williams's *Refuge: An Unnatural History of Family and Place* (1991) tells of three generations of women all afflicted with breast cancer as a result of America's nuclear tests, and their search for solace on the banks of the Great Salt Lake. Contemporary American nature writer Scott Russell Sanders, in *Staying Put: Making a Home in a Restless World* (1993), brings up a common theme explored by nature writers: How do I find my own piece of land in this turbulent world?

The twentieth-century nature writers emphasized coexistence of humanity and nature, and particularly a "sense of place." For them, without a geographical anchoring point, there can be no spiritual anchoring point. Nature writing in the twentieth century, therefore, revolved around a common theme in different geographical environments, and it attracted a large number of writers, although each of them has their own unique writing style. Muir, who set his works against the backdrop of Yosemite Valley in the western United States, was known as "John o' Mountains"; while Burroughs—whose works include essays inspired by visits to his native Catskill Mountains in the eastern United States, including his first collection *Wake-Robin*—was nicknamed "John o' Birds." Austin and Abbey are known for writing about deserts, and their works are considered desert classics. Dillard wrote about her creek-side nature pilgrimage, and Williams about the solace she found on the banks of a salt lake. These are twentieth-century nature writers, yet they remain inextricably linked to the pioneers of the genre: Muir has been described as the last disciple of the Concord transcendentalists and Abbey as the Thoreau of modern times, while Dillard's *Pilgrim at Tinker Creek* is viewed as a bold *Walden*. At the same time, they have also gone beyond their predecessors in their view of the natural world.

In terms of breadth, in the twentieth century nature writing was no longer limited to America. The Association for the Study of Literature and Environment, founded in America in 1992, now has affiliated associations in the UK, Australia, Japan and many other countries. At the same time, works focusing on the exploration of the relationship between nature and humanity began to emerge in countries around the world. In terms of depth, as many twentieth century nature writers had a good knowledge of the natural sciences and human ecology, they possessed a deeper understanding of their subject than their predecessors. Leopold proposed the idea of a "land ethic" in an abandoned Wisconsin farm and asked people to cultivate an ecological conscience; Abbey proposed a new model of harmonious coexistence between humanity and nature in the desolate western deserts, based on opposition, compromise and balance; Williams, on the side of the salt lake, asked that humanity see the wilderness as a feeling and love it as though it were a person. For modern nature writers, the relationship between humanity and nature is no longer one of "me and it," but of "me and you." They believe that there is no longer a pure self, but a self-in-place, as we are one with the environment in which we live. They no longer believe in the survival of the fittest, but in coexistence.

Nature writing has not only had new writers and new works appearing all the time, but has also matured as an area for teaching and research. In the last few decades of the twentieth century, many good anthologies of nature writing have been published, including Daniel Halpern's *On Nature: Nature, Landscape, and Natural History* (1987); Thomas Lyon's *This Incomperable Lande* (1989); and Nina Baym and Ronald Gottsman's *The Norton Book of Nature Writing* (1990). At the same time authoritative criticisms of nature writing have also appeared, such as Buell's *The Environmental Imagination* (1995); Scheese's *Nature Writing* (1996); Fritzell's *Nature Writing and America* (1990); American literature professor and literary critic Sherman Paul's *For the Love of the World: Essays on Nature Writers* (1992); and Scott Slovic's *Seeking Awareness in American Nature Writing: Henry Thoreau, Annie Dillard, Edward Abbey, Wendell Berry, Barry Lopez* (1992), as well as many critical works on key individual nature writers.

More than a century and a half ago, in his eulogy for Thoreau, Emerson ("Thoreau", [1862] 1989 in Baym & Gottsman, p. 1037) wrote that "wherever there is knowledge, wherever there is virtue, wherever there is beauty, he will find a home." In *Down the River* (p. 48), Abbey echoes Emerson: "Wherever there are deer and hawks, wherever there is liberty and danger, wherever there is wilderness, wherever there is a living river, Henry Thoreau will find his eternal home." These contrasting comments about Thoreau reveal the differing views of nature writers in the nineteenth and twentieth centuries, demonstrating the shift from a self-centered to an ecology-centered view.

Today, nature writing is still developing, with all kinds of new terms appearing. Related genres of environmental literature and ecological criticism have also established themselves, but the three have similar themes and can be regarded as coming from the same school of thought. Literary critics still debate whether nature writing has become part of the literary mainstream, but no one can deny that this genre of literature boasts a rich history, and judging from its current popularity, can look forward to an equally rich future.

## Concepts and Features of American Nature Writing

It can be said that contemporary American nature writing not only carries on the Romantic and transcendentalist traditions, but also exudes a strong flavour of modernity. In contemporary American nature writing, the influences of Romanticism and transcendentalism are easily found—in, for example, its promotion and praise of nature, its disregard and contempt for materialism, and its pursuit of and yearning for spiritualism. On the other hand, as contemporary nature writing is occurring in an era when technology is developing at a great speed, there are also some clear modernist and postmodernist characteristics. These are shown, for example, in its iconoclastic attitude and its contempt for authority, in its ambiguous feelings of awe and suspicion towards nature, and in the jigsaw puzzle images and blurred boundaries presented in some works. However, contemporary American nature writing is neither a reproduction of Romanticism nor the equivalent of postmodernism. It transcends tradition. It is of its own time and does not need further verification.

Conceptually, nature writing has abandoned the literary tradition of focusing on humanity, and in its place puts forward the idea of a land ethic which promotes harmonious coexistence between humanity and nature. The term "land ethic" comes from Leopold's *A Sand County Almanac* (1949). Leopold, an ecologist, explains the principle of the land pyramid and the food chain. He explains that humans are only part of the land community, while other members include soils, waters, plants and animals. In this land community, every member has its own position and is in an interdependent relationship with all other members. On the long journey of evolution, humanity is only a traveler in company with other biological organisms. No matter what intentions we have, nature will always go its own way. To be in step with nature, we must be in harmony with it. To prove that we are part of the land community, we must learn to respect and accommodate it. Leopold called for a rethinking of ecology and for an ethical duty to the natural environment that we rely on for survival. He asked us to think like a mountain—that is, to think from an ecological perspective, and in terms of the relationship between humanity and nature, and of the need to maintain the health of the land. He wanted humanity to cultivate a kind of "ecological conscience." Leopold called for us not only to practise our ethical duties toward the land we rely on for our survival, but also to put forward behavioral criteria for a land ethic, saying: "A thing is right when it tends to preserve the integrity, stability and beauty of the biotic community. It is wrong when it tends otherwise" (in Finch & Elder, pp. 418–421). Paul regarded the natural scenic composition of the wilderness as the measure of success for contemporary human achievement. He says in *For Love of the World:*

> When I pause in my shoveling and look out over the lake, now dark with scud, and note the reeds (the natural beauty of natural things, this landscape), I know that what I build must conform to this, belong to this composition. (p. 12)

In works of nature writing, we have seen the cycle of love: human love extends to love for the land, and the calm and meditative concentration of nature gives back to us spiritual comfort. It can be said that nature writing commingles the love between human beings and humanity's love for nature, extending social ethics to the land ethic and extending a duty to society to the duty to land, and promoting a great and boundless cycle of love.

In addition, nature writing exudes a strong sense of wilderness, and views and understanding of the wilderness are constants. Emerson states that "i[I]n the woods, we return to reason and faith," and Thoreau that "in wildness is the preservation of the world"; Muir believed that "in God's wilderness lies the hope of the world" and Leopold considered the spiritual satisfaction gained from the wilderness to be superior to that gained from material things, an inalienable human right like that of free speech. On the other hand, Abbey saw the wilderness as a buffer zone that could slow down the frenzied pace of cultural change. Another American nature writer, Sigurd F. Olson, believed that deep down in each of our hearts is a primal disposition and a passionate longing for the wilderness. In nature writing the notion of "wilderness awareness" includes both rationality and sensibility. In terms of rationality, the wilderness is the root of human existence, something that can remind us of our connection with the natural world, and also that we can only maintain the health of our culture by keeping our land healthy. In terms of sensibility, the wilderness fosters a unique kind of emotion and is the home to human spirituality, as our souls in particular need to be nourished by the untamed wilderness of nature. The notion that tranquillity is invaluable is perhaps the most succinct interpretation of wilderness awareness by nature writers living in a materialistic, unstable and characteristically noisy modern society.

It is precisely because of the aforementioned ideas that nature writing has, for the most part, rejected the anthropocentric literary traditions and abandoned the ever-popular subjects of war, romance and death, to cast its gaze on nature and focus on the description and exploration of the harmonious relationship between humanity and nature as its own literary domain.

We often say that to study literature is to study humanity, but nature writing does not follow the same format as other literary genres. The role of humanity has changed, the locale has shifted away from densely populated towns and cities, and the focus has turned to the depiction of the wilderness, parks and the countryside. In a sense this is similar to the paintings of the Hudson River School and their philosophy of "nature on the canvas." In Buell, in a chapter dedicated to "The Aesthetics of Relinquishment" in *The Environmental Imagination*, describes the diminishing role of humanity as the protagonist. He uses Austin's *The Land of Little Rain* to show that she places more importance on the natural environment than on stories about human characters.

> Her protagonist is the land, more particularly the geography of its watercourses and the patterns of life created by water scarcity: the habits of life, the routes of travel, the visible marks on the landscape all being shown as shaped by the need of both animals and humans to adapt themselves to this basic necessity." (p. 80)

What nature writing presents to us is not a pure and simple view of nature or a complicated view of humankind, but a co-dependent community of animals, plants and humans that have the same needs and adapt to the same living environment. Through this community we can see the whole world, and even the world in the future.

Some scholars have examined nature writing through rhetorical analysis. In the introduction to *Green Culture: Environmental Rhetoric in Contemporary America* (Herndl & Brown 1996), the editors use classic rhetorical theory—in particular, the three modes of persuasion: ethos, pathos and logos—to articulate a rhetorical model of environmental discourse. Poetic discourse is assigned to the category of pathos. Poetic discourse is the type of language we use to talk about the beauty, value and emotional impact of nature, whereby nature is portrayed as a spiritual or transcendental entity. It follows that this type of language will tend to view humanity as part of nature, and will often try to define humanity in the context of a harmonious existence with nature, which is why it is also called ecocentric discourse. The power this kind of language wields lies in its ability to elicit pathos; in other words, it appeals to the emotions of the audience to elicit aesthetic and spiritual responses from them. Herndl and Brown (p. 12) cite works such as Thoreau's *Walden* and Beston's *The Outermost House* as good examples of pathos. It could be said that "poetic discourse" is a term used by scholars of nature writing to describe and summarize the rhetorical form of nature writing, which emphasizes the personification and emotive aspects of nature writing. In all the important nature writers who are examined in this book, especially Thoreau and Beston, we see manifestations of poetic discourse in nature writing.

The structure and style of nature writing have revealed new literary possibilities, as the subjects that nature writers visit are different. As Thoreau said, "instead of calling on some scholars, I paid many a visit to particular trees" (quoted in Scheese, p. 10). In the works of nature writing are not just the writing, but also the natural world along the author's journey. Consequently, we can see that almost every nature writer bases their work on a particular natural location with which they are familiar. This is the sense of place that they emphasize. Sanders (p. 150) writes: "I have been thinking about the story of the land and tried to understand how the map of the heart is attached to the map of the geography," while in *Birds and Man* (1930, p. 117), English author W. H. Hudson states: "We have but to listen to the human tones in wind and water, and in animal voices; and to recognize the human shape in plant, and rock, and cloud, and in the round heads of certain mammals, like the seal." Humanity and nature are mutually interdependent and have a symbiotic and complementary existence, an inseparable relationship. Works of nature writing invariably document intimate encounters of their authors with nature and on some level can be seen as autobiographical in style, but they also differ in style from traditional autobiographies. American scholar Mark Allister, in *Refiguring the Map of Sorrow: Nature Writing and Autobiography* (2001, p. 32), explains that nature writers transform features of autobiographies such as "self, life, writing" into "land, memory, story." This means that, when writing about nature and landscapes, the author can also be considered to be writing about their own life, an attribute that has integrated natural history writing and autobiographies. This type of nonfiction not only records but also describes the author's feelings, bringing together their experiences and emotions, a method which combines both realism and aesthetics. This kind of literature, based on intimate encounters with nature, is bound to have an extraordinary appeal. At the same time a literary genre that is so connected with the land will use a distinct language, a kind of earthy language that is as simple as the soil and as refreshing as morning dew. Nature writers want their writing to be natural and new, with freshness and vividness, rather than dead. Due to the use of this novel language, new style and structure, nature writing—like flourishing nature with its varied forms and shapes—has a unique charm.

Art and literature are closely related and complementary, and through this the notion of beauty has evolved. As mentioned earlier, the Hudson River School artists actually attempted in their paintings—using a brush instead of a pen and depiction instead of narration—to express the same ideas as those expressed by nature writers in words. In Cole's painting *The Voyage of Life* (1840), the scenes depicted recreate the spiritual journey described in Emerson's *Nature*. Another of Cole's paintings, *The Oxbow* (1836), seems to hint at Thoreau's view: humanity's optimistic view toward the environment is a composite of the wild and the civilized. The optimism of humanity towards the environment comes from a form of integration of wilderness and civilization. Barton Levi St. Armand, a professor of American Civilization and English at Brown University, once analyzed Thoreau's *Walden* and his *Cape Cod* (1865) and found that their portrayal of landscapes resonated most closely with the luminist landscapes of the Hudson River School (St. Armand 1980). Readers of other works of nature writing find similarities with the "word painting" in the writings of the English author John Ruskin, while Austin's *The Land of Little Rain* has been compared to a series of small landscape paintings. Some nature writers are themselves artists: Alexander Wilson, author of *American Ornithology* (published between 1808 and 1814); William Bartram; and in more modern times mother and daughter Ann and Susan Zwinger have all provided illustrations for their own books. This close relationship between literature and art resembles the Chinese artistic concept of "poetry in the painting" and "painting in the poem" described by Su Shi (1037–1101), a well-known Chinese poet of the Song Dynasty, of the works of Wang Wei (a. 699–761), a poet of the Tang Dynasty. American nature writers go further. For them, "beauty springs from a sense of continuity, an all's-right-with-the-world, a deep sense of ecology, the elegant symbiosis of whole. Beauty is not a luxury but a necessity" (Zwinger 1991, p. 148).

In summary of the above, we could say that in nature writing, the authors are not only writing, but also listening to nature with their ears and experiencing it with their heart. They present to their readers a multidimensional scene which contains landscape, soundscape and soulscape, and it is this multidimensional depiction that makes nature writing unique.

Although nature writing has shown a diversity and individuality, it also leaves an impression that it is a large collection of loosely associated works. It could also be understood as a field of study that emerged from a crossover of the natural and human sciences. Readers are able to interpret and appreciate nature writing from a number of perspectives; and although this is a source of endless pleasure for them, it presents an immense challenge for researchers who wish to analyze the genre systematically. When editing *This Incomperable Lande*, Lyon painstakingly classifies works of nature writing into seven different categories according to prose structure, but at the same time reminds the reader that

> "the types I have listed tend to integrate, and with great frequency. This may be somewhat irritating to lovers of neatness who would like their categories to be immutable, but nature writing is not in truth a neat and orderly field." (Lyon, p. 3)

In America, nature writing has been acknowledged as a distinct school of literature, but some are still trying to interpret nature-themed literary works more generally. They believe that such works do not necessarily fall under the heading of nature writing, as there are many related literary schools. Patrick D. Murphy, for example, in *Farther Farther Afield in the Study of Nature Oriented Literature* (2000, p. 10), talks of "Nature Oriented Literature," of which nature writing is simply one part. This type of classification is more conducive to the adoption of such literature internationally.

In view of the above, it is clear that nature writing is a literary form which seeks to guide us to a state of mind that promotes healthy body and soul through immersion in the natural world. It emphasizes the importance of physical contact and spiritual connection with nature, and seeks to find a cultural and spiritual path that will increase awareness of the need for humanity to coexist harmoniously with its natural environment, so that we embark on a new lifestyle. Aesthetically, nature writing presents a fresh and unique beauty. Reading nature writing gives us access to a fluid beauty, a spiritual pleasure. Exploration of this field is like the European migrants exploring the New World, filled with excitement by encountering seemingly endless new discoveries.

## References

Baym, Nina, and Ronald Gottsman, eds. 1989. *The Norton Anthology of American Literature*. 3rd ed. Vol. I. New York: Norton.

Branch, Michael, ed. 2004. *Reading the Roots: American Nature Writing before Walden*. Athens, GA: University of Georgia Press.

Cole, Thomas. (1836) 2011. "Essay on American Scenery," in Grese, Robert ed. *The Native Landscape Reader*. Amherst: University of Massachusetts Press, pp. 27–36.

Edwards, Jonathon. 1948. *Images or Shadows of Divine Things*. Ed. Perry Miller. New Haven: Yale Univ. Press.

Emerson, Ralph Waldo. (1903) 1979. Centenary Edition, *The Complete Works of Ralph Waldo Emerson*. Vol. I: Nature Addresses and Lectures, 2nd edition of the Centenary Edition. New York: AMS.

Finch, Robert, and John Elder, eds. 1990. *The Norton Book of Nature Writing*. New York: Norton.

Fritzell, Peter A. 1990. *Nature Writing and America: Essays upon a Cultural Type*. Iowa City, IA: Iowa State University Press.

Herndl, Carl G., and Brown, Stuart C. 1996. *Green Culture: Environmental Rhetoric in Contemporary America*. Madison: University of Wisconsin Press.

Lyon, Thomas J., ed. 1989. *This Incomperable Lande: A Book of American Nature Writing*. Boston: Houghton Mifflin Company.

Marshall, Ian, and David Taylor. 2006. "A Catskills Dialogue: Looking for John Burroughs, from Wake Robin to Slabsides." *ISLE* 13.1: 167–181.

Matthiessen, F. O. 1941. *American Renaissance: Art and Expression in the Age of Emerson and Whitman*. Oxford: Oxford University Press.

Scheese, Don. 1996. *Nature Writing: The Pastoral Impulse in America*. New York: Twayne Publishers.

Smith, Henry Nash. (1950) 1970. *Virgin Land: The American West as Symbol and Myth*. New ed. Cambridge, Mass.: Harvard University Press.

St. Armand, Barton Levi. 1980. "Luminism in the Work of Henry David Thoreau: The Dark and the Light." *The Canadian Review of American Studies* 11.1, pp. 13–30.

Thoreau, Henry David. (1947) 1977. *The Portable Thoreau*, ed. and with introduction by Carl Bode, New York: The Viking Press; Penguin Books.

Zwinger, Ann H., ed. 1994. *Mary Austin and John Muir: Writing the Western Landscape*. Boston: Beacon Press.

Zwinger, Susan. 1991. *Stalking the Ice Dragon: An Alaskan Journey*. Tucson, AZ: University of Arizona Press.

2

# The New Continent in Nature Writing

## Introduction

After Columbus "discovered" the "New World," the image of this land flashed through the European consciousness like a ray of light. As an imagined ideal world and a fertile and unexplored virgin land, the New World attracted wave after wave of explorers, travelers and settlers. Not until the seventeenth century, however, did the English explorer Captain John Smith set up the first permanent English settlement, Jamestown in Virginia. The New World deeply impressed the new settlers, first and foremost with its dawn-like freshness. Unlike the much-abused ancient land of Europe, the New World was wild and full of vitality, heralding opportunities for a new life. At the same time, it was also a foreign and silent land. Although there were endless ranges of high mountains and an incomparable sea of dense forests, these mountains told no legends and tales of ancient knights, and the forests sang no songs of human history. Few ideas from the geographical map of the Old World were of much use here, and the settlers had to start anew: to get to know the natural environment in order to survive and prosper in the lonely wilderness.

John Smith and William Bradford were not only pioneering explorers of the New World, but also the first writers to describe its natural scenery. They were the first to cast their gazes towards nature, making nature the focus of the concerns of the inhabitants of the New World and preparing the ground for the theme of American nature writing. In the eighteenth century, the American theologian and writer Jonathan Edwards (1703–1758) and the naturalist John Bartram (1699–1777) sought to approach nature from spiritual and scientific perspectives respectively, and began to write about nature on the basis of their own understandings of the subject, further underpinning the genre. In 1791, when John Bartram's son William Bartram (1739–1823) published *Travels Through North and South Carolina*, he became the first New World writer to receive acclaim in the European literary world. With this book William Bartram established the genre of American nature writing, and he was later recognized as its founder. William Bartram's protégé Alexander Wilson, with his vast volume *American Ornithology* (published 1808–1814), expanded the scope of the meaning of nature in the New World. He represented nature as an organic composition of the land, the animals and the living organisms, thus enriching the meaning and broadening the vision of later American nature writing. Through the efforts of these pioneers, the land of the New World was able to communicate with humankind via a variety of sounds, such as the rustling of moving leaves, gurgling of flowing water and soft chirping of birds. It is this pioneering work that formed the foundations of later American nature writing.

## Lecture 3. Smith and Bradford: The Image of Nature

When people trace the origins of American history and literature, they usually cast their eyes back to John Smith (1580–1631) and William Bradford (1590–1657). From a literary perspective their writings may not be of very high value, but they were nonetheless the pioneers of nature writing, and their importance cannot be overstated. First of all, the act of traveling to an unexplored land from a mature civilization to become pioneers and chroniclers of this new land was itself creative. Fritzell, in *Nature Writing and America: Essays upon a Cultural Type* (1990, p. 146), writes about the special significance of these pioneers in the New World and compares them with nature lovers in Europe during the same period. He believed that people in Europe faced a natural environment that was familiar and well-studied and that had undergone many changes and sufferings. Moreover, they could easily return to their civilized cities should they tire of nature. On the other hand, Americans of the early period faced a wilderness that was foreign and unfamiliar, and they did not have anywhere to retreat to. Perhaps precisely because of this, early pioneers like Smith and Bradford, when faced with the desolate wilderness, developed an unwavering determination—they had no choice but to live with nature and follow its ways.

Matthiessen, in *American Renaissance: Art and Expression in the Age of Emerson and Whitman* (1941), discusses the close connection between Henry Thoreau of the American Renaissance period and John Smith and William Bradford of the early colonial period. Thoreau once commented that Smith's works took him "back to a wilder country and a little nearer to primitive times." When describing the first winter endured by English puritans at Plymouth, Bradford's unforgettable sentence "For summer being ended, all things stood in appearance with a weather-beaten face" prompted Thoreau to declare that only from these emotional words can "standard English" be found, and that this language did not spring from those who "gazed upon nature from the gentlemanly vista of a countryseat," but "from the peasant's horn windows" (p. 116). Smith and Bradford also had a unique writing style that was straightforward in form, which continues to highlight the characteristics of nature writing to this day. In this respect, the American scholar Perry Miller explains the unique cultural significance of America being known as a "Nature's Nation" and as the "New World." In the chapter titled "An American Language" of his *Nature's Nation* (1967, pp. 208–216), he argues that the American language evolved from "the plain style" used by Puritans who had migrated from Europe to the New World in the seventeenth century. It is a style that came from the wilderness. As the settlers were "Protestants in the wilderness," this plain style was not learnt from the schools of Europe but from the necessities of surviving in the American wilderness. Miller explains that should people wish to directly understand what this plain style means, they would only need to read any part of Bradford's 1630–1650 work *Of Plymouth Plantation* (in Miller 1967, pp. 208–216). The reason why Thoreau and modern American nature writers frequently look to Smith and Bradford is because they were the first to highlight the close relationship between humanity and nature. They were the first to focus on nature as the subject of their descriptions, presenting scenes of natural landscape of the New World to later generations. When describing nature they used a straightforward style of expression, creating a unique literary style for the genre of nature writing that later came into being. Their words revealed a rough and wild spirit that was full of vitality, a spirit that is absent in modern society.

Smith's *A Description of New England* (1616) and Bradford's *Of Plymouth Plantation* (1630–1650) both vividly depict images of the New World through their authors' personal experiences, assigning marks of knowledge to this unnamed land. The former truthfully detailed the practicalities of travel, settlement and cultivation in the new land, and also drew detailed depictions of the coastline, rock formations, forests and the climate, all of which presented an image of a pure, sacred land, a plentiful paradise. The latter narrated the entire history of the English puritans who migrated to America in 1620, a story of risking their lives to search for paradise—leaving their homeland behind, enduring a long voyage, building a settlement and being self-sufficient. This book, which has been described as having "the epic nature," lays the cornerstone for the eternal American mythology (James E. Miller in his *Heritage of American Literature*, 1991, p. 10). Through different perspectives, these two books showed the various images of the New World and its natural environment: a pure virgin land, a bountiful Garden of Eden, a land of terrifying forests and a roaring wilderness.

Smith often described the New World as a "virgin land" or a "Garden of Eden." "First, the ground is so fertile, that questionless it is capable of producing any Grain, Fruits or Seeds you will sow or plant.... All sorts of cattle may here be bred and fed in the Iles or Peninsulas, securely for nothing.... The waters are most pure, proceeding from the entrails of rocky mountains; the herbs and fruits are of many sorts and kinds." He went on to describe the richness of this Garden of Eden in detail: There are many kinds of herbs and fruits such as currants, mulberry, grape, spices, gooseberry, plum, walnut, chestnut, pumpkin, gourd, strawberry, bean, pea and maize; many kinds of trees such as oak, fir, birch, elm, cedar, cypress, hazel, ash and pine; a variety of birds and beasts such as eagles, cranes, geese, brants, gulls, turkeys, moose, otters, wolves, foxes, raccoons, bears, wild-cats and martens; fishes such as whales, sharks, sturgeon, cod, eel, mackerel, herring and crustaceans such as lobsters, muscles, whelks, crabs and oysters, in short, a plentiful world. Smith called for Europeans to come to this bountiful paradise that God had bestowed and to begin innocent work like Adam and Eve to cultivate this Garden of Eden for the next generation (in Miller 1991, pp. 33–41). *A Description of New England* was written in descriptive prose that was fresh and simple, full of life and energy—a unique writing style that was copied by later nature writers. As "author of the first English book written in America" (Ruland and Bradbury, *From Puritanism to Postmodernism: A History of American Literature*, 1991, p. 8), Smith personally visited each island and each harbour to study their natural features so that he could gather the information first hand. It was in an old tent in the wilderness or by a campfire in a pine forest that he noted down what he saw, heard and experienced. This practice of personally experiencing nature in order to observe and describe it faithfully and objectively has become one of the distinct characteristics of all modern nature writers.

Roderick Nash, an American writer and literary critic, feels "While the word [wilderness] is a noun it acts like an adjective. The term designates a quality (as the '-ness' suggests) that produces a certain mood or feeling in a given individual and, as a consequence, may be assigned by that person to a specific place" (Nash, *Wilderness and the American Mind*, 1968, p. 1). Such mood and feeling is expressed by Bradford in *Of Plymouth Plantation*. "…Being thus passed the vast ocean, and a sea of troubles before in their preparation (as may be remembered by that which went before), they had now no friends to welcome them, nor inns to entertain or refresh their weather-beaten bodies, no houses or much less towns to repair to, to seek for succour…besides, what could they see but a hideous and desolate wilderness, full of wild beasts and wild men…. For Summer being done, all things stand upon them with a weather-beaten face; and the whole country, full of woods and thickets, represented a wild and savage hue. If they looked behind them, there was the mighty ocean which they had passed, and was now as a main bar and gulf to separate them from all the civil parts of the world" (in Miller 1991, p. 62).

In Bradford's book, it is hard to find the charming Edenic scenes filled with Adam's beautifully sweet melodies described by Smith. Instead, there is confusion and fear experienced by the puritans, who feel isolated or lost in the roaring wilderness. From these different images of the New World and its natural environment presented by Smith and Bradford, we can see that from the very beginning, American nature writing contained unavoidable contradictions: love intermingled with fear of nature, and the pursuit for material civilization conflicting with a longing to preserve the purity of the land—contradictions which have remained to this day. *Of Plymouth Plantation* was based on Bradford's personal experiences and was written in the form of a diary that was succinct and honest in style, without any embellishments or grandeur—a literary style that also later became one of the distinct characteristics of nature writing.

Smith's and Bradford's aforementioned works both contain portrayals of Native Americans, describing them respectively as good people who were naive, kind, shy and honest; and as monsters who were uncivilized, ruthless and wild. If one were to argue that when the first European settlers reached the New World they had set foot upon a land that was no longer a pure wilderness, that this land already had signs of human history, then it was the Native Americans who had left their cultural mark upon it. For these people, the heavens were their father and the land their mother, and everything that walked, flew or took root were their children; to them, animals were also humans and had souls. They had a brotherly bond with nature, and anthropomorphized every part of it: the land was a lifeform, rocks had souls and animals were able to communicate. This attitude that treated humanity and nature as being on equal terms, and the pursuit of intimate bond with nature, had an impact on later nature writers, and still does to this day, so we must acknowledge the special mark that Native Americans have also left on the genre of American nature writing.

Another seventeenth century figure worth mentioning is America's first female poet, Anne Bradstreet (1612–1672). Her poem "Contemplations" (1645), which describes her solitary stroll in the woods on a sunny autumn afternoon, contemplating the imageries of nature and life, is regarded as America's first poem about nature.

No matter how the seventeenth century European settlers perceived and portrayed nature in the New World, they all dedicated themselves passionately to this land: either engaging in "innocent labour" on the pure land of their Garden of Eden, or undertaking their great mission of building God's city in the midst of a roaring wilderness. In solitude they learnt how to find their place in the New World. In writing diaries, essays, autobiographies and chronicles, while seeking to find their inner sense of self-worth through the external activities of exploration of the land, they infused the land under their feet with special feelings. Therefore, unlike any literature from other countries, "American nature writing contains…the early and classic American's quest both for a coherent sense of self and a coherent sense of place" (Fritzell, p. 154).

## Lecture 4. Edwards and Bartram: A Sacred Landscape

In the eighteenth century, more than a century after Europeans first began to settle in the New World, most Protestants still viewed the natural environment as inherently dangerous. Although the wilderness and the forests signified freedom and liberation, they also meant getting lost and taking risks, as they were—after all—the domain of wild beasts and people who were considered by some to be monstrous savages. Under the influence of the century's literary mainstream, nature was still seen as rough, wild and uncivilized.

Jonathan Edwards (1703–1758), according to St. Armand, "might be considered the Mozart of American Puritanism, taking old forms and infusing them with a rhapsodic élan that threatened to break down the rigid dualistic barriers that heretofore had held firm between nature and culture, creature and creator, the world and the soul" (St. Armand 1997, p. 32).

In his autobiographical essay "Personal Narrative" (c. 1739–1742), Edwards departs from what was then the norm and boldly associates inner spiritual experiences with the natural features of the outside world, imbuing the natural landscape with human spirit. At the time, nature had taken on a different meaning for him: it was a landscape of the soul. He describes an unforgettable moment he experienced on his father's peaceful pasture as a boy: "And as I was walking there, and looking up on the sky and clouds; there came into my mind, a sweet sense of the glorious majesty and grace of God, that I know not how to express" (in Miller 1991, p. 343). Edwards would later call this his "Sixth Sense," a kind of supernatural sense like being bathed in the light of God. However, only God's chosen people—when in communion with nature—could feel this new sense and could, through this, commune with God. In Edwards's works we see the divinity of nature: if nature is a book, then God must be the author who wrote every letter in every page of this book. He says in "Personal Narrative" that "God's Excellency, His wisdom, His purity and love, seemed to appear in everything: in the sun, moon and stars; in the clouds and blue sky; in the grass, flowers, trees; in the water, and all nature" (p. 343).

In the journal he kept over the years which he referred to as "The Images of Divine Things," which was later edited by Perry Miller and published under the title of *The Images or Shadows of Divine Things* by Yale University Press in 1948, Edwards uses a metaphorical approach to illustrate that God made the physical world a shadow or representation of the spiritual world. He compares the rose to the "crown of glory"; wind and rain, light and warmth to "effusions of His Holy Spirit"; a river to the endless flow of God's love; trees on the river bank to the saints who have been nurtured by the Holy Spirit; and the occasional shadow of nature, the crow—manifestation of the Devil who feeds on the soul of the dead (in Perry Miller 1948).

Of course, Edwards's works had both bright and dark themes. Both "Personal Narrative" and *The Images or Shadows of Divine Things*, in which he worships nature like a child and describes a vivid and beautiful natural world, show a bright side. Some of his other works, such as his sermon "Sinners in the Hands of an Angry God" (1741), show a darker side as he talks about terrifying topics such as calls of death and the fire from hell. Although the themes of his works alternate between bright and dark, when he wrote about nature and the soul his tone was predominantly joyous: even the widely feared phenomenon of thunder and lightning often elicited feelings of happiness and prompted him to ponder the greatness of God.

The way that Edwards boldly associated nature with God and the human soul, his miraculous imaginations of the natural world and his use of metaphors to describe nature has continued to influence later generations. It is not difficult for us to see Edwards's legacy, for example in Emerson's *Nature*, and in many other works of nature writing.

Unlike Edwards, John Bartram (1699–1777) was not a theologian. He is widely regarded as the "father of American botany" and was appointed as King's Botanist for the American colonies by King George III. Bartram founded the first internationally renowned botanic garden in North America, and was also one of the founding members of the American Philosophical Society and a friend of both Benjamin Franklin and Thomas Jefferson. He was very well read, and was even able to find a copy of Confucius's *Analects* in North America in 1757, although it was titled *The Morals of Confucius.*[1] According to his biographer Thomas P. Slaughter in *The Natures of John and William Bartram: Two Pioneering Naturalists, Father and Son, in the Wilderness of 18th-Century America* (1997), Bartram was not only aware of Confucius and his philosophies, but also wrote three pages of notes after reading the *Analects*. These notes did not copy Confucius's words, but showed that Bartram sought to understand Confucius from his own perspective and admired him for his peaceful and unshakeable moral character (p. 16). As a botanist and explorer, Bartram visited regions in the Catskill Mountains in Carolina and Florida to conduct extensive studies, collect plant specimens and explore the wonders of the natural world, his overall aim being to uncover the meaning of life. His botanic garden, his collection of plants, diaries and letters, and the inspirational and educational nature of his ways all contributed to him being the most beloved naturalist of his generation. He and his son William Bartram set the original standard for American nature writing, and their perspectives on this subject served as a guide for later generations of nature writers (in Ann Ronald *Words for the Wild*, 1987, p. 369).

John Bartram's line of work was studying nature. He wanted to conduct scientific explorations and to realize his self-worth in nature. He was a very accurate

1 The edition of *Analects* that Bartram read was probably the 1691 English translation published in London and commonly regarded as the first ever English translation. However, the American sinologist Jonathan D. Spence believed that the 1691 edition of *The Morals of Confucius* was simply a collection of Confucius's aphorisms on morality and cannot be regarded as a full translation of the *Analects*. Further details can be found in Spence's article "What Confucius Said"—a review of Simon Leys's translation of the *Analects* (The New York Books Review, 10 April 1997).

observer, and liked to watch the rejuvenation of nature through the passing of the seasons. The wonders of the natural world always surprised him, and this caused him to be forever interested in and curious about nature. For him, "Nature and Bible are both open books...can be read in two directions: down, as in the passages just quoted, as an expression of God's plan for us all, and up, as a means through which to know God. It was through the 'telescope' of nature that John saw 'God in his glory'" (Slaughter, pp. 58–63). Because of this, Bartram devoted his life to the observation and study of nature, working tirelessly to maintain his botanic garden, as his work in the garden was his explorations, his collection of specimens, and also his dream.

Bartram saw order, balance and similarities in the organic world of nature. In the connections, he found an intelligence that is shared by all animals and plants, believing that even plants were sentient and had a will to live. He once gave as example the story of a blooming daisy. When people approached to admire its flower, it would close up. In a letter to a friend, he wrote about animals, "They have love, hatred, fear, and revenge if not foreknowledge, thought, will, and desire, faculties of the soul, memory, and reason in a great degree and some in a greater than some of our species" (quoted in Slaughter, p. 65). It is because of his belief that all organisms of the natural world shared a kinship that Bartram adopted what was, at the time, a somewhat unique stance: anti-hunting. His idea that a form of intelligence is shared by all living organisms led him to romanticise nature, and this is commonly thought to be the origin of the Romantic and transcendental movements of the following century (p. 134).

John Bartram, however, viewed nature through two different perspectives: practical and aesthetic—he saw nature's uses and also its beauty. Of course, as the King's Botanist of the North American colonies he had to serve the empire, discover new plants, and provide useful information in order to facilitate economic growth in the colonies. With this social responsibility, Bartram delved into nature to discover things that would be of use to his fellow colonists. He made many trips into the wilderness, all closely associated with his interest in the practical use of nature. It is not difficult for us to find evidence of this in his diaries: "good soil measures 'three feet deep'…a grassy plain is as large as '20 acres'; a creek is 'big enough to turn two mills'" (quoted in Slaughter, p. 12). Bartram saw the practical value of soil, grasslands and creeks, which reflected the classic view of nature in the eighteenth century. While discovering the practical uses of nature, he also appreciated the beauty all around him: pines are "lofty and noble," creeks and valleys are "pleasant," weather "truly charming," mountain ranges "lovely," and wolves' howling "musical." In appreciating nature, as Slaughter explains, he used his five senses to the full to experience the spectrum of natural colors and forms. Taking just fresh flowers as an example, "some glitter like a silver lace, and others shine as if embroidered with gold." Some are emerald green, others deep violet, others a pure white from top to bottom, and still others scarlet. In terms of shape, some are like fantastic cups, others overhanging bells, others small open umbrellas, and still others like twinkling star clusters. He would dig up the earth and examine the soil layers. He would immerse his hand in streams to feel the temperature of the water, and in one creek deemed it "loathsome and warm, but not so hot as one's blood" (quoted in Slaughter, pp. 67–75).

Unlike people today, Bartram did not feel a conflict between nature's practical value and its aesthetic value. The ideas of a "pleasant" creek that is "big enough to turn two mills" and "lofty and noble" pine and cypress trees that can be cut to become timber sat harmoniously in his mind, each complementary to the other. His unique way of seeing and writing about nature through both practical and aesthetic perspectives challenged the traditional segregation of science and poetry and gave rise to a new literary style.

John Bartram is generally thought of as someone who observed nature through a scientific perspective and used simple prose to describe it. However, his writings about nature are often only summary observations of natural phenomena which lack finesse and emotional appeal. The consensus view at the time was best described by the words of one of Bartram's contemporaries: "John Bartram left the esthetic appreciation of the…wilderness to his son William" (quoted in Slaughter, p. 70).

## Lecture 5. Bartram Junior: Lonesome Pilgrimage

As the puritans of the New World cast their gaze towards God in the eighteenth century, John Bartram's son William Bartram (1739–1823) cast his gaze towards nature and undertook a pilgrimage to the forests. William was America's first native-born naturalist and botanist, while at the same time an artist and writer. Like Edwards, William had foresight and was not afraid to innovate. From 1773 to 1777, William Bartram traversed through the then wild south-eastern region of America to study its flora and fauna and observe its natural scenery and indigenous people. He recorded what he saw in his diary, and later used his diary entries as the basis of his renowned book about nature and his personal experiences, *Travels Through North and South Carolina* (1791), written in a romanticized style.

Buell (1996, pp. 63, 398), writes about William Bartram's "excitement by the sense that the country in which he finds himself really seems like a pastoral Eden" and describes him as "the first to achieve transcontinental literary fame." Lyon (1989, p. 24) feels that "When William Bartram published his *Travels* in 1791, he became the first American nature essayist."

*Travels Through North and South Carolina* not only describes the flora and fauna and natural scenery, but also expressed the author's philosophical musings as a man of nature: William Bartram felt that a kind of kinship existed between all living organisms, and that each was sentient and had a soul. The appeal of the book stems from the narrative skill of the writer, the richness of his imagination, and the poetic beauty of his language. It has been translated into many European languages including German, French and Dutch, which attracted the interest of Europeans in the Old World and had some influence on the development of European Romanticism. English Lake Poets William Wordsworth and Samuel T. Coleridge and French Romantic writer F. R. Chateaubriand all found inspiration in this book, with Coleridge calling it "a work of high merit every way." English author Thomas Carlyle was also highly complementary in a letter to Emerson dated 8 July 1951: "All American libraries ought to provide themselves with that kind of book; and keep them as a kind of biblical article" (quoted in Miller 1991, p. 531).

The influence of Bartram's *Travels Through North and South Carolina* in America was even more far reaching. At the time, the people of the New World needed to learn not only about themselves but also about the unfamiliar natural world. In coping with these double necessities, *Travels Through North and South Carolina* no doubt provided them with guidance. What's more, as a naturalist Bartram was the first to use the word "sublime" to describe nature that is characteristically American, and this served to distinguish the sublime beauty of American nature and the graceful beauty of the European continent. William Bartram was an educated and cultured person with a scientific mind who had also endured the harshness of living in the wilderness, having ventured beyond the frontier of civilization into the domain of the wilderness many times. He observed nature from a scientific perspective, experienced it with a passion and described it in a romanticized style, blending natural history, literature, science and spirituality together into one, making *Travels Through North and South Carolina* "a landmark accomplishment in American literature" and "widely recognized as both a scientific and a literary classic of the period" (Branch in Cheryll Glotfelty and Harold Fromm, eds. *The Ecocriticism Reader: Landmarks in Literary Ecology* 1996, p. 287).

William Bartram died on 22 July 1823, at the age of 84. Unlike others coming to the end of their lives, when death came he was not lying on his bed, but under an ash tree in the garden. There were no children around him (he never married), but he was surrounded by the familiar plants and flowers of his garden and passed away peacefully in the world of plants and animals that he loved. He was buried in a cemetery, although many would prefer that he had been buried under that ash tree where he would be able to watch over his plants, the river and his home, the place where his soul had taken root. Bartram had renounced fame and fortune throughout his life, and apart from accompanying his father to the south-eastern region of America in his youth and going back there several times on his own as an adult, he remained living in the house and gardens that his father had left him. Thus, he was able to undertake the spiritual work of observing nature and documenting his observations in peace.

Of course, William's love of nature and his keen sense of observation came from his father, and he also inherited his father's knowledge of nature and philosophies. John Bartram affectionately called William "my little botanist" when he was small (Slaughter, p. 112). From the age of 14, William started accompanying his father on field trips to places such as the Catskill Mountains to study plants. On these trips, he developed a keen eye for capturing and observing nature, and became obsessed with sketching birds, animals and plants (more than a century later the nature writer John Burroughs, another bird lover, also ventured into the Catskill Mountains—to bring the exquisite feathered kingdom to life in words). It was William's father who taught him to see the kinship between humanity and nature, and it was also his father who encouraged him to add color and infuse energy into his sketches. Bartram junior soon became known as an "admirable painter of plants," and his father in a letter to a friend wrote proudly, "My son William hath drawn most of our real species of oaks and all our real species of birches with an exact description of their particular characters, not according to grammar rule, or science, but nature" (quoted in Slaughter, p. 115). Later on, British Royal Society Fellow John Fothergill asked William to freely sketch the natural beauty of the New World and send the drawings to him. Both the London Natural History Museum and the American Philosophical Society hold collections of his drawings, some of which were even used as illustrations in the renowned English botanist George Edwards's momentous works *A Natural History of Uncommon Birds* (1743–1751) and *Gleanings of Natural History* (1758–1764).

At the time, however, one could not earn a living by just drawing birds and flowers. William's father and his friends (including Benjamin Franklin) came up with many different potential careers for William, hoping that he would become a merchant, farmer, sculptor or even doctor, but none of these held his interest and he remained "an unambitious lover of nature" (Slaughter, p. 256). William would probably not have guessed that a century later, following in his footsteps, another nature lover and life-long bachelor Henry David Thoreau would also be described as having "no ambition" and was given the title of "the captain of a huckleberry party" (Baym & Gottsman 1989, p. 1044). Both William Bartram and Henry Thoreau possessed the sensitivities of an artist, talents of a writer and soul of a poet, but they were not entirely suited to their time in history, a time when only people of great worldly achievement were respected. Bartram and Thoreau were men of refinement who were dedicated to the pursuit of inner spiritual fulfilment. William's life journey came to an end under a tree in his beloved garden, while Thoreau uttered the two dying words: "Moose" and "Indians" (Channing 1873, p. 319). Both men spent their lives in search of a kind of inner peace, kinship with nature and personal spiritual enjoyment, rather than the fame which worldly people coveted. This is perhaps the root of their success, and it is perhaps precisely their solitary lives that led to the recognition and great respect they both received after their deaths.

Although William Bartram's interest in nature was fostered by his father, who taught him how to observe nature and passed on a wealth of knowledge to him, there were obvious differences in their ways of life and their styles of writing. As a subject of the British crown, John embraced the traditional notions of fame and fortune, and had an insatiable desire to succeed; whereas William chose a simpler, quieter life and paid little regard to worldly affairs. He turned down a professorship in botany at the University of Pennsylvania and declined an appointment as a naturalist on Thomas Jefferson's Red River expedition, preferring total seclusion in his garden, drawing, writing and communing with nature. John was a very rational and pragmatic man who was very much in tune with the times in which he lived, whereas William acted according to his feelings despite living in a world in which rationality dominated. John harboured hatred and hostility toward the native Americans (who were said to have killed his father), while William respected and admired them, as is evident in his *Travels Through North and South Carolina.* Both father and son were very fond of nature but expressed this in different ways. John's works showcased his blueprint for developing the wilderness, whereas William would describe every one of his exciting discoveries in the untouched wilderness. Most of John's writings of nature can be found in his published diaries and letters, which mainly document specific natural phenomena, data and the practical value of nature. In contrast, William's writings and drawings are infused with his emotions and imagination, and evoke Gothic and Romantic images. European Romantic writer Charles B. Brown and American novelist James F. Cooper both found inspiration and creative background knowledge in William's *Travels Through North and South Carolina*. Writers such as Emerson and Thoreau learned from William's philosophies on nature and from his writing style, which helped them to create their own famous works on humanity and nature.

William Bartram's reputation as "the first American nature essayist" is also due to the importance of his book *Travels Through North and South Carolina* in laying the foundations for the establishment of the American nature essay as a new literary form. His thoughts and opinions about the connection between humanity and nature, the kinship and equality between nature and humanity, the characteristics of American nature, and the duality of beauty and terror in nature were far-reaching. They would later become topics of study, and form the thematic basis of nature writing.

In addition to the depiction of the wondrous sights of a "primitive" land and its hitherto unknown flora and fauna, what impresses people most about *Travels Through North and South Carolina* is the richness of the author's spiritual experiences and the dreamlike feelings he evokes: Bartram gently floated from one scenic spot to another, welcoming each wonderful surprise with passion, transforming natural landscapes into spiritual landscapes. Just strolling through the forest, watching the flight of a butterfly, hearing the buzz of an insect, witnessing the torrents of a river or capturing a wondrous sight of the ocean would elicit a spiritual awakening within him: "O thou Creator supreme, almighty! How infinite and incomprehensible thy works! Most perfect, and every way astonishing!" (Bartram 1996, p. 71) Unlike his father's explorations, which served the interests of the British Empire, William's travels through America's Southeast from 1773 to 1777 served not to aid European settlers to control this land, but as a spiritual journey, a spiritual homage to Mother Nature. He completely immersed himself in "the magnificent forest" and the "solemn shades." When he hiked into the mountains and paid homage to the "god of the forest" we can feel his sincere belief in nature's innate powers of creation and his reverence for the supreme creator. When he was in the mountains, we can feel his awe and wonder at the spectacles of the setting sun and his praise of it as a "minister of the Most High in the rule and government of this earth" as he watched "with gratitude his departing smiles, tinging the fleecy roseate clouds, now riding far away on the eastern horizon; behold they vanish from sight in the azure skies" (Bartram, p. 144).

William Bartram chose to dedicate his life to nature, not for fame or profit, but to lose himself in nature and then rediscover himself. For him, the closer we are to nature, the closer we come to realizing self-fulfilment. For him, self-fulfilment did not mean achieving some common notion of success or accomplishment, but a kind of inner self-realization, a journey to spiritual peace, a spiritual state akin to witnessing nature's sunset.

> It was drawing on towards the close of day, the skies serene and calm, the air temperately cool, and gentle zephyrs breathing through the fragrant pines; the prospect around enchantingly varied and beautiful; endless green savannas, chequered with coppices of fragrant shrubs, filled the air with the richest perfume. (Bartram, p. 43)

This descriptive section from *Travels Through North and South Carolina* reveals the uniqueness of Bartram's pursuits. His fulfilment comes from simple spiritual and natural experiences, such as watching a sunset or the flight of a bird, or the growth of plants and trees. This is because he believed that: "The works of a person that builds begin immediately to decay; while those of him who plants begin directly to improve. In this planting promises a more lasting pleasure than building" (in Slaughter, p. 248).

As Bartram's journey in America's Southeast was drawing to its end, what he was seeking was no longer the exploration and love of nature, but inner peace. After this journey and until his death, Bartram almost never left the home and garden surrounded by nature that had been left to him by his father. While working in the garden alone, Bartram sought a kind of spiritual state that would allow his mind to wander freely. For him, solitude was not just a preference, but a necessity that would lead to inner peace and happiness. The connection he established between nature and human spirituality pointed the way for American literature to find spiritual inspiration in the wilderness, and also influenced subsequent generations of nature writers such as David Thoreau and Barry Lopez. They later put forward similar views like "I went to the woods because I wished to live deliberately, to front only the essential facts of life" (Thoreau, *Walden* 1910, p. 118) and "I think of two landscapes—one outside the self, the other within" (Lopez 1989, p. 64).

Branch believes that "Bartram's appreciation for the wonderful intricacy of natural systems and his belief that every living thing manifests 'the divine and inimitable workmanship' combined to produce a sensibility that may be described as proto-ecological" (1996, p. 288). Throughout *Travels Through North and South Carolina* the book shows the interrelationship between nature and humanity that Bartram identified through his observations in the wilderness. He believed that the natural world existed as a single entity, filled with the spirit of the creator—that the spirit of God exists in all natural things, and that this connected every living organism. Bartram's depiction of the Altamaha River and the scenery along its riverbanks presents such an image of all things being interdependent and living in cooperation:

> How gently flow thy peaceful floods, O Alatamaha! How sublimely rise to view, on thy elevated shores, yon Magnolian groves, from whose tops the surrounding expanse is perfumed, by clouds of incense, blended with the exhaling balm of the Liquid-amber, and odours continually arising from circumambient aromatic groves of Illicium, Myrica, Laurus, and Bignonia. …the peaceful stream, on whose polished surface were depicted the mutable shadows from its pensile banks; whilst myriads of finny inhabitants sported in its pellucid floods. (Bartram, p. 63)

Of all the natural things, what first piqued William Bartram's interest was the plant kingdom. As he explains, "Perhaps there is not any part of creation, within the reach of our observations, which exhibits a more glorious display of the Almighty hand, than the vegetable world" (Bartram, p. 13). This, he believed, is because plants follow changes in the seasons by transforming into a variety of different forms and colors to form countless numbers of visual spectacles. He praised the stately dignity of the Florida palm, the harmonious beauty and gracefulness of the Carica papaya, the vivacious loveliness of *Kalmia latifolia* and *Azalea coccinea*, and the beauty and fragrance of *Crinum floridanum*. The flower-speckled meadows and valleys were where he longed to be. The Native Americans called him the *Puc-Puggy* (Flower Hunter) because he was forever engrossed in the collection of all sorts of flower specimens. Even in his last letter to his sweetheart, at the end of his only love affair, his parting words were: "Fresh blown flowers are and ever have been the symbol of joy" (quoted in Slaughter, p. 165).

Bartram's passion for plants stemmed from his belief that they were sentient, like humans. In the introduction to *Travels Through North and South Carolina*, he used the insect-catching properties of the yellow pitcher plant and the Venus flytrap to illustrate his claim that plants are gifted with human-like senses and attributes and are organic, living and self-moving bodies, for in certain plants we can see motion and volition. He then presented evidence that animals also possessed emotions and described a scene he witnessed during his journey: after a female bear was shot, a cub approached the dead body. It sniffed and touched the slain bear, and with an agonizing expression "cried out like a child" (quoted in Slaughter, p. 20). When observing the sentient conduct of plants and the emotions of the animals, Bartram realized that he was not only a witness of the natural order and balance, but also part of the natural world: he was keenly aware that when he was watching a spider with excitement, the spider was also watching him.

William Bartram did not have as much faith in science, commerce and humanity as did his father John, and he had concerns for the future. "Though admitting that human inventions, arts and science to be a part in the progress of nature, yet [they] are perpetually productive of innovations, and events, that show the defects of human policy" (quoted in Slaughter, p. 184). He preferred that nature would remain unspoilt and continue to be the realm of the plant kingdom, hoping that mankind's violent hand would never disturb the freshness of nature. Bartram's thoughts and views were different from those of his contemporaries and more in line with those of nature writers toward the end of the nineteenth and throughout the twentieth century, who opposed anthropocentric views and wished to place plants and animals on an equal footing with humanity. John Muir would later call trees his "botanical brethren," Mary Austin would view the world as a net that humanity and nature wove together, and Aldo Leopold would propose his theory of "land ethic" and call for people to cultivate an ecological conscience. It is not difficult to see Bartram's influence on these later writers' thoughts and views, and it is also very hard not to admire his visions and foresight from two centuries ago.

Before Bartram, botanists were often too engrossed in their studies to pay more than cursory attention to the wilderness. Bartram was different. Although he made detailed descriptions of the fauna and flora of the American Southeast and listed 215 species of birds in *Travels Through North and South Carolina*, he was so deeply fascinated by the wilderness as a whole that he often did not notice the existence of a particular plant specimen. On one occasion, in the mountains of northern Georgia, his imagination was so "wholly engaged in the contemplation of this magnificent landscape" that he "was almost insensible…of…a new species of Rhododendron" (quoted in Nash 1968, p. 54). Nash believed that "What made William Bartram forget the rhododendron and rejoice in wilderness was its sublimity. His descriptions mark the first extensive use of that term in American letters" (1968, p. 54).

The Merriam-Webster dictionary defines the word "sublime" as: (1) lofty, grand or exalted and (2) tending to inspire awe usually because of elevated quality. Edmund Burke, William Bartram's English contemporary, explains in his book *Philosophical Enquiry into the Origin of Our Ideas of the Sublime and*

*Beautiful* (1757) that "the sublime, specifying the strongest emotion of which the mind is capable, should be ranged in a category coordinated with the beautiful." This "strongest emotion" was close to terror because he believed that "the ideas of pain are much more powerful than those which enter on the part of pleasure." Therefore, among the properties of the sublime he lists "objects of great dimension," "vastness" and "darkness" (quoted in Hans Huth's *Nature and the American*, 1957, p. 11). On some level, these properties accurately describe the natural landscape of early America and the astonishment of those who witnessed its beauty. In using these terms of expression, Bartram artfully defines the unique natural and spiritual American landscapes. It is precisely from America's natural landscape that he validates Burke's definition of the sublime: "astonishment… is the effect of the sublime in its highest degree" (Scheese, *Nature Writing*, p. 1).

Unlike people in Europe or other parts of what was then thought of as "the civilized world," when early American settlers went to nature, they encountered a wilderness that was almost devoid of human presence and they had no home of civilization to return to. Faced with seemingly virgin and desolate mountains, rivers and plains, they felt a solemn inner melancholy and a sense of astonishing beauty. Bartram called his journey to the American Southeast from 1773 to 1777 a "lonesome pilgrimage," and his book *Travels Through North and South Carolina* was regarded as "an ode to unspoiled natural beauty" (Slaughter, p. 178). Throughout his book, the New World is bathed in the sublime beauty of rays of divine light. For him, the sublime beauty of nature and the majesty of the creator are intertwined, because he believes that the wisdom and power of God, the "supreme author of nature," are evident throughout the wilderness. Bartram confesses that he is "seduced by these sublime enchanting scenes of primitive nature." In the wilderness of the Carolinas, he "beheld with rapture and astonishment a sublimely awful scene of power and magnificence, a world of mountains piled upon mountains" (Bartram, pp. 106, 295). When he was on the Altamaha River, he was overwhelmed by the sight of the lovely bushes that lined the riverbanks, the pleasant meadows, and the tall forest in the distance as they "unfolded fresh scenes of grandeur and sublimity." He called the Isle of Palms a "delightful spot, planted by nature…[a] blessed unviolated spot of earth," and described groves that resonated with the "divine hymns of the feathered songsters" (pp. 63–65).

We should not forget that Bartram's journey was a "lonesome pilgrimage."

> At midnight I awake; when, raising my head erect, I find myself alone in the wilderness of Florida, on the shores of Lake George. Alone indeed, but under the

> care of the Almighty, and protected by the invisible hand of my guardian angel. (Bartram, p. 144)

In describing his experience of a storm on the bank of the Altamaha River, he paints a sublime picture of nature:

> The livid purple clouds thickened on the frowning brows of the morning; the tumultuous winds from the east, now exerted their power. O peaceful Alatamaha! Gentle by nature! How thou wert ruffled! Thy wavy surface disfigured every object, presenting them obscurely to the sight, and they at length totally disappeared, whilst the furious winds and sweeping rains bent the lofty groves, and prostrated the quaking grass, driving the affrighted creatures to their dens and caverns. (Bartram, p. 65)

Bartram's views of nature which link the American natural world with the concept of the sublime also help to explain the artistic style of a school of American landscape paintings—the Hudson River School, founded by Thomas Cole. It should be pointed out that this art movement had a close relationship with American nature writing.

Although *Travels Through North and South Carolina* was regarded as "An ode to unspoiled natural beauty" (Slaughter, p. 178), Bartram did not present a view of nature that was completely idealized and romanticized; he understood that it had two sides. As he appreciated the tranquil summer's day with its blue sky and plains dotted with flowers, he was also naturally aware of the ruthlessness and harshness of nature and accepted its violent and cruel side. In him rationalism and romanticism coexisted, and in his works the harmony and order of nature appears alternately with its heartlessness and cruelty. In Chapter 6 of *Travels Through North and South Carolina* where Bartram describes the "paradise of fish," he presents a picture of perfect harmony: with the fragrant air filled with the singing of merry birds, with reflections of the fruit trees and flowering trees in the crystal water, innumerable fish of all kinds swim around freely, each politely making room to let others pass by. The tree-trunk-like crocodile, the greedy garfish, the hostile trout, the barbed catfish, all seem to live peacefully side by side, with "no sign of enmity" or any "attempt to devour each other," "no covert, no ambush; here the trout freely passes by the very nose of the alligator, and laughs in his face, and the bream by the trout" (Bartram, pp. 150–151). What surprised Bartram was that the river creatures being "consciousness of each other's safety, or some other latent cause, should so absolutely alter their conduct, for here is not the least attempt made to injure or disturb one another" (Bartram, p.151).

In contrast, in the part about "the alligators," Bartram paints a completely different picture.

> I have seen an alligator take up out of the water several great fish at a time, and just squeeze them betwixt his jaws, while the tails of the great trout flapped about his eyes and lips, ere he had swallowed them. The horrid noise of the closing jaws, their plunging amidst the broken banks of fish, and rising with their prey some feet upright above the water, the floods of water and blood rushing out of their mouths, and the clouds of vapour issuing from their wide nostrils, were truly frightful. (Bartram, p. 118)

Here Bartram presents the dark side of nature, filled with terrifying images. At the same time, he also points out nature's cruel order: it is a world in which the strong prey on the weak, you are either predator or victim, and you either conquer or be conquered. From this, we can appreciate the complexity and mysteriousness of the natural world. In the 1970s, the American nature writer Annie Dillard painted an almost identical picture of nature in her most famous work *Pilgrim at Tinker Creek*, the only difference being that her point of view and style of writing had more of a modern flair.

William Bartram's accomplishments were not limited to his renowned book *Travels Through North and South Carolina*; he also became a mentor for the next generation of naturalists in his later life. From 1777 until he died 46 years later, Bartram did not leave the city of Philadelphia and his life revolved about the eight-acre botanic garden left by his father. It was in that botanic garden that he received two young European immigrants—Alexander Wilson from Scotland and Thomas Nuttall from England—and became their mentor. These two individuals would later become famous figures in American natural history and would exert considerable influence upon American nature writing, with Wilson considered Bartram's most accomplished student.

## Lecture 6. Wilson: A Eulogy to Birds

Alexander Wilson (1766–1813) was born in Scotland. He was a weaver and a poet, and after migrating to America in 1794, he became immensely interested in the wildlife of his adopted country. He soon started venturing into the forests of Pennsylvania to observe and paint the birds there, and when he took up a teaching post in a school near Bartram's garden, he met Bartram and decided to take his instruction and guidance in natural history and painting. Wilson's decision was wise, as his tutor was not only a great painter who taught him painting techniques, but also an experienced naturalist who taught him the ways of nature, ways to observe and appreciate nature, and ways to see from nature's perspective and to grasp what was truly important in the world. This is reflected in a letter Wilson wrote to Bartram sometime later:

> I confess that I was always an enthusiast in my admiration of the rural scenery of Nature; but, since your example and encouragement have set me to attempt to imitate her productions, I see new beauties in every bird, plant, or flower, I contemplate.... I sometimes smile to think that while others are immersed in deep schemes of speculation and aggrandizement—in building towns and purchasing plantations, I am entranced in contemplation over the plumage of a lark, or gazing like a despairing lover, on the lineaments of an owl.... While others are hoarding up their bags of money, without the power of enjoying it, I am collecting, without injuring my conscience, or wounding my peace of mind, those beautiful specimens of Nature's works that are for ever pleasing. (quoted in Slaughter, pp. 252–253)

Dillard, in *Pilgrim at Tinker Creek*, invents a romantic fable explaining the origin of a species of bird found in America—the starling. In her fable, starlings were introduced into America from Europe by a rich New York businessman whose curious hobby was to introduce into America all the birds mentioned by William Shakespeare (p. 35). Wilson's interest in birds was not just a hobby. Under Bartram's influence, his interest in the birds of his adopted land became his life passion and pursuit. Like his tutor, Wilson viewed the study of natural history as paying homage to God and his ornithological studies as a devout mission. One year after meeting Bartram, Wilson proclaims that "I am now about to make a collection of all our finest birds" (quoted in Lyon 1989, p. 41).

Wilson soon started venturing further into the American wilderness, traveling over 10,000 miles in search of new species of birds. He despised the "indoor naturalists and armchair travelers," claiming that he only painted birds living in their natural habitat. He firmly believed that his study of birds was inspired by God, and his explorations and research would therefore inspire American culture. He combined his scientific and literary talents in order to record the national treasure of America's birds. When he died, 11 years after meeting Bartram, he left behind a rich spiritual legacy: nine volumes of *American Ornithology*. This was the first set of books about birds in America, and its publication made Wilson America's "foremost authority on birds" (Branch 1996, p. 289). *American Ornithology* is regarded as "Wilson's efforts toward an emotionally resonant yet empirically accurate prose, a style that would in effect poeticize science represent.… His ten year book amounted to a hymn of love to America nature. It should be regarded as an important American document" (Lyon 1989, pp. 42–43).

Wilson's love of the wilderness, his poetic insight and penmanship enabled him to capture the characteristics of wild animals and plants. His experiences of making detailed observations in the wild also moulded his writing style to create a precise style of essay, a literary form that combined poeticism with science, similar to but surpassing the literary style of Bartram. In Wilson's writings, his descriptions of birds resemble paintings. Take that of the Ivory-billed Woodpecker as an example:

> He may be called the king or chief of his tribe; and nature seems to have designed him a distinguished characteristic in the superb carmine crest and bill of polished ivory with which she has ornamented him. His eye is brilliant and daring; and his whole frame so admirably adapted for his mode of life, and method of procuring subsistence, as to impress on the mind of the examiner the most reverential ideas of the Creator. His manners have also a dignity in them superior to the common herd of woodpeckers. Trees, shrubbery, orchards, rails, fence posts, and old prostrate logs, are alike interesting to those, in the humble and indefatigable search for prey; but the royal hunter now before us, scorns the humility of such situations, and seeks the most towering trees of the forest; seeming particularly attached to those prodigious cypress swamps, whose crowded giant sons stretch their bare and blasted or moss-hung arms midway to the skies. In these almost inaccessible recesses, amid ruinous piles of impending timber, his trumpet-like note and loud strokes resound through the solitary savage wilds, of which he seems the sole lord and inhabitant. (quoted in Finch & Elder 1990, p. 78).

In the above text, Wilson's words not only paint a vivid image of nature, but also contain rich connotations. From it we can appreciate the sanctity of nature in the majesty and ingenuity of the Creator; the similitude of animals and humans from the woodpecker's conscious choice of habitat; and the awe-inspiring sublime beauty of American nature, from the sound of the woodpecker resounding through the solitary forest. Could this sublime image of nature conjure up another image in people's minds—is the implacable woodpecker relentless searching for food in the wilderness not a reflection of explorers in the New World? Is "sole lord" not an apt description of the early American settlers, proud of their chosen path and yet saddened by their spiritual solitude?

In *American Ornithology*, we can also find evidence of Wilson's initial efforts to find America's cultural identity in the land he explored, and to protect the natural environment. Like his predecessors Thomas Jefferson and William Bartram, Wilson believed that his natural history research did not only contribute to science, but also to America's cultural identity. In the words of the historian Robert Elman, Wilson expressed a "somewhat mystical belief…that the living riches of America's wilderness formed a common heritage—a kind of unifying fabric—linking all the peoples of diverse ancestry and background to a single destiny in a young, vigorous nation" (quoted in Branch 1996, p. 290). Wilson also once promised that if "the generous hand of patriotism be stretched forth to assist and cherish the rising arts and literature of our country…[they will] increase and flourish with a vigor, a splendor and usefulness inferior to no other on earth" (quoted in Branch 1996, p. 290). Wilson also tried to anthropomorphize the birds that he found in the wild, relating them to poets and American cultural identity. He once described several varieties of wood thrushes, observing in detail their birdsong, birdcall, habits and habitats. Such efforts also influenced a later naturalist—the renowned ornithologist John J. Audubon.

In Wilson's works, we can see that he was a budding environmentalist in his criticism of humanity's efforts to conquer the natural world. He conducted pioneering studies into cataloguing varieties and numbers of wild creatures and foresaw that human habitation would decimate the breeding grounds of certain species. As Wilson adored the song of each and every bird, he deeply deplored anyone who would harm or hunt them. He was deeply hurt and saddened by the untimely death of birds around his home as "one of the victims was a cardinal he knew well: it would come and sing at his window every morning" (quoted in Finch and Elder 1990, p. 78).

From the works of the important figures of the early colonial period and the eighteenth century, it is easy for us to see the foundations from which American nature writing emerged. In finding ways to survive in the wilderness, learn about the land and understand themselves, it was inevitable that the ways of the seventeenth century settlers, so intimately linked with nature, would be reflected in early American literary works. It can be said that the depictions of the New World and its natural environment by early European settlers such as Smith and Bradford cemented a new literary genre that was themed upon nature and written in essay form. Going into the eighteenth century, the settlers began to look at the natural environment around them in two different ways: how to make use of it, and how to appreciate and understand it. Writers such as Jonathan Edwards and John and William Bartram combined their personal experiences with the study of nature and brought together science and literature to create unique nature essays. In the words of Lyon, it "is probably not coincidental that the nature essay developed as a distinct genre only toward the close of the 18th century," and the publication of William Bartram's *Travels Through North and South Carolina* marked "the first fully developed nature essay in this country" (Lyon 1989, pp. 19, 38).

Although the seventeenth and eighteenth centuries witnessed the appearance of writers who defined the theme, form and style of nature writing, those who would define its philosophical ideas would not emerge until the following century. This is because it was only in the nineteenth century that people began to seriously consider ways to find the origins of American spirit and culture. Works such as Cole's "Essay on American Scenery" and Emerson's *Nature* (both published in 1836) show a new kind of passion that people at the time had developed for the land.

## References

Bartram, William. (1791) 1996. *Travels and Other Writings*. New York: Library of America.

Baym, Nina, and Ronald Gottsman, eds. 1989. *The Norton Anthology of American Literature*. 3rd ed. Vol. I. New York: Norton.

Branch, Michael. 1996. "Indexing American Possibilities: The Natural History Writings of Bartram, Wilson and Audubon." In Cheryll Glotfelty and Harold Fromm, eds. *The Ecocriticism Reader: Landmarks in Literary Ecology*. Athens, GA: University of Georgia Press.

Buell, Lawrence. 1996. *The Environmental Imagination: Thoreau, Nature Writing, and the Formation of American Culture*. Second printing. Cambridge, MA: Harvard University Press.

Channing, William Ellery. 1873. *Thoreau, The Poet-Naturalist: With Memorial Verses*. Boston: Roberts Brothers.

Finch, Robert, and John Elder, eds. 1990. *The Norton Book of Nature Writing*. New York: Norton.

Fritzell, Peter A. 1990. *Nature Writing and America: Essays upon a Cultural Type*. Iowa City, IA: Iowa State University Press.

Huth, Hans. 1957. *Nature and the American: Three Centuries of Changing Attitudes*. Berkeley: University of California Press.

Lopez, B. 1989. *Crossing Open Ground*. London: Picador.

Lyon, Thomas J., ed. 1989. *This Incomperable Lande: A Book of American Nature Writing*. Boston: Houghton Mifflin Company.

Matthiessen, F. O. 1941. *American Renaissance: Art and Expression in the Age of Emerson and Whitman*. Oxford: Oxford University Press.

Miller, James E., Jr., ed. 1991. *Heritage of American Literature: Beginnings to the Civil War*. Vol. I. San Diego, CA: Harcourt Brace Jovanovich.

Miller, Perry, ed. 1948. *Images or Shadows of Divine Things by Jonathan Edwards*. Conn: Yale University Press.
Miller, Perry. 1967. *Nature's Nation*. Cambridge, MA: Belknap Press of Harvard University Press.
Nash, Roderick. 1968. *Wilderness and the American Mind*. Second printing. New Haven, CT: Yale University Press.
Ronald, Ann, ed. 1987. *Words for the Wild*. San Francisco: Sierra Club Books.
Ruland, Richard, and Malcolm Bradbury. 1991. *From Puritanism to Postmodernism: A History of American Literature*. New York: Penguin Books.
Slaughter, Thomas P. 1997. *The Natures of John and William Bartram: Two Pioneering Naturalists, Father and Son, in the Wilderness of 18th-Century America*. New York: First Vintage Books.
St. Armand, Barton Levi. 1997. "The Book of Nature and American Nature Writing: Codex, Index, Contexts, Prospects." *ISLE* 4.1 (1997): 29–42.

3

# The Influence of Transcendentalism on Nature Writing

## Introduction

For America, the nineteenth century was a time of glory, whether from a historic or a literary perspective. As society developed, people's attitudes toward nature also underwent important changes. The imagery that associated the wilderness with the devil depicted by early Puritans began to fade, along with the eighteenth-century perception that nature was uncivilized and dangerous. People began to see the natural world as their spiritual temple, giving rise to an unprecedented optimism and an eagerness to experience nature. According to Buell (1996, p. 56), during this period "the link between American cultural identity and exurban and preindustrial spaces became one of the enabling myths of American literary nationalism. The American environment became one of its most distinctive cultural resources." This sentiment is shared by Nash (1967, p. 67): "Americans sought something uniquely 'American,' yet valuable enough to transform embarrassed provincials into proud and confident citizens.... But in at least one respect Americans sensed that their country was different: wilderness had no counterpart in the Old World."

This period also marked the birth of transcendentalism, a movement that had far-reaching effects on American culture and ideology. Those who studied transcendentalism understood and defined the movement in different ways. Octavius Brooks Frothingham, author of *Transcendentalism in New England: A History* (1972, pp. 136, 134), saw it as a "distinct philosophical system" and "essentially poetical." Perry Miller, editor of *The Transcendentalists: An Anthology* (1971, p. 8), believed that "the Transcendental movement is most accurately to be defined as a religious demonstration."[1] Donald Koster, in *Transcendentalism in America* (1975, p. 2), reports that H. C. Goddard considered transcendentalism to be "a literary movement, a philosophy and a religion all in one" while as a scholar of transcendentalism himself, Koster (p. 25) summarizes it thus: "Whatever else the transcendentalists were, they were persons of an intellectual, an artistic, a literary inclination." It is reasonable to say that at the time transcendentalism was a contentious movement,[2] and that it is still defined and interpreted in a variety of ways even today. In *The Republic*, Plato mentions "an ancient quarrel between philosophy and poetry" ("Book 10"). Literature and philosophy are a contradictory pair, both inseparable and difficult to delineate, linked by an inseparable, logical and yet irrational connection. For the purpose of this book, I would like to use the definition of "Transcendentalism in America" from Mayer Howard Abrams's *A Glossary of Literary Terms* (2009) as a basis for a simple discussion on transcendentalism and its influence on nature writing. The term is defined as: "A philosophical and literary movement, centered in Concord and Boston, which was prominent in the intellectual and cultural life of New England from 1836 until just before the Civil War….Transcendentalism was neither a

1 It is worth mentioning that in the introduction to the 1959 edition of Frothingham's *Transcendentalism in New England*, Yale professor Sydney E. Ahlstrom gives high praise to both Frothingham's work and Miller's *The Transcendentalists: An Anthology*, describing the latter as voluminous and of extraordinary value. He makes it clear that apart from Miller, Frothingham—with his detailed descriptions of transcendentalism's philosophical origins, ideological purpose, status and contributions, and also its important figures—is the only other writer whose work comprehensively covers the whole of the transcendentalist movement. As this book is mainly about American nature writing, and this chapter a discussion on the influence of transcendentalism on nature writing and not transcendentalism itself, I have not described the movement in detail.

2 At the time, the man leading the attack on the transcendentalist movement was Harvard Divinity School Professor Andrew Norton. For details see Miller, *The Transcendentalists*, p. 6 and Frothingham, *Transcendentalism in New England*, pp. 123–125.

systematic nor a sharply definable philosophy, but rather an intellectual mode and emotional mood that was expressed by diverse, and in some instances rather eccentric, voices" (Abrams, pp. 375–376). Modern historians of the movement tend to take Emerson and Thoreau as its central exponents and take Emerson's *Nature* and lectures such as "The American Scholar," and Thoreau's *Walden* and his journals as the texts of the focus of their study. These historians believe that Emerson's transcendentalism was influenced by German philosopher Immanuel Kant, English thinkers Samuel Taylor Coleridge and Thomas Carlyle, Plato and Neoplatonists, Swedish theologian Emanuel Swedenborg and various Asian philosophies (Abrams, p. 376). Most scholars believe that the term "transcendentalism" came from Kant's expression "transcendental knowledge" but before Kant there was Descartes's "rationalism" in the seventeenth century and Locke's "empiricism" that spanned the seventeenth and eighteenth. To a certain extent, Kant believed that knowledge is gleaned from experience, but he also stated that not all knowledge has to be derived empirically. As intuition is part of human experience, which determines human knowledge, "apriorism" is the natural conclusion. Emerson's transcendentalism and Kant's apriorism are similar in their emphasis on the strength of the mind and the belief that humans are intuitively capable of independent thought. In fact, as one of Emerson's biographers Robert Richardson quotes in *Emerson* (1995, p. 351), Emerson once said: "The one thing of value in the universe is the active soul." The essence of transcendentalism is deeply connected with nature. On this subject, Richardson also mentions that a trait clearly shared by American transcendentalism and German apriorism is their emphasis on a connection between the soul and nature (Richardson, p. 199). However, Emerson was not constrained by apriorism. He extended the concept of transcendental knowledge, believing it to include an intuitive cognizance of moral and other truths that transcend the limits of human sense-experience. Emerson's transcendentalism affirmed "an ethics of individualism that stressed self-trust, self-reliance, and self-sufficiency; a turn away from modern society, with its getting and spending, to the scenes and objects of the natural world, which were regarded both as physical facts and as correspondences to aspects of the human spirit." It also had "faith in a divine 'Principle,' or 'Spirit,' or 'Soul' in which both humanity and the cosmos participate." Emerson said that this omnipresent spirit "constitutes the 'Unity within which every man's particular being is contained and made one with all other'" (Abrams, pp. 653–655).

In 1842, six years after the founding of the Transcendental Club, Emerson delivered his lecture "The Transcendentalist." This signified that in America, patriotism had become the more inclusive transcendentalism, and it had led to the appearance of transcendentalists. Groups of people who wanted to practice transcendentalism formed and created communities such as Brook Farm (1841–1847). These transcendentalists wished to be away from the crowd and the noise in order to be close to nature and to live a way of life that is self-fulfilling, self-sufficient and mutually beneficial within their community. At the time this was in stark contrast to the ways of commercialism and materialism that were becoming widespread as a result of the acceleration of industrialization in America. Seeing the continuous decline of values and social morals, these transcendentalists felt they could no longer allow their minds to be corrupted. They wanted to understand the power of nature, to rediscover themselves in nature and live in harmony with their surrounding environment. Although Brook Farm ended in failure, another transcendentalist, Thoreau, demonstrated the values of transcendentalism through his personal experiences at Walden Pond. American literary historian Robert E. Spiller, in his *The Cycle of American Literature* (1967, p. 66), writes: "In Thoreau the affirmative argument had reached the stage of demonstration; the spirit of independence had become a way of life." Therefore, although Emerson's transcendentalism followed the tradition of Kant's apriorism, it had a broader meaning in the New World, and he was able to transform it into a concept that could be put into practice. Frothingham points out that in Germany and France the influence of apriorism is limited within schools and among scholars and has never gone beyond philosophy, literature and art while in New English the seed of this European idea has grown locally and blossomed in various forms of practical social life (1972, p. 105). Joel Myerson further explained, "Transcendentalism is not a concept that operates in isolation from the rest of its world; it is not navel gazing. There is a strong sense of social commitment in Transcendentalism" (Myerson 2000, p. xxxiv). Although some ideas and practices of transcendentalism were controversial at the time, Emerson placed his faith in the next generation. He firmly believed that transcendentalism would retain its beauty and strength, and that transcendentalists would "reorganize themselves in nature, to invest themselves anew in other, perhaps higher endowed and happier mixed clay than ours, in fuller union with the surrounding system" (Emerson, Centenary Edition, *The Complete Works of Ralph Waldo Emerson*. Vol. I, [1903] 1979, p. 227). As nature writing soared in popularity towards the end of the last century, it was clear that the nature writers were fulfilling Emerson's prophecy.

In looking back at the origins of transcendentalism in America it is not difficult to see that ideologically, transcendentalism and nature writing emerged around the same time. They had the same aspirations, each influenced the other, and the spiritual qualities promoted by transcendentalism also went hand in hand with those expounded by nature writers. In analyzing the origins of nature writing, Scheese (*Nature Writing*, pp. 12, 22) saw transcendentalism as one of its main sources and declared that the important transcendentalists were also pioneers of nature writing. It is worth noting that the Transcendental Club was founded in New England in 1836, also the year in which Emerson's *Nature* was published. Emerson has been called "the spokesman of transcendentalism," as well as being a pioneering nature writer. His book *Nature* is regarded as a text that not only "would have served better than any other as the manifesto of the Transcendentalists" (Koster, pp. 1, 14), but also contained important ideas about nature writing. Perhaps Buell summarizes the important points of transcendentalism most succinctly: "spirit, nature and man" (1973, p. 19). We may wish to expand on this and say that the purpose of transcendentalism is to encourage humanity to return to nature, to elevate human spirituality through direct contact with nature and communing with God. This process, through which humanity can be one with nature or the universe, is similar in premise to the eastern and Chinese philosophical concept of heaven and humanity as one. It is also a point of interest in nature writing. Buell felt that the spirit of transcendentalism can be better understood from a literary perspective, and he pointed out that most works written by transcendentalists are nonfictional prose. He noted the characteristics of transcendentalist works, such as the form of their prose tradition, diarized structure and colloquial style (Buell 1973, pp. 9, 1, 94–96). Myerson also noted that transcendentalists wrote primarily nonfiction prose or poetry, and they utilized private experiences and materials to a great extent, so that much of the best of Transcendentalist literature was taken from journals and letters of the period (Myerson, p. xxx). Scheese put it more specifically: "The literature of transcendentalism represented a confluence of autobiography and natural history, combining to form a new genre" (Scheese, p. 23). It could, therefore, be said that transcendentalism and nature writing, whether in terms of a spiritual perspective or of literary form, shared a familial bond. Unlike other American literary trends, transcendentalism "wended its abstract way out of the college classroom and into the pathless wilderness" (Ronald, p. 1). In this chapter we will discuss the works of Emerson, Thoreau and Whitman, writers who were all strongly associated with transcendentalism, and also the Hudson River School painter Thomas Cole, not only because works of this art movement are indicative of transcendentalism, but also because Cole's essays are considered fine examples of nature writing. Turner in *Spirit of Place: The Making of an American Literary Landscape* (1992, p. x) even says that "before Thoreau there wasn't any such thing as an American literary landscape. And before his contemporary, Thomas Cole, there wasn't a tradition of American landscape painting." This shows the important positions of Thoreau and Cole—heavily influenced by transcendentalism, and their most famous works all centered on the theme of nature—in the history of American literature and art.

One of the main concepts of transcendentalism is that humanity can find spiritual and artistic inspiration from nature. As leaders of their respective fields, Thomas Cole—American romantic landscape painter and founder of the Hudson River School—and Ralph Waldo Emerson—the transcendentalist writer whose speech "The American Scholar" became known as "America's Intellectual Declaration of Independence"—were the first to look for spiritual enrichment from nature. We can probably say that only after Cole and Emerson both published their respective works "Essay on American Scenery" and *Nature* in 1836 did American writers and artists start to look to the landscapes of the New World for literary and artistic inspiration. Before this point, they often looked towards Europe to find the cultural basis for their literary and artistic works.

An important example in this respect is Washington Irving (1783–1859), who is known as the "father of American literature." In *The Sketch Book* (1819–1820), Irving writes that unlike Europe, America lacked "all the charms of storied and poetical association" and that his motivation for visiting Europe was "to escape, in short, from the commonplace realities of the present, and lose myself among the shadowy grandeurs of the past" (Irving, new ed. 1821, p. 3). Cole and Emerson represented a different and new attitude. They disagreed with Irving's view that America lacked an ancient European culture which prevented its cultural and artistic development. Cole felt that America was an unspoiled land whose inexhaustible cultural treasures existed in its virgin landscapes. In "Essay on American Scenery and Nature" and in his paintings, Cole extolled the sublime and vast beauty of the American landscape and showed that nature was not only a source of light, darkness and color, but also the source of cultural, emotional and artistic inspirations. As the founder of the Hudson River School, Cole expounded the concept of transcendentalism, which meant that artists of the movement, through their paintings, visualized the ideas that their literary counterparts such as Emerson, Thoreau, Whitman and other writers expressed in words. As a transcendentalist nature was, for Emerson, a window to the soul, a symbol of human spirituality and a place to commune with God. He deeply believed that "In the woods, we return to reason and faith" (Emerson, "Nature" in Centenary Edition, *The Complete Works of Ralph Waldo Emerson*. Vol. I: *Nature Addresses and Lectures*, [1903] 1979, p. 10). Moreover, he also demonstrated that nature served as a literary template, encouraging writers and scholars to become observers of nature (Buell 1973, pp. 143–144).

In the course of American literary history, Thoreau and Whitman are perhaps the two people who benefited the most from Emerson, and the relationship between these three was quite special. In *From Puritanism to Postmodernism*, its editor Richard Ruland describes their close relationship thus: "Emerson the thinker, Thoreau the bold experimenter in life's living, Whitman the singer and seer." Then Ruland goes on to describe their attitudes toward nature: "Emerson—despite his insistence that the world was real and not the shadow of ideas—remained closest to philosophical idealism.… Thoreau expressed a far more direct contact with the natural world.… Whitman always retained a transcendentalist mysticism and a belief in nature as force and source" (Ruland and Bradbury, p. 129). Their relationship existed on many levels. In a traditional sense, it could be seen as a relationship between mentor and disciples, but it could also be regarded as one between theorist and practitioners. Koster, for example, also states that "if Ralph Waldo Emerson was the chief expositor of Transcendentalism, Henry David Thoreau was surely its chief practitioner" (p. 48). Thoreau and Whitman were devout followers of Emerson's transcendentalism and their thinking and views on nature were heavily influenced by Emerson, who to varying degrees helped them to develop their literary careers. Living by Walden Pond, Thoreau quietly practiced Emerson's theoretical concept of enjoying "an original relation to the universe" (Emerson, "Nature", p. 4). Whitman, on the other hand through poems and essays in *Leaves of Grass* (1855) and *Specimen Days and Collect* (1882)—"incarnates its geography and natural life and rivers and lakes" (Whitman, "Preface to *Leaves of Grass* [1855], in Baym and Gottsman, eds. *The Norton Anthology of American Literature*, p. 1960). From a modern perspective, it would seem that Thoreau and Whitman were more influential than Emerson in America and in the world, as both garnered world renown with their unique thinking, personality and writing. It could be said that Thoreau and Whitman, through their respective means, brought to life Emerson's somewhat abstract image of nature and introduced Emerson's transcendentalist concepts to ordinary people. Through this process, they gradually demonstrated their uniqueness of thought and distinctive creativity, and established their own philosophies and literary styles, eventually becoming immortal figures in the literary world and in people's hearts.

## Lecture 7. Cole: An Eternal Gallery

The Hudson River School (1825–1870) was America's first ever independently formed landscape art movement. It followed transcendentalist principles while retaining romantic traditions and enjoyed nearly 50 years of popularity in America. The artists of this movement sought to move away from the European schools of art and discover their own identity. It is characterized by artists all painting outdoors and creating works through close observation of the American wilderness, especially around the revered region of primal wilderness by the Hudson River. These paintings not only reveal the characteristically desolate and mysterious American landscape, but also reflect a new concept: in a sublime and awe-inspiring landscape, humanity is insignificant and only nature is eternal.

One well-known painting from the Hudson River School, *Kindred Spirits* (1849), was painted in remembrance of the passing of Thomas Cole (1801–1848), the founder of the movement. It was the work of Asher B. Durand (1796–1886), Cole's lifelong friend and an important figure in the Hudson River School. Set in the Catskill Mountains where Cole had lived and painted, Durand's painting depicts Cole and his poet friend William Cullen Bryant[3] as a pair of kindred spirits. In the painting, the two men are standing on a rocky ledge over a deep valley admiring the spectacular views. In the background there are misty mountains, from which comes a stream that rushes forward through the dark valley floor amid large rock formations, while prominent in the foreground is a broken tree trunk, a motif often found in paintings of the Hudson River School, which Cole called a "reminder." It alludes to the fragility and fleeting nature of human existence, suggesting that only nature and our spiritual beliefs can be everlasting. It is interesting to note that the ideas found in this nineteenth century landscape painting coincide with the views of modern nature writers. When faced with the barren desert, the twentieth century nature writer Edward Abbey laments: "Men come and go, cities rise and fall, whole civilizations appear and disappear—the earth remains" (1971, p. 194). We could say that Hudson River School painters are in essence using imagery to interpret ideas that nature writers seek to put into words, yet its founder Thomas Cole was not only a painter but also regarded as a nature writer. His famous "Essay on American Scenery" laid the ideological

3 Bryant is also known as the "father of American poetry" and a "Poet of Nature." For more details, see: Begiebing and Grumbing (eds.), *The Literature of Nature*, p. 272 and Hart, *The Oxford Companion to American Literature*, p. 95.

and connotative foundations for nature writing, so in some respects the Hudson River School and American nature writing were kindred spirits.

## *Wilderness Culture*

Thomas Cole, born in England in 1801, had two dreams when he was young: (1) become an artist, and (2) sail across the ocean to America and settle in the New World. What he did not expect was that years later his childhood dream would come true and he would become a famous landscape painter in America, where his passion for the wilderness would give rise to America's first art movement, the Hudson River School, earning him recognition as the "father" of the movement. Cole's major works consisted of sets of paintings such as *The Course of Empire* (1836), *The Voyage of Life* (1840)[4] and *The Cross and the World* (1846–1848), as well as many other natural landscape paintings set in the Catskill Mountains, and White Mountains and along the Hudson River.

Cole immigrated to America with his family in 1818, when he was 17. They lived in rural areas of Ohio and then Pennsylvania. As a young man Cole soon began exploring nearby mountains and, like William Bartram before him, trained himself to be a keen and sharp observer. He carefully observed the living trees, the dead tree trunks and the leaves whose color changed with the seasons and recorded their various states in his sketch book. As he drew his sketches in the wilderness, just as William Bartram had during his journey through America's Southeast, Cole was developing the skills that would later shape the style of his paintings and that of the Hudson River School. Cole's early explorative sketches and his later sketching trips in the Catskill Mountains led him to form the view that cultural and artistic inspiration could be found in the wilderness and helped to create the distinctive nature-themed Hudson River School art movement.

Cole visited the village of Catskill in the Catskill Mountains for the first time in 1825, and eventually settled at Cedar Grove Farm in the village. Cedar Grove was situated on the west bank of the Hudson River, and its position on a slope enables the property to command a complete vista of the Catskill Mountains. It

4 There are two sets of *The Voyage of Life*, one painted in 1840 and the other in 1842. The 1840 set was commissioned by New York banker and philanthropist Samuel Ward Snr and is currently kept at the Museum of Art at the Munson-Williams-Proctor Arts Institute in New York State. Due to the death of Ward and difficulties in dealing with his heirs, Cole completed an identical set of the paintings in 1842. This later set is in the collection of the National Gallery of Art, Washington, D.C.

is thought that the farm may have been named after a local forest of cedar trees. From the beginning, Cedar Grove and the Catskill Mountains nurtured Cole's artistic spirit. The undulating peaks of the Catskill viewed from the valley of the Hudson were like the waves of the ocean after a storm both enchanted and excited Cole. The Hudson River and the Catskill became the theme of Cole's artistic creations and gave purpose to his life. Many of Cole's ideas for "Essay on American Scenery" came from his time at Catskill, and many of his Hudson River School masterpieces were created at Cedar Grove.

There are two reasons why I want to discuss Cole's work in a book about American nature writing: (1) His "Essay on American Scenery" is intimately connected in theme, creative background and style of expression with American nature writing, and (2) the Hudson River School art movement and American nature writing share a similar purpose and complement each other.

Throughout the first half of the nineteenth century, questions such as "Is America's natural scenery comparable with that of Europe?" and "Should American writers and artists follow the European Romantic movement or follow a uniquely American form of expression instead?" continued to be asked. Europe's long history and rich cultural heritage lend its natural scenery a golden luster and provide sources of inspiration for writers and artists. In contrast, American writers and artists faced a wilderness that had little history or cultural heritage known to them. How then should American writers and artists view America's natural landscape and what path should they follow? These are the questions that Cole sought to answer in "Essay on American Scenery." As Nash comments: over the long period of several decades, "Thomas Cole attracted wide attention as a celebrant of the American wilderness. His landscapes added art to poetry and fiction as a medium through which his countrymen could be instructed in the glories of the native landscape" (Nash 1967, p. 78).

In "Essay on American Scenery" Cole did not seek to denounce the Old World, whose "mountains, woods, and streams, made sacred in our minds by heroic deeds and immortal song—over which time and genius have suspended an imperishable halo." However, he believed that the American landscape was in no way less beautiful than that of Europe and, if anything, perhaps even more sublime and pure, and closer to God. The most impressive characteristic of the American scenery was its wilderness, where nature was still predominant, where "these scenes of solitude from which the hand of nature has never been lifted, affect the mind with a more deep-toned emotion than aught which the hand of man has touched." Cole goes on to detail the sublime beauty and characteristics of the American natural landscape, praising America's mountains as its primary element, water, like the eye in the human countenance, as a most expressive feature; waterfalls as the voice of the landscape and the sky as the soul of all scenery. He concludes that: "American associations are not so much of the past as of the present and the future" (Cole, in *The Native Landscape Reader*, 2011, p. 36). If Europe represents culture, then America nature and those born in nature's nation should find inspiration from nature.

The value of "Essay on American Scenery" lies not only in its breakthrough argument that American writers and artists should seek cultural inspiration from the American landscape, and a way was shown for America to develop its own literature and art, but also in the fact that the essay itself is an excellent piece of nature writing, as Cole uses a fresh and graceful style of writing to praise the unique American landscape and its effect on the American people.

Let us first look at America's mountains: In the mountains of New Hampshire there is a union of the picturesque, the sublime, and the magnificent; there the bare peaks of granite, broken and desolate, cradle the clouds; while the valleys and broad bases of the mountains rest under the shadow of noble and varied forests…although in some regions of the globe nature has wrought on a more stupendous scale, yet she has nowhere so completely married together grandeur and loveliness—there he sees the sublime melting into the beautiful, the savage tempered by the magnificent…[in autumn] then every hill and dale is riant in the luxury of color—every hue is there, from the liveliest green to deepest purple from the most golden yellow to the intensest crimson. The artist looks despairingly upon the glowing landscape, and in the old world his truest imitations of the American forest, at this season, are called falsely bright, and scenes in Fairy Land. (Cole, "Essay on American Scenery," in Grese ed. *The Native Landscape Reader*, 2011, p. 35)

Cole also quotes the words of an American poet to show the superiority of American forested mountains over those in Europe: "when the woods have put their glory on, the purple heath and yellow furze of Europe's mountains are in comparison but as the faint secondary rainbow to the primal one" (Cole in *The Native Landscape Reader*, 2011, p. 31).

Then let us look at America's waters: "the unrippled lake, which mirrors all surrounding objects, we have the expression of tranquility and peace—in the rapid stream, the headlong cataract, that of turbulence and impetuosity." Cole goes on to describe the lakes of the northern states:

> In character they are truly and peculiarly American. I know nothing in Europe which they resemble.… Embosomed in the primitive forest, and sometimes

> overshadowed by huge mountains, they are the chosen places of tranquility; and when the deer issues from the surrounding woods to drink the cool waters, he beholds his own image as in a polished mirror,—the flight of the eagle can be seen in the lower sky; and if a leaf falls, the circling undulations chase each other to the shores unvexed by contending tides. (Cole in *The Native Landscape Reader*, 2011, p. 32)

As a lonely traveler, when Cole finds himself shut in by stupendous mountains in a deep and secluded valley, he is

> overwhelmed with an emotion of the sublime, such as I have rarely felt, it was not that the jagged precipices were lofty, that the encircling woods were of the dimmest shade, or that the waters were profoundly deep; but that over all, rocks, wood and water, brooded the spirit of repose, and silent energy of nature stirred the soul to its inmost depths. (Cole in *The Native Landscape Reader*, 2011, p. 32)

Then let us hear the waterfall, the voice of the landscape:

> The waterfall strikes its own chords, and rocks and mountains re-echo in rich unison.… In gazing on it [Niagara] we feel as though a great void had been filled in our minds—our conceptions expand—we become a part of what we behold!…In its volume we conceive immensity; in its course, everlasting duration; in its impetuosity, uncontrollable power. (Cole in *The Native Landscape Reader*, 2011, p. 33)

And lastly, let us feel the American skyscape as the soul of all scenery.

> Whether it be the serenity of the summer's blue, or the dark tumult of the storm. It is the sky that makes the earth so lovely at sunrise, and so splendid at sunset." (Cole in *The Native Landscape Reader*, 2011, p. 35)

Cole is passionate in his praise of American sunsets, claiming that their beauty can vie with those in Italy while many are more gorgeous:

> Look at the heavens when the thunder shower has passed, and the sun stoops behind the western mountains—there the low purple clouds hang in festoons around the steeps—in the higher heaven are crimson bans interwoven with feathers of gold, fit for the wings of angels—and still above is spread that interminable field of ether, whose color is too beautiful to have a name. (Cole in *The Native Landscape Reader*, 2011, p. 35)

For Cole, the evening sky under the sunset of the American winter is even more moving.

> In the westering sun the iced groves glitter as beneath a shower of diamonds—and through the twilight heaven innumerable stars shine with a purer light than summer ever knows. (Cole in *The Native Landscape Reader*, 2011, p. 35)

"Essay on American Scenery" not only allows us to appreciate Cole's artistic visual depictions of America's unique natural scenery, it also introduces us to a unique awareness of nature: a new experience that is completely undisrupted by any human element and untouched by the hand of man. Unlike the polluted and even decaying natural landscape of Europe that has been marred by man, the primal, desolate, fresh and honest natural landscape of America is indicative of its vitality, giving confidence, strength and charm to this new country.

Despite Cole's love of the American wilderness, he also occasionally had clouds of doubt hanging over him or felt the pull of an antipodal attraction of civilization. Therefore, "Essay of American Scenery" also shows mixed feelings: his love tinged with a fear of nature, and the pursuit for material civilization conflicting with the desire to preserve unspoiled nature. Like his predecessors John Smith and William Bradford, Cole saw nature as both a bountiful Eden and a fearsome wilderness; and like John Muir, who struggled with the dilemma of nature beckoning but civilization preventing one from going forward, Cole also could not escape the dilemma that continues to plague the American people. Although Cole wrote that "pleasures spring like flowers within the bosom of the wilderness" and declared that he always returned from the mountains to the city "with a presentiment of evil," he at times "dreaded the wilds" (quoted in Nash 1967, p. 79). For him, the elated pleasure experienced in enjoying the sublime beauty of nature was only a step away from real fear.

According to Nash, Cole was once caught in a thunderstorm in the Catskill Mountains. At first, he gave himself up to the wilderness of the elements and pronounced the situation "romantic," but as the fury increased, his ecstasy changed to apprehension, and when the storm ended, Cole was relieved to see "in a neighboring dell, the blue smoke curling up quietly from a cottage chimney" (quoted in Nash, p. 79). During a trip to the White Mountains, Cole recognized conflicting emotions in the presence of wilderness, generalizing in his journal that "man may seek such scenes and find pleasure in the discovery, but there is a mysterious fear [that] comes over him and hurries him away. The Sublime features of nature are too severe for a lone man to look upon and be happy" (quoted in Nash, p.

79). Cole's longing for solitude and wilderness together with his interest in society and culture, his love and fear of nature, and his antipodal mentality towards civilization all point to a trait of American nature writing that has existed since its inception. Fritzell describes this trait as "a tolerance for ambiguity that is very difficult to sustain—a dedication to paradox, and even an occasional delight in uncertainty, ..." (Fritzell, p. 16).

It is evident in Cole's 1836 painting *The Oxbow*[5] that he recognized the conflict between nature and civilization, and sought to reflect this in the composition. *The Oxbow* depicts the Connecticut River as viewed from Mount Holyoke after a thunderstorm. The left half of the painting symbolizes the wilderness with rugged cliffs, dark forests, tree trunks split by thunder and lightning and violent clouds. The right half is a peaceful and harmonious country scene-the Connecticut River makes an "oxbow" turn here. On the riverbanks are pieces of level farmland and among the green groves are small and neat cottages. Sunshine adds a tone of joyfulness. Nash (p. 80) comments that: "Cole's divided canvas implied the idea Henry David Thoreau accepted as axiomatic: man's optimum environment is a blend of wildness and civilization."

## *Nature as the Canvas*

Cole's theory that culture and creative inspiration could be found in the American wilderness came from personal experiences. He founded the Hudson River School in 1825, and even though he and other artists of the movement were to some extent influenced by the European Romantic movement, from the beginning they followed Emerson's call to disregard "the courtly muses of Europe" (Emerson, "The American Scholar" [1837], p. 114) and established a uniquely American form of art that presented the fairytale that was the American landscape. The Hudson River School was able to capture what was new in the New World, and they considered America's primitive natural scenery a gift from God that paralleled the Garden of Eden. As the creation of a culture that was unique to America was essentially the main idea of transcendentalism, we can say that the Hudson River School showcased the movement.

Cole believed that if nature had not been touched by the hand of man, it would be easier for man to touch the hand of God. The Hudson River School

5 The complete name of this painting is: *View from Mount Holyoke, Northampton, Massachusetts, after a Thunderstorm—The Oxbow*. It was later shortened to *The Oxbow*, which refers to the shape of the depicted part of the river.

saw the wilderness as a place where spiritual enlightenment could be achieved and art as a medium could bring together man and nature. They firmly believed that if America was a new Garden of Eden, then they as artists possessed the keys to its gates. Unlike traditional painters who confined themselves indoors, the Hudson River School proposed nature as the canvas and attracted a group of nature painters who Emerson would call "to look at the world with new eyes" (Emerson, "Nature," p. 75). They used paintings to represent similar ideas that Emerson, Thoreau and Whitman expressed in words.

Like the emerging American nature writing, the Hudson River School sought to depict the relationship between man and nature. Cole's paintings are enchanting landscapes characterized by steep cliffs, dark valleys, and stormy clouds with a ray of sun shining through. A common trait found in his paintings is the absence of human figures or their extremely small stature, with the wilderness dominating the composition, as it is nature that artists of the Hudson River School sought to promote, and they believed that only when man is in harmony with nature and only with the support of a magnificent landscape can human nobility be seen.

American nature writing and the Hudson River School share the same creative background, so if American nature writing is a form of outdoor writing, then the Hudson River School is simply a form of outdoor painting. Not only did Cole revolutionize the theme and style of painting in America, he also created a new way of painting—directly sketching in nature. He frequently ventured into the Catskill and the White Mountains to sketch their scenery, and would then transfer his drawings onto a large canvas after he returned to his studio. Drawing inspiration from nature and painting in nature would not only become a trait of the Hudson River School, but also influence successive generations of writers.

As Huth notes (p. 84), this form of close connection between man and the wilderness "began to conform to a pattern" in the first half of the nineteenth century. Painters, poets and writers all looked to the wilderness for inspiration in creating a New World culture, and therefore a unique air of culture existed at the time. The nineteenth century American literary critic and essayist Thomas Wentworth Higginson named his collection of nature essays *Out-Door Papers*; Walt Whitman, the author of *Leaves of Grass*, was called "poet of the wilderness"; Thoreau was obsessed with the wilderness and called for writers to use a language that is connected to the land. In addition, we can see people's interest in the wilderness from the titles of books and essays on man and nature, such as the "Summer Books" that were popular in the 1840s and 1850s; *Forest Life* by Caroline Kirkland published in 1842; and *Life in the Open Air*, co-written by

Theodore Winthrop and Hudson River School painter Frederick E. Church in the early 1840s (Huth, pp. 82–84).

This style of creativity based on the wilderness that began in the nineteenth century continued to enjoy popularity in the twentieth. Sanders explained that his 1993 book *Staying Put: Making a Home in a Restless World* came from his "own tasting of the dirt" (p. xiii), while Paul saw the combination of simple natural scenery in the wilderness as a successful feature of modern society (*For Love of the World,* 1992). It is not difficult for us to see their close link with the Hudson River School, as their works were complementary to the art movement.

The Hudson River School is known for its treatment of light and shade through clever juxtaposition of sunlight and shadows. This technique is able to realistically depict the sublime features of the American landscape and demonstrates the movement's aesthetics. Emerson considered light to be the incarnation of the soul, and Thoreau viewed the sunset as a glorious map where imagination can gallop among the incomparable isles, continents and cities. Like Emerson and Thoreau, the Hudson River School believed that nature and spirituality shared the same qualities and that one's spiritual landscape mirrored the natural landscape, ever changing and diverse. In the words of Durand, "Sunshine is the joyous expression of nature, the lovely smile that lights up all her beauty, so changing and adoring all it rests upon, as to seem itself creative" (quoted in St. Armand 1989, p. 273). In the Hudson River School's landscape paintings, sunlight is used to confer emotion and movement to the scene, turning every shadow into a thing of beauty and adding a strong contrasting effect to the whole painting. For example, Cole's early work *The Clove, Catskills* (1827)[6] depicts a V-shaped valley flanked by two hills, one in the light and the other in the shade. The shaded slope on the right is covered by a dark forest in a solemn gloomy green, while the one on the left has autumn leaves in lively shades of green, red and yellow glistening with sunlight, and the V shaped valley in the middle is almost completely in a cradle of the sun, enjoying its warm caress. In the foreground are rugged rocks, small twisted trees and a shiny waterfall descending from the clifftop. In the distant sky roll dark clouds.

In the late Hudson River School period, this style of painting of Cole's influenced his followers to develop "luminism," a painting style that emphasized the effect of light. The light radiating from the clear and pure sky clearly outlines the

6 Details of the three Hudson River School paintings (*The Oxbow*; *The Clove, Catskills*; and *Kindred Spirits*) described in this chapter are based on illustrations from: Yaeger, *The Hudson River School*, pp. 24, 6, 35.

dark scenes, emphasizing the solemnity and grandeur of nature, involving man in a silent and mysterious artistic world. St. Armand obtained inspiration from this painting style of the Hudson River School. In "Luminism in the Work of Henry David Thoreau's: The Dark and the Light" (1980) he analyzes Thoreau's two works *Walden* and *Cape Cod*, and demonstrates the similarities between the Hudson River School and the literary works from the same period. St. Armand feels that in *Walden* and *Cape Cod* readers are able to find landscapes that most closely resemble the luminist landscapes of the Hudson River School. In *Walden*, the land Thoreau describes is a natural paradise and the scenes that he depicts are mostly not dark and scary, but in the form of soft and changing bodies of light. For Thoreau, a body of water is able to manifest the essence of its surroundings, so Walden Pond became a mirror of light that showed Thoreau the illusory existence of two heavens, one in the water and the other in the sky. In contrast, the first chapter of Cape Cod tells a depressing story of a shipwreck, and in that context Thoreau sees the ocean as watery wilderness, a black mirror that reflects an empty and foreboding image rather than a personal, pure and luminous image. The expansive ocean and wild coastline form a deeply lonely seascape. Unlike the spiritual solitude of *Walden* that benefited the soul, this is unfamiliar and ever so empty. "Even a thousand people could not break this emptiness and they will be lost in this vast scene like their footprints disappearing in the sand" (St. Armand, pp. 13–30).

From the perspective of the Hudson River School's contrast of light and dark in their painting to look at the interpretation of nature in American nature writing, we could perhaps argue that a new vista has presented itself to those studying American nature writing. Thoreau's expressive techniques used in the two aforementioned works also show the influence of the Hudson River School on American literature as a self-contained and enduring cultural phenomenon.

For Cole and other Hudson River School painters, aside from the contrast of light and dark, their works have always been covered by a thin mysterious mist—nature's veil. On this, in his book *Emily Dickinson and Her Culture* (1989), St. Armand says, "Above all, the artist's highest goal was to capture the effect of that atmosphere which hung like a celestial veil between the heavenly throne of the deity and the earthly realm of the artist" (p. 228). Ian Marshall, another American English professor, commenting on the American poet Robert Frost's love of the mountains, mentions that in the paintings of the Hudson River School, clouds and mist caress the shape of the mountains, expressing in visual terms the transcendental belief that the natural and the spiritual worlds intermingle and communicate with each other. He thinks that "the clouds representing the intangible realm of the spirit" (Marshall 1998, p. 195). The Hudson River School needed not only to faithfully capture firsthand experience of the wilderness and its regional characteristics, but also to incorporate a moment of personal interaction, especially the moment in which the painter attempts to capture the essence of that layer of floating mist. It is perhaps the diffusive quality of the mist that gives the light-and-dark composition of the Hudson River School a distant and hazy allure that is intensely charming. In my opinion, this light-and-dark treatment of nature, and the inclusion of a pervasive mist, accurately reflect how nature writers see nature; these qualities embody the opposing forces of nature against humans and wilderness against civilization, a conflict that humans have been unable to avoid since the dawn of civilization. That layer of mist also reflects Emerson's other interpretation of nature: the mystery of nature. The relationship between humans and nature also seems to share similar qualities with this responsive layer of mist, sometimes bright and close, at other times gloomy and distant. Nature writers would later come to value this layer of mist, as it became a mechanism that would alleviate the conflict between man and nature.

Although the Hudson River School painters, like nature writers, used a variety of ways to depict it, they never faltered in their enthusiasm for nature. They were artists who wholeheartedly embraced nature to portray its beauty; they were deeply idealistic and searched passionately for divine imagery through the natural scenery to achieve spiritual fulfilment. In contrast to their attention-grabbing works of art, these artists were mostly sensitive introverts who were above worldly considerations. Cole was such a shy and melancholy individual, with poor social skills. He avoided the busy cities and refused to be interviewed, living exclusively at Cedar Grove by the Hudson River for the rest of his life. Durand, another prominent artist of the school, was familiar with the mountains of Connecticut from a young age. For much of his life he explored those nearby mountains, and according to his own description would "hide behind a tree or bush at the approach of a person or vehicle" (quoted in Yaeger 1996, p. 32).

The Hudson River School artists' enthusiasm for nature, however, was unmatched. This can be seen in the romanticized death of another well-known artist of the school George Inness, which marked the end of a life filled with passion for nature. According to his son, George Inness brought together appreciation of natural scenery and that of the holy spirit and the human soul. He was especially interested in the shades of light and dark and their pulsating vitality in the sky and on land. A sunset would deeply move him. One autumn evening in 1894, while waiting for his wife to get dressed to go out for a drive with him, Inness walked alone outside to view the setting sun. The sunset that day was unusually splendid, and as the fiery red mass went down below the horizon Inness, mesmerized by the sight, threw his hands into the air, exclaimed to the setting sun, "My God! Ah, how beautiful!" and fell to the ground, never to get up again (St. Armand 1989, p. 294). It can be imagined that it was in a moment of profound happiness and with an indescribable sense of beauty that he embarked on his last journey to the heavens he had longed for, and that he was accompanied on his journey by the sun that he had admired throughout his life. This story is similar in theme to a beautiful Hudson River School painting, in which man becomes one with nature thereby achieving immortality.

In February 1848, Cole died at Cedar Grove, in a bedroom that overlooked the Catskill Mountains, and he was later buried at Cedar Grove. Cole once wrote a short poem in which he expressed his desire to be buried in a spot where he had lived and adored, somewhere near the mountains so that they could keep each other company. In 2001, on the bicentenary of Cole's birth, Cedar Grove opened to the public as a National Historic Site. Today, within the Cedar Grove grounds, visitors can still see the magnificent mountain scenery that Cole enjoyed. In the summer, flowers in the Cedar Grove garden still bloom in splendor, and the cedar trees are still lush and green. When he was alive, Cole was an artist who embraced nature and depicted its beauty with all his heart; after he died, he returned to nature's embrace and achieved a sense of immortality in human history.

Whether they were praising the colors of autumn sunsets or were crossing the lonely winter wilderness, Hudson River School painters always kept their inner eyes persistently and gently open and maintained clear heads and sensitive and accepting hearts. Perhaps this was the secret to their success and the secret of the greatness and loftiness of their art. St. Armand (1989, p. 204) states that "literature...would outstrip both music and painting as the queen of the arts only when perfection in nature writing had been achieved." Today when people are looking at the perfect or great American nature writing, they would naturally think of Ralph Waldo Emerson first.

## Lecture 8. Emerson: Nature Always Wears the Color of the Spirit

In 2003, on the bicentenary of Emerson's birth, Buell published *Emerson*, a book about Emerson's life, thoughts and works. In its introduction (p. 6), Buell says of the book: "I finished the first crude version when I was 26. I complete this latest at age 62. If it persuades others that Emerson is worth pondering for so long a time, I shall be very glad." Emerson's character and his humanistic philosophies have for a long time been important subjects of research in the fields of American culture and literature. When Henry James visited the writer's former Concord home in 1904, he said in tribute that "not a russet leaf fell for me, while I was there, but fell with an Emersonian drop" (Buell 1996, p. 321). The exploration and discussion of Emerson's character and his connection to nature is, however, a relatively new focal point in worldwide literary circles. When looking at his rich life, his transcendentalist beliefs and his written works, we cannot fail to notice his special passion for nature and the intimate connection that exists between him and nature.

## *Nature as a Textbook*

The nineteenth-century American writer and critic James Russel Lowell (1819–1891) had a unique understanding of transcendentalism, describing it as "simply a struggle for fresh air, in which, if the windows could not be opened, there was danger that panes would be broken, though painted with images of saints and martyrs" (quoted in St. Armand 1989, p. 205). As a transcendentalist, Emerson boldly refined the traditional beliefs passed down from Edwards, who believed that only those chosen by God could, through nature, commune with God. In contrast, Emerson declared that not only those chosen by God but anyone with sight could, with a little concentration, see that "nature always wears the colors of the spirit" (Emerson, "Nature," p. 11). Unlike others of his generation who saw only the practical value of nature's bounty, Emerson was able to see its spiritual value. In his mind, nature not only represented the spirit, but was at the same time a teacher of noble morals, a textbook from which people could search for knowledge, and also the source of American culture and art.

Of Emerson's most famous works, his first was *Nature*. Before its publication he had considered giving his first collection of essays a title that related to the woods, as many of the essays contained within evolved from ideas that came to him during his walks in the woods of Concord. In *The Life of Ralph Waldo Emerson* (1957, p. 365), its author Ralph L. Rusk describes vividly how "on a fair day he (Emerson's friend Amos Bronson Alcott) and Emerson might walk and talk as far as Walden Pond. On a rainy day they merely talked, ideas pouring inside the house, rain pouring outside." In fact, before Emerson called for his fellow countrymen to seek the sources of American culture in nature, he had already found his spiritual nourishment in the wilderness of Concord, and his woodland walks became a prerequisite factor in his search for literary inspiration.

His walking companion was first Alcott, then Thoreau, and later Thoreau's poet friend William Ellery Channing.

Walter Harding, in his *The Days of Henry Thoreau: A Biography* (1982), describes an interesting incident: in the autumn of 1844, Emerson decided to organize a club to meet on Monday evenings in his library as he liked to have interesting, stimulating young men around him. Alcott, Hawthorne, Thoreau, Channing and others were all invited. However, it is said that the first Monday evening was spent in awkward silence. The host asked everyone to say something interesting. Alcott made an effort and said something, then there was silence.

> Thoreau made a brief observation. More silence. Emerson beamed and said nothing. Hawthorne shrunk further into the shadowed corner of the room. Finally a bowl of apples was brought in and each munched in silence. When the end of the long evening came, each disappeared his own way into the darkness. (Harding, p. 174)

Two such meetings were held and they were not very successful. Harding comments: "Their place of conversation, they had discovered, was not the parlor but the woods" (p. 174). No wonder it is said that, unlike many any other literary movements, transcendentalism "wended its abstract way out of the college classroom and into the pathless wilderness" (Ronald, p. 1).

Emerson believed that an intimate connection existed between man and nature, and he would often refer to natural phenomena when talking about his personal life. For example, in his journal entry of 27 July 1837, he wrote: "Many trees bear only on alternate years. Why should you write a book every year?" In another journal entry, he talked about Webster's oration at Cambridge: "Mr. Webster never loses sight of his relation to Nature. The Day is always part of him. 'But, Mr. President, the shags of evening which close around us, admonish me to conclude,' he said at Cambridge" (in Porte ed. *Emerson in His Journals*, 1982, pp. 165, 151). Emerson's daughter Edith remembered that when she was young, the family orchard and woods were the classrooms in which her father introduced his children to nature, patiently telling them the names of each kind of tree. Before bed, he would sing them an improvised lullaby that was fanciful or absurd and usually began with "Goodnight to Mr. Minot's cow, Goodnight to Mr. Minot's barn." His youthfulness and his poetic moods were unforgettable to his children. Edith remembered that she "entirely recognized Father as a poet," and connected his blue eyes and fresh appearance with Concord's skies and woods, and with

Walden Pond. She "had an idea that Waldo and Walden had some relation to each other" (quoted in Rusk, p. 363).

If Cole's fantastical visions of the wilderness led him to establish the Hudson River School, then Emerson's devotion to nature and the intimate connection he shared with it enabled him to shed the influences of English culture and become the father of American literature. If Edwards is said to have injected a sense of fervor into conservative Puritanism in the eighteenth century and to have broken down the barriers between nature and culture, creation and creator, the material and spiritual, then Emerson was even more bold as he broke away from traditional puritanism and cultivated a group of new and independent-minded Americans while building a new understanding and perception of nature. Ruland and Bradbury (in Ruland & Bradbury, p. 119), correctly conclude that "Emerson's statement of the innocent vision at work in a natural world was meant as a new beginning for himself and all Americans, and that is what it eventually came to seem." In his first book *Nature*, Emerson called for people to look at nature from a different perspective, and he believed that the independence of the American scholar and creation of an American culture depended upon a new understanding of nature. "There are new lands, new men, new thoughts. Let us demand our own works and laws and worship" (quoted in Miller 1991, p. 874). However, for people to develop a new understanding of nature, they need to awaken their spirituality. Emerson felt that tradition and fear of authority had become obstacles in allowing people to understand the universe, so he encouraged people to free themselves from fear and old perceptions in order to connect with the joyous and exciting spiritual energy that permeates nature. He believed that humans' relationship with the universe is harmonious, our existence is integral to the cosmos, and so every person is able to maintain a relationship with the universe on a most basic level.

Thomas J. Lyon (1989, p. 50) states that "perhaps more persuasively than any other thinkers of his time in America, Emerson made his audience feel truly modern—liberated, and able to conceptualize the universe not as mystery but as something they could experience." This was an emotional and spiritual experience, something that no one had experienced before. Like Cole, Emerson was filled with confidence in America. Like the Hudson River School, Emerson believed that "the river is a perpetual gala, and boasts each month a new ornament." His understanding, however, was much deeper, and he linked the beauty of nature with the thoughts of humans. "Nature stretches out her arms to embrace man, only let his thoughts be of equal greatness.... Only let his thoughts be of equal scope, and the frame will suit the picture" (quoted in Miller 1991, pp. 878–879).

Emerson's disciple Thoreau, in his essay "Walking," describes this connection between the thoughts and the new world in more romantic terms.

> I trust that we shall be more imaginative, that our thoughts will be clearer, fresher, and more ethereal, as our sky—our understanding more comprehensive and broader, like our plains—our intellect generally on a grander scale, like our thunder and lightning, our rivers and mountains and forests,—and our hearts shall even correspond in breadth and depth and grandeur to our inland seas. (Thoreau 1975, p. 533)

It was precisely on the basis of this firm belief that ideas and culture can be found in nature. Emerson in his speech "The American Scholar," which would become known as "America's Intellectual Declaration of Independence," put forward an idea that would have far-reaching influence, "the ancient precept, 'Know thyself,' and the modern precept, 'Study nature,' become at last one maxim" (Emerson, "The American Scholar," p. 87). This idea would not only be seen as the sacred life-long mission of the American scholar, but also the theoretical basis for the subsequent blossoming of American nature writing. In works of nature writing from the time of Emerson right up to the modern day, the study of nature and the understanding of self have been enduring subjects of interest.

On a wide grassy bank of the Concord River sits a beautiful old English-style house—The Old Manse. The American people are very fond of this old house, as it is a witness to the start of two American revolutions, one political and the other cultural. In 1775, Emerson's grandfather William Emerson left the Old Manse and his family to fight in what became the American War of Independence. It was from the windows of this house that William's family watched the retreat of the British army from the North Bridge. Sixty years later, in the same house, William's grandson Ralph started a second war of independence: a historical struggle for independence from the constraints of European culture. From a small room on the second floor of the Old Manse with a view of the North Bridge, where his grandfather and the local militia had once engaged with the British army, Emerson wrote his first published work *Nature*—an essay that would lead to another "shot heard around the world" (Emerson, "Concord Hymn" [1837], in Vol. IX: Poems, 1979, p. 159).

According to Rusk, Emerson began to contemplate writing an essay about nature towards the end of 1833, on return from his travels around Europe. When he moved into the Old Manse in 1834, he was finally in the right state of mind and began work on the essay. During this time, the waves of the Atlantic seemed to sound in his ears again. He wrote:

> The great willow tree over my roof is the trumpet and accompaniment of the storm and gives due importance to every caprice of the gale, and the trees in the avenue announce the same facts with equal din to the front tenants. Hoarse concert: they roar like the rigging of a ship in a tempest. (quoted in Rusk, p. 209)

It is perhaps this symphony of nature that gave Emerson the same passion as Thomas Jefferson had when he wrote the Declaration of Independence. In the introduction to *Nature* Emerson wrote:

> Why should not we also enjoy an original relation to the universe?...Embosomed for a season in nature, whose floods of life stream around and through us, and invite us by the powers they supply, to action proportioned to nature, why should we grope among the dry bones of the past, or put the living generation into masquerade out of its faded wardrobe? The sun shines today also. (Emerson [1836] [1903] 1979, p. 4)

*Nature* was first published in Boston in September 1836, by James Munroe and Company. It was a small booklet of 95 pages, and the cover page did not even have the name of the author. However, it went on to become the book that has the greatest influence on the development of American literature. It was the highest expression of Emerson's ideals, and would form the basis for all his later works. It was also an important book that revealed Emerson's understanding of nature through his study of the natural world.

The initial reception to *Nature* was varied. Orestes A. Brownson, reviewing the book in the *Boston Reformer*, fully recognized the value of Emerson's work, calling it "the forerunner of a new class of books, the harbinger of a new Literature as much superior to whatever has been, as our political institutions are superior to those of the Old World." However, another literary critic, Francis Bowen, found the misty and vague atmosphere in the book frustrating:

> The reader feels as in a disturbed dream, in which shows of surpassing beauty are around him, and he is conversant with disembodied spirits, yet all the time he is harassed by an uneasy sort of consciousness, that the whole combination of phenomena is fantastic and unreal. (quoted in Myerson, p. 3)

but felt that

> the purpose of the book, so far as it may be said to have a purpose, is, to invite us to the observation of nature, and to point out manifestations of spirit in material existences and external events. (quoted in Myerson, p. 5)

It is undeniable that the book has a vague and mysterious air, but through this alluring mist we can see the shimmering light of truth. In *Nature* we can see Emerson's understanding of nature, his attitude toward it, and what he strives for through his relationship with it.

## Rational Nature

For Emerson, nature is the symbol of spirit, it's a classroom in which we can learn and a book which we can study. Following Edwards's view that God made the physical world "a shadow of the spiritual world," Emerson states in *Nature* that "Nature is the symbol of spirit." However, he also had a more specific view that "every natural fact is a symbol of some spiritual fact. Every appearance in nature corresponds to some state of mind, and that state of the mind can only be described by presenting that natural appearance as its picture." For someone who believed that "it is sown a natural body; it is raised a spiritual body," Emerson saw nature itself as a map of the soul. To him, the surging river symbolized humans' state of mind and the silence of the night sky a manifestation of reason, while the boundless wilderness was indicative of an unrestricted imagination. However, "this relation between the mind and matter is not fancied by some poet, but stands in the will of God, and so is free to be known by all men" (Emerson [1836] [1903] 1979, pp. 33–34). He believed that "we know more from nature than we can at will communicate" and warned that we "shall not lose their lesson altogether, in the roar of cities or the broil of politics" (p. 31). He also believed that "a life in harmony with nature, the love of truth and of virtue, will purge the eyes to understand her text. By degrees we may come to know the primitive sense of the permanent objects of nature, so that the world shall be to us an open book, and every form significant of its hidden life and final cause" (p. 35).

Ruland and Bradbury (p. 123) comment that "it is Emerson's notion of earth as text and sacrament that marks the place he holds in American literature." Or perhaps it is the metaphorical way Emerson associated spiritual symbols and meaning with the kaleidoscopic forms of natural scenery that prompted later American nature writers to realize that human culture should be derived from nature.

Emerson's pursuit of nature was essentially still a pursuit of spirituality. For him, both reason and soul needed to be nurtured by views of the wilderness and the countryside. In *Nature* he wrote: "In the woods, is perpetual youth.... In the woods, we return to reason and faith" (Emerson, [1836] [1903] 1979, pp. 9, 10). The woods had always had deep meanings in the New World. They were a symbol of freedom, and also indicative of danger. However, Emerson saw the woods through a new and unprecedentedly optimistic perspective. He told people that the woods were not always the wilderness feared by the puritans, and that they could instinctively and naturally go from the darkness to the light without the guidance of a pastor.

Furthermore, man could also find virtue and selfhood in the woods. Emerson felt that "beauty is the mark God sets upon virtue" (p. 19) and that "every natural process is a version of a moral sentence" (p. 41). He taught people to grasp the "moral sentiment which thus scents the air, grows in the grain, and impregnates the waters" (p. 42), learn from the firmness of the "sea-beaten rock," and discover the secret of attaining tranquility from the azure sky. More importantly, we are told that "In their eternal calm, he finds himself" (p. 16). We could say that finding one's self in nature was Emerson's tenet. He found beauty in nature, but it was the beauty of wisdom. He equated beauty with achieving completeness and harmony within the universe. "The standard of beauty is the entire circuit of natural forms—the totality of nature" (pp. 23–24): a single object is not particularly beautiful as there is only beauty in the whole. "Beauty," he surmised, "in its largest and profoundest sense, is one expression for the universe" (p. 24). The characterization and pursuit of overall beauty in the universe in the nineteenth century reflected people's yearning at the time for the wild and unspoiled natural world as their spiritual sanctuary.

In regard to understanding language, we are also able to see Emerson's vision of a rational nature. For him, nature is a vessel that holds ideas. Words are signs of natural facts. Words originate from natural phenomena, and nature itself is a form of metaphorical essay.

> Every natural fact is a symbol of some spiritual fact. Every appearance in nature corresponds to some state of the mind, and that state of the mind can only be described by presenting that natural appearance as its picture. An enraged man is a lion, a cunning man is a fox, a firm man is a rock, a learned man is a torch...the blue sky in which the private earth is buried, the sky with its eternal calm, and full of everlasting orbs, is the type of Reason. (p. 26)

Therefore, Emerson deeply admired

> The poet, the orator, bred in the woods, whose senses have been nourished by their fair and appeasing changes, year after year, without design and without heed,—shall not lose their lesson altogether, in the roar of cities or the broil of politics. (p. 31)

Perhaps by looking at American nature writers, we will find these traits so admired by Emerson.

Those who have read *Nature* are all likely to have been moved by the first paragraph of its first chapter.

> But if a man would be alone, let him look at the stars. The rays that come from those heavenly worlds, will separate between him and what he touches. One might think the atmosphere was made transparent with this design, to give man, in the heavenly bodies, the perpetual presence of the sublime. (p. 7)

Kenneth Burke (1966, p. 57) once remarked that "in keeping with the quality of Emersonian individualism, he equates the stars with a desire to be alone." For Emerson, stars became the twinkling physical embodiment of thought, a symbol of a desire to be alone. In *American Renaissance*, Matthiessen references T. S. Eliot's description of four ways of thinking: to talk to others, or to one other, or to talk to oneself, or to talk to God. He puts Emerson and Thoreau in the category of shifting between the last two (Matthiessen 1941, p. 67). Perhaps this explains why Emerson was able to enter a transcendental state as he gazed alone at the starry sky and reveled in the time he alone shared with the heavenly stars.

Emerson valued individuality above intellect, and therefore always associated self with nature. He believed that the passing of time in nature and the changing of our emotions are in unison, and that nature's four seasons go hand in hand with the seasons of our soul. In *Nature*, Emerson presents us with a map of the soul that reflects the passing of a day: "The dawn is my Assyria;[7] the sunset and moon-rise my Paphos,[8] and unimaginable realms of faerie; broad noon shall be my England of the senses and the understanding; the night shall be my

7 Assyria—an ancient and once powerful empire located in what is now northern Iraq and southeastern Turkey.

8 Paphos—An old Cypriot city which Greeks inhabited during the Mycenaean era. The city is known for Aphrodite's temple, and according to legend it is the site where she rose from the sea.

Germany of mystic philosophy and dreams" (p. 17). His map, exposing the link between the soul and nature, demonstrates why soul and nature—or culture and nature—cannot be separated from one another.

In following Emerson's map, we are reminded of Thomas Cole's series of paintings *The Voyage of Life* (1840). The paintings drew upon the four seasons to symbolize life's journey from childhood to youth, adulthood and old age. The journey through childhood feels like the coming of dawn or the arrival of spring: a single boat carrying an infant emerges from a cave under the guidance of an angel, as rays of the rising sun shine over the boat in the middle of the river and the flowers and plants on the river banks. The cave is "emblematic of our earthly origin, and the mysterious Past." In the age of youth, the solo traveler navigates the "River of Life"; the sun shines brightly upon the lushly green landscape where flowers bloom, and the clouds form a palace in the sky filled with the hopes of youth. Reaching adulthood, the optimistic and romantic scenery is being left behind. In its place come rapids and banks of gnarled rocks, a scene of gloomy autumn, but the rays of the setting sun through the mountain peaks show a supernatural power, giving people something to hope and dream for. In contrast, the painting of old age imparts a dream-like stillness, a winter's dream: against an almost night-time backdrop, an angel descends upon a lonely old man on a boat as a ray of divine and everlasting light pierces through the rolling dark clouds, revealing a staggeringly sublime beauty.

Cole's *The Voyage of Life* shows in visual form the spiritual journey Emerson describes in *Nature*. Like the man looking at the star alone described by Emerson, Cole's *The Voyage of Life* is also, in a certain sense, a lonely journey, one of seeking the self and individuality. In nineteenth-century America, when people had not been able to free themselves from the European cultural mode, the efforts Emerson and Cole made to seek the self and individuality are of particular significance.

While Emerson called on people to look at nature with a new perspective and to seek the spiritual and the self in nature, his attitude toward nature was not a blind worship. He had a controlled attitude and kept an observer's distance. As Ronald points out (p. 9), Emerson's attitude toward the wilderness was inclined to cultivate human intelligence rather than direct participation. "The true center of Emerson's world is man." In fact, nature was, in Emerson's eye, subservient to the needs of the man. In his eyes the relationship was that of master and servant. The importance of nature lay in its use as a means to communicate with God or to understand morals. Emerson believed that Nature was silent and that "man" was "his brother" who could speak as Nature's spokesperson, therefore his attitude toward nature was also a condescending one. At the conclusion of the first chapter of his *Nature*, also titled "Nature," Emerson advises people to use the pleasures from nature with temperance, as to him going to nature was not for pleasure but for virtue. In another essay titled "The Method of Nature," he explains more clearly that "there is a certain infatuating air in woods and mountains which draws on the idler to want and misery.... Therefore man must be on his guard against this cup of enchantments" (Emerson, pp. 212, 213).

We can say that Emerson's nature had a strong didactic and moral color. Nature serves humans so that we can carry out our duties and abide by our morals, and it is not for us to wallow in pleasure. For him, we humans going to nature is a temporary spiritual trip, soon to be assisted by reason to return to human society. Therefore, in Emerson's attitude toward nature we also see a conflict, a close yet detached relationship, an intriguing, misty ambiguity similar to that of the Hudson River School.

Emerson's works are known for their relaxed structure and poetic descriptions. *Nature* has been called a long prose poem. Matthiessen (p. 64) comments: "All of Emerson's books can be reduced to the same underlying pattern. They are hardly constructed as wholes." He then gives examples to show that most of Emerson's works were originally written as lectures before they were published as essays, and the lectures were based on his diary entries. James E. Miller points out in *Heritage of American Literature* (1991) that "in any event, with Emerson it is a mistake to separate too deeply the prose and poetry: the essays may be read on one level as imaginative poems; the poems on one level as distillations of the ideas gleaming in the prose" (p. 873).

Emerson has been called 'the father of American literature' because the river of American literature has carried his insight and attitudes to today. Looking at the American literary works of past and present, I believe nature writing best reflects the literary characteristics of Emerson as discussed above. Contemporary American nature writing, as far as its essence is concerned, started in the period represented by Emerson. Even though later nature writers describe the particulars of their respective times, their eyes frequently refer back to Emerson, Thoreau and other representative writers of that period.

More than 150 years later, when we look at "the sage of Concord" from the modern perspective, we see not only a great thinker representative of the American spirit of that time and a Washington-like intellectual leader, but also a pioneer of American nature writing. Ann Ronald explains in *Words for the Wild* that Emerson's

> *Nature*, and his later collections of essays stirred a number of his contemporaries into fresh ways of thinking and writing about the American landscape. It was Emerson who first described the correspondences between man, nature and God. It was Emerson who saw in nature signs analogous to human and spiritual facts... And it was Emerson who confirmed, "Nature is made to conspire with spirit to emancipate us." Of course, it was Henry David Thoreau who attached these precepts to an actual way of life, but it was Emerson among the Transcendentalists who first saw the possibilities. Moreover, it was Emerson who brought the essay about nature into the American literary marketplace. (Ronald, p. 9)

Buell calls *Nature* "an American classic," and in his book *Emerson* (2004, p. 143) comments, "Though he did not write very concretely about the natural environment, more than anyone else he inspired the first waves of American nature writing: Thoreau, John Burroughs, John Muir, Mary Austin." This is because they are all followers of Emerson's spirit.

Of course, when reading Emerson's works on nature, the reader is not engaged directly with nature because under Emerson's pen it is a rational and an abstract nature. William Henry Channing, a writer and a contemporary of Emerson, says in an article commenting on Emerson's essays that there are four kinds of writers: the first kind writers are chasing the setting sun in the twilight, in an opposite direction to the motion of the earth, determined to follow with longing admiration the descending glory of the past; the second kind are midnight thinkers, lone and abstracted, focusing on the truths of eternity in the distant dark sky. The third kind are, in morning wakefulness and filled with hope, watching the reddening east and welcoming the rise of the sun with religious enthusiasm. The fourth kind are in the noonday and full of sunny cheerfulness, focusing on the sunny present. Channing assigns Emerson to the second class (he does not deny that Emerson sometimes also showed the passions of the third kind of writers) (quoted in Myerson 1992, p. 25). Comparatively speaking, Emerson's nature was philosophical and filled with the brilliance of reason, beautiful and distant like stars in the night sky. Fortunately, his disciple Thoreau spent his life putting Emerson's theory into practice, and we readers can experience tangible nature.

## Lecture 9. Thoreau: The Story of Walden

In the general view, Henry David Thoreau (1817–1862) is closely associated with Walden. Although he only lived there for two years, he seems to have stayed there forever and as a result become a mythological figure. However, when his book *Walden* was first published in 1854, the reception was not that warm. Only 2,000 copies were printed, and they took five years to sell. It was not reprinted until 1862, the year Thoreau passed away. Since then, it has undergone many and frequent reprints. In 1985, *Walden* appeared as the first in a list of the "Ten Books That Shaped the American Character" in *American Heritage* magazine. In 2004, Houghton Mifflin published a new edition of *Walden* in commemoration of the 150th anniversary of its first edition. *Walden* has been translated into many languages, and is a well-known and influential book around the world. Buell explains the reason for its eternal charm, to some extent, in *The Environmental Imagination*, calling *Walden* a "canonical" and "aspiring literary classic" and describing "*Walden* as actual life rather than *Walden* as immortal book" (1996, pp. 370, 371, 379).

With the modern trend of returning to nature and returning to the simple and truthful, the fascination with Thoreau has increased. If we say that he is regarded as a saintly idol in American culture and his spirit as a part of American cultural heritage, then when we discuss American nature writing we must not fail to mention Thoreau. We can even say that Thoreau is the most influential writer in American nature writing, and that Walden has become the holy land which many Thoreau followers long for.

## The Nourishment of the Outdoors

Unlike the Concord sage Emerson, Thoreau did not enjoy fame and glory when he was alive. On the contrary, he was always in an awkward social position. After graduating from the then Harvard College, Thoreau taught at schools, worked in a pencil factory and as a surveyor and a carpenter, but none of these ways of making a living attracted him. Like the British naturalist Gilbert White (1720–1793), Thoreau found a fulfilling and meaningful lifelong career in studying the vigorous life of nature and in pursuing the art of living well. However, New England in the nineteenth century could not tolerate the leisurely lifestyle of a jobless person and did not regard wandering in the mountains and on the ranches as a virtue. It was difficult to experience inner growth in industrial times. Thoreau wrote in his diary, "I must confess I have felt mean enough when asked how I was to act on society—what errand I had to mankind—undoubtedly I did not feel mean without a reason—and yet my loitering is not without defence" (quoted in Robert Milder's *Reimagining Thoreau*, 1995, p. 3).

As early as when he was at university, Thoreau had read Emerson's *Nature*. In it, he felt he was hearing a sacred call to abandon worldly ambitions and set out on a road to nature study, to the experience of exploring truth in nature and in seeking a spiritual sublimation. Emerson makes a vivid comparison with Thoreau's choice of life in his eulogy at Thoreau's funeral: there grows on the cliffs of the Tyrolese Mountains in Europe a flower. Many brave men would go for it at the risk of their lives in pursuit of beauty or love. Emerson exclaims: "Thoreau seemed to me living in the hope to gather this plant, which belonged to him of right" (quoted in Baym & Gottsman 1989, p. 1045). Emerson's use of this romantic metaphor to summarize Thoreau's pursuit is very appropriate, except that the love Thoreau pursued is spiritual. He did not gather the flower for a sweetheart, for he had long ago declared that "all nature is my bride" (quoted in Buell 1996, p. 211).

All his life Thoreau appeared to be seeking a most simple and direct connection with nature. He gave up everything worldly that was sought by others: wealth, fame and comfort. In fact, he probably understood little of routine pleasures and enjoyment, for he was, as Emerson pointed out, seeking the beauty that ordinary people would not dare to seek. The image of Thoreau wandering in the countryside around Concord, an old book of music scores and a cane in his hands, seems to have stayed in people's memory forever. The book of scores was used to keep the plant samples, the cane was for measuring land, and his clothes were always grey or a combination of green and brown, the colors closest to nature and to animals. Merging nature into oneself and at the same time merging oneself in nature was the extraordinary life pursuit of Thoreau.

When Thoreau went into nature he used all his senses to fully engage in the experience. "I see, smell, taste, hear, feel that everlasting something to which we are allied...the actual glory of the universe" (quotes in Matthiessen, p. 76).

Thoreau's "see" is not the casual look, rather a kind of gazing and an intense observation through which he found "red snow" and "green sun" (quoted in Miller 1991, pp. 1260–1261). In *Walden* he describes what he saw when spring came.

> The grass flames up on the hillsides like a spring fire...as if the earth sent forth an inward heat to greet the returning sun; not yellow but green is the color of its flame;—the symbol of perpetual youth, the grass-blade, like a long green ribbon, streams from the sod into the summer. (Thoreau, *Walden*, [1854] 2004, p. 283)

He felt that his life also took on the color of nature, for example when:

> once it chanced that I stood in the very abutment of a rainbow's arch, which filled the lower stratum of the atmosphere, tinging the grass and leaves around.... If it

> had lasted longer it might have tinged my employments and life. (Thoreau, [1854] 2004, p. 181)

Thoreau believed that smell was the most primal and reliable way of exploring nature. He could detect the presence of a certain species of azalea more than half a mile away by smell. He loved the energetic lilac because it told its story to the lonely traveler. From the smells of nature Thoreau also deduced a life philosophy, "If the day and the night are such that you greet them with joy, and life emits a fragrance like flowers and sweet-scented herbs, is more elastic, more starry, more immortal,—that is your success" (Thoreau [1854] 2004, p. 195).

Thoreau loved to take children out to pick berries and he liked to picnic in nature. To him, picking berries was an art like that of Mozart's composing and Michelangelo's painting, and eating berries was like Holy Communion. When nature was full of fruits and inviting him to a picnic, he was often intoxicated with the heavenly taste of the fruit. In the cold winter he would dig a hole in the frozen pond of Walden, taste the cold water from the pond, and admire images of the two paradises, one in the water and the other in the heaven.

Thoreau's hearing was extraordinary. A lot of pleasures in life were conveyed to him via sound. The thrushes' singing at night in the woods, the roosters' crowing in the morning, the frogs' croaking at dusk beside Walden Pond, and even the soft sound of a leaf falling to the ground all fascinated him. In *Walden*, there is a chapter that describes "sounds." Sounds from nature and the sound of the church bells are all parts of the sounds of Concord.

Touch was also Thoreau's assistant in his exploration of nature. In *Walden*, he describes his experience of walking at night:

> Sometimes, after coming home thus late in a dark and muggy night, when my feet felt the path which my eyes could not see, dreaming and absent-minded all the way, until I was aroused by having to raise my hand to lift the latch, I have not been able to recall a single step of my walk, and I have thought that perhaps my body would find its way home if its master should forsake it, as the hand finds its way to the mouth without assistance. (Thoreau *Walden*, [1854] 2004, p. 153)

This sensitive touch, like that of a blind person, was not innate but a result of long experience. Perhaps it is precisely because all his senses were completely awake and open when he was observing nature that he was full of extraordinary energy and sensitivity and that he became the greatest observer of American scenery.

The nature that Thoreau loved was almost a wild nature. To him, going to nature was going inward. Observing the four seasons of nature was also observing the four seasons of the heart. He wrote in his diary, "the seasons and all their changes are in me.... The perfect correspondence of Nature to man, so that he is at home in her!" Therefore in March, in spring, he felt the spirit of spring. "My life partakes of infinity," he wrote in his diary in March 1853. In contrast, writing in winter in November, he said, "whose name [November] sounds so bleak and cheerless," even "men are more serious." His description of nature is all tinted with the colors of the seasons. "I pick up a white oak leaf, dry and stiff, but yet mingled red and green, October-like" (quoted in Slovic 1992, pp. 26, 28). In fact, Thoreau tried throughout his life to be in step with nature and to grow like the seasons, becoming green in spring and maturing to yellow in autumn. And he was, like the leaf of a tree or plant, part of nature, whether in life or in death.

Thoreau, with his pursuit for and fascination with nature, tried to awaken people to reinvigorate their lives with the seasons and give life vitality and hope. Perhaps we can see his life in the description of the wild apple tree in his later work titled *Wild Apples*: The wild apple is a native fruit that grows in the wilderness and generally forgotten and ignored by man. This tree grows in difficult environment and has resilient and strong life force. Though it bears fruit very late and unknown to most people, it never forgets its own noble mission. (Thoreau [1862] 2002, p. 152) Throughout his life, Thoreau was seeking a hidden inner sweetness, an intimate connection with nature and a spiritual pleasure of an individual rather than worldly fame. This may have been where his success lies.

## *The Call of the Green*

Thoreau wrote four books about nature: *A Week on the Concord and Merrimack Rivers* (1849), *Walden* (1854), *The Maine Woods* (1864) and *Cape Cod* (1865). There are also the diaries he wrote from 1837 to 1861, recording his observations of nature in the wilderness. These diaries were first selected and published toward the end of the nineteenth century in four volumes titled respectively *Spring*, *Summer*, *Autumn* and *Winter*. Later more volumes were added, making up 20 volumes in total. Apart from these he also wrote many essays and poems.

Although both Emerson and Thoreau loved nature, they saw and wrote about different natures. The nature in Emerson's eyes was rational, didactic, abstracted and sublime. He believed that "in the woods, we return to reason and faith" (Emerson, "Nature" p. 10). His attitude was that of an observer. He wanted to participate in nature, but he did not find a way into it. The nature under his pen was philosophical, and it glistened with reason like stars in the night sky, beautiful but remote. The nature that Thoreau loved was almost a wild nature, a nature that relaxed the body and mind, and it had nothing to do with moral teaching. In nature Thoreau was looking for a child-like, pastoral pleasure, a freedom from constraints, a medicine that was good for the physical and mental health, an externally simple but internally rich way of life. He therefore regarded nature as his bride, and threw himself into the study of nature with wholehearted passion.

In *Walden*, Thoreau used his own life as a model to teach people how to live wisely and think vividly. His pursuit in life can be summarized with one word: "simplify." He gave up almost any pursuit for material things in life, and succeeded in realizing his goal of living rather than being burdened with making a living. In his diaries, Thoreau gave a very clear definition of what he meant, "There are two kinds of simplicity, one that is akin to foolishness, the other to wisdom. The philosopher's style of living is only outwardly simple, but inwardly complex. The savage's style is both outwardly and inwardly simple" (quoted in Paul 1962, p. 104). Thoreau believed that once people inherited the farm, the house, the animals and farming tools, they became slaves of the land, burdened with the material life. Emerson pointed out the direction in his *Nature*, asking "why should we grope among the dry bones of the past, or put the living generation into masquerade out of its faded wardrobe? The sun shines today also" (Emerson [1836] [1903] 1979, p. 3), and in *Walden*, Thoreau went further:

> Why should they begin digging their graves as soon as they are born?...let us first be as simple and well as Nature ourselves, dispel the clouds which hang over our own brows, and take up a little life into our pores. ([1854] 2004, pp. 3, 65)

If we say that Emerson wanted to awaken the American people to shake off the cultural shadow of the old world and seek a spiritual independence, what Thoreau wanted was for people to get rid of their old way of life and seek freedom in life. He wanted to awaken people like a rooster from their dreams, from their old way of life, and for them to adopt a new way of life that was in step with nature. He wanted people to minimize their material demands in order to free their bodies and hearts and enrich their spirits. His practice by the side of Walden Pond and his description of his life in the woods are the kind of new way of life he advocated.

People often think that *Walden* marks the peak, and the latter part of his life was the winter season of his writing. However, if we read carefully the essays Thoreau wrote in his later life, we will find that it is precisely in these later works that Thoreau made the call that "I wish to speak a word for Nature" (Thoreau [1947] 1977, p. 592), put forward the idea that "In wildness is the preservation of the world" (p. 609), and expressed the relationship between wild nature and healthy human civilization, showing his insight and spirit before his time.

It is well known that Thoreau's other way of experiencing nature was through walking. Walking has also become the most symbolic aspect of his life. As a fervent nature observer, Thoreau needed clear and simple natural scenery, full wilderness, and freedom to record the grandeur of nature. His walks were leisurely, and his goal had nothing to do with profit. To him, this was a real and sacred occupation, a work that could not stand disruptions. Harding describes Thoreau's frustration when neighbors asked if they could join him on his walks. Thoreau felt that he could not allow this, and he explained in his diary:

> They do not consider that the wood-path and the boat are my studio, where I maintain a sacred solitude and cannot admit promiscuous company...ask me for a certain number of dollars if you will, but do not ask me for my afternoons. (quoted in Harding, p. 292)

Emerson, in his eulogy of Thoreau, said that Thoreau's walking was something ordinary people cannot match, and his body and his mind formed a surprisingly harmonious relationship. "The length of his walk uniformly made the length of his writing. If shut up in the house, he did not write at all" ("Thoreau," in Baym and Gottsman, p. 1037). It can be said that in his walking Thoreau merged time, space and thought into one, and his physical footsteps were in correspondence with those of nature and of the spirit, thus enriching his walking and enabling him to write the famous essay titled "Walking" with the prescient sense of environmental protection. The greatest gain of his walks was his discovery of the value and allure of the wilderness: "The West of which I speak is but another name for the Wild; and what I have been preparing to say is, that in Wildness is the preservation of the World" (Thoreau [1947] 1977, p. 609).

Where Thoreau went further than his contemporary transcendentalist writers was in his discovery of the value of the wilderness, and this has become the focus of American nature writing today. As "Walking" was his last essay, his view on the wilderness is also the greatest achievement of his short life. The significance of Thoreau's discovery lies in that he broke away from the old, traditional views of the wilderness. Going to the wilderness is not going to the primitive and to the past and is not going backward historically; on the contrary, wilderness signifies future and hope. He concluded that

> we go eastward to realize history and study the works of art and literature, retracing the steps of the race; we go westward as into the future, with a spirit of

> enterprise and adventure.... Hope and the future for me are not in lawns and cultivated fields, not in towns and cities, but in the impervious and quaking swamps. (Thoreau [1947] 1977, pp. 604, 611)

Wilderness contains an un-awakened force and vitality, and also signifies beauty and health. Thoreau believed that life was full of wilderness. The most alive was the wildest and the closest to wilderness, and this was also closest to goodness and beauty. He also saw the damage that industrialization does to wilderness. "Nowadays almost all man's improvements, so called, as the building of houses and the cutting down of the forest and of all large trees, simply deform the landscape, and make it more and more tame and cheap." He hated the cruel cutter and was glad that it could not cut off the clouds from the sky. Therefore, he called: "Give me the ocean, the desert, or the wilderness!" (Thoreau [1947] 1977, p. 613) "I would have every man so much like a wild antelope, so much a part and parcel of Nature, that his very person should thus sweetly advertise our senses of his presence, and remind us of those parts of nature which he most haunts" (pp. 610–611). He already foresaw the consequences of industrialization in blind pursuit of development, with no regard for the natural environment. He believed that an individual or a whole culture, once separated from wilderness, would become weak and stupid. Therefore, preserving a small oasis of wilderness in the desert of civilization became the most persistent pursuit of Thoreau (Thoreau 1975, p. 535).

For Thoreau, wilderness signified hope not only for the world, but also for culture and literature. "In literature it is only the wild that attracts us." He cites *Hamlet* and the *Iliad* as examples and states that, whether in *Hamlet* or in the *Iliad*, the most charming part is "the uncivilized free and wild thinking" (Thoreau 1975, p. 539). It can be said that Thoreau's view on the value of the wilderness has another significance: it directs the eyes of literary artists to the wilderness. He calls on writers to "Go to grass" (Thoreau, "Walking", [1947] 1977, p. 623) and use a language that is closely connected with the land. This call of the wild in literature set a new horizon for later American nature writing and attracted generations after generations of American nature writers. As the contemporary American nature writer Gary Snyder says, "Consciousness, mind, imagination and language are fundamentally wild. "Wild" as in wild ecosystems—richly interconnected, interdependent, and incredibly complex. Diverse, ancient, and full of information" (quoted in Gifford "The Social Construction of Nature," 1996, p. 33). Although more than 150 years have passed since the death of

Thoreau, Snyder's comment shows that Thoreau's views on the value of the wilderness are continuing to develop and expand in our time.

In another essay titled "Huckleberries," Thoreau specifies the spiritual value of nature.

> A river with water-falls, meadows, lakes, hills, cliffs or individual rocks, a forest and single ancient trees. Such things are beautiful. They have a high use which dollars and cents never represent. If the inhabitants of a town were wise, they would seek to preserve these things though at a considerable expense. For such things educate far more than any hired teachers or preachers or any at present recognized system of school education. (Thoreau [1862] 2002, p. 196)

However, what pains Thoreau most is that

> most men, it appears to me, do not care for nature, and would sell their share in all her beauty, for as long as they may live, for a stated and not very large sum. Thank god they cannot yet fly and lay waste the sky as well as the earth. (Thoreau [1862] 2002, p. 198)

He then uses a vivid analogy to further explain. He says that we ask people to teach botany while permitting others to cut down the few precious specimen trees that are left. "It is like teaching children Latin and Greek while we burn the books printed in those languages" (Thoreau [1862] 2002, p. 200). Preservation of wild nature is to preserve human culture; saving animals and plants that are on the verge of extinction is to save humanity itself. Thoreau already had this awareness of ecology well over a century and a half ago, and that is why he could see clearly the close connections between nature and culture.

*Autumnal Tints* (1862) is another masterpiece by Thoreau. Some critics believe that this work "operates with the density and symbolic richness of poetry, which it comes nearest to approaching" (Milder, p. 186). In *Autumnal Tints* Thoreau further explains the relationship between nature and literary art. He praises the beauty of the autumn leaves of New England and exclaims that the sunny October day gives them brilliant colors and the sun makes them shine. He compares the forest to a garden and calls the autumn leaves "the forest-flower," saying: "Flowers are but colored leaves, fruits but ripe ones." He exclaims that, in comparison with the autumn scenes, "our paint-box is very imperfectly filled.... Think how much the eyes of painters of all kinds, and of manufacturers of cloth and paper, and paper-stainers, and countless others, are to be educated by these autumnal colors." He compares the autumn scenery to galleries along the side of the road, and shows that indoor galleries could not compare with these roadside ones. "Let us have Willows for spring, Elms for summer, Maples and Walnuts and Tupeloes for autumn, Evergreens for winter, and Oaks for all seasons" (Thoreau [1862] 2002, pp. 127, 129). These are the galleries of the four seasons of nature in Thoreau's heart.

*Autumnal Tints* also promotes an optimistic attitude toward life and lofty spiritual pursuit. Thoreau believed that, just like humans who have markets and meetings, Nature also has an annual October festival. Looking at red maple leaves, golden elm leaves and dark red oak leaves joining the celebration of nature, Thoreau could not help but ask: "Did not all these suggest that man's spirits should rise as high as Nature's,—should hang out their flag, and the routine of his life be interrupted by an analogous expression of joy and hilarity?" He especially admired the autumn leaves at the top of tall trees, because they grew higher and higher with the trees and received more influence from the heavens, and were therefore nobler and more refined. Besides, they had the sacred sky as their dancing stage and the light as their dancing companion. When leaves danced together with light, "so intimately mingled are they with it, that, what with their slenderness and their glossy surfaces, you can hardly tell at last what in the dance is leaf and what is light" (Thoreau [1862] 2002, pp. 129, 131).

People usually regard autumn as the season of harvesting, and at the same time of things dying. However, in *Autumnal Tints* we do not see any of the sadness of autumn that is usually written about. For Thoreau, coming to the end of the journey, returning the gratitude of the land and the sun, and having given the best of one's life, leaves die very elegantly. "How beautifully they go to their graves!" He exclaims:

> How many flutterings before they rest quietly in their graves! They that soared so loftily, how contentedly they return to dust again, and are laid low, resigned to lie and decay at the foot of the tree, and afford nourishment to new generations of their kind, as well as to flutter on high! They teach us how to die. (Thoreau [1862] 2002, pp. 124–125)

It can be said that *Autumnal Tints* shows the maturity of Thoreau's thinking and writing style and marks the peak of the author's life and literary achievement. He learned to obtain sublimity and grandeur in a hard life and also spread the spiritual seeds and offered the fruit of his spiritual pursuit to humankind for the benefit of generations after.

## *Late Glory*

In the eyes of Thoreau's contemporaries, he was Emerson's disciple or even his imitator, and never left Emerson's shadow. After his death, however, Thoreau has won numerous followers and exerted great influence, and is looked at and appreciated in a fresh light. Matthiessen, in *American Renaissance* (1941), considers Thoreau to be "the actual glory" of American Renaissance, and an even more brilliant writer than Emerson. Buell, in *The Environmental Imagination* (1995), focuses on Thoreau and says that even in the American Renaissance of the mid-nineteenth century the main figure was not Emerson but Thoreau. He gives five reasons why Thoreau was the symbol of American culture: (1) that his originality relative to Emerson's was greater than had been supposed, (2) that Thoreau was America's best and most influential nature writer, (3) that he soared above mere nature writing into higher (more spiritual) realms, (4) that he was a good conscientious citizen and person, contrary to the charges of civic dereliction and misanthropy, and (5) that he was great in the courage and the character of his dissent (pp. 351–359).

In the field of American nature writing, Thoreau's influence also far exceeds that of Emerson. Thomas J. Lyon believed that Thoreau had his own unique understanding and experience and consequently formed views different from those of Emerson's. Even though Emerson outlined the new nature view and elaborated on the relationship between man and nature, he never had the same passion as Thoreau to engage in the wilderness. Evidently in the early period of his study of nature Thoreau had already formed a view of nature that was more radical than that of Emerson's. He emphasized the personal experience of participating in nature and practice nature, and he showed that direct experience of nature could produce real and tangible wisdoms. Lyon considered "the possibilities of the nature essay as a modern literary form were already first outlined in Thoreau's first essay," "The Natural History of Massachusetts," which was published in *The Dial* in 1842. (Lyon 1989, pp. 51–52)

Thoreau said in a diary entry:

> Men commonly exaggerate the theme. Some themes they think are significant, and others insignificant. I feel that my life is very homely, my pleasures very cheap; …I see that they look with compassion on me, that they think it is a mean and unfortunate destiny which makes me walk in these fields and woods so much, and sail on this river alone. But so long as I find here the only real Elysium, I cannot hesitate in my choice. (quoted in Milder, p. 300)

Thoreau did not receive much recognition when he was alive. Perhaps he only regarded his pursuit of the Elysium as his personal goal. However, unexpectedly, many years after his passing his views would become classics of human civilization and his choice of life the model for many people. Like the wild apple tree he described, his fruits ripened very late. However, time and reality has proved the late glory of Thoreau.

Nowadays, Thoreau's attitude toward life and his writings have become the focus of attention of the American and even the international literary world. As mentioned above, Buell's *The Environmental Imagination,* generally regarded as a seminal work on nature writing, focuses on Thoreau and explores the close connection between Thoreau and nature writing, and the formation of American culture. Milder, in *Reimagining Thoreau* (p. xi), reviews Thoreau's life, writings and their influence on human civilization from a modern perspective. He sees Thoreau's career path as "neither a parabola whose vertex is *Walden* nor a continuous line, but a rising arc with periodic disruptions and recommencements, constant only in its impulse toward ascent." Thoreau has also been studied as an important figure in the development of ecological thought, and he was, of course, also a very popular writer. Toward the end of the twentieth century, an American scholar saw on the shelves of a Barnes & Noble chain bookstore eight books by or about Emerson, 19 by or about Hawthorne and 23 by or about Thoreau.

Thoreau has been called a "pioneer of nature writing," influencing and guiding a large group of nature writers in the twentieth century. American nature writer and once professor of literature at Columbia University Joseph Wood Krutch (1893–1970) read Thoreau's works repeatedly and, adopting his advice, moved to Arizona, where he wrote about the desert there, publishing *The Desert Year* (1952), *Grand Canyon* (1958), *Best Nature Writing of Joseph Wood Krutch* (1970) and many other works (Finch & Elder 1990, pp. 429–430). Henry Beston built a little wooden cottage and wrote the classic work *The Outmost House: A Year of Life on the Great Beach of Cape Cod.* Edward Abbey first published his *Desert Solitaire,* based on his personal life experience in the Grand Canyon areas in the American West, in 1968, and later wrote *Down the River* (1982) based on traveling downstream for many days on the Green River in southwestern Utah. One of the chapters in this last book is titled "Down the River with Thoreau," because he took a copy of the 33rd edition of Thoreau's *Walden* on the journey. Annie Dillard's *Pilgrim at Tinker Creek* was also written based on the author's year-long experience of observing nature along Tinker Creek in Virginian, and was described by some as a "kind of gutsy *Walden.*" There are many other similar examples.

Thoreau left behind a different cultural tradition for later American literature: seeing the natural world from a specific location. Emerson, in his tribute to Thoreau, told of when Thoreau had finished reading a book titled "Arctic Voyage" and returned it to a friend, commenting that "most of the phenomena noted might be observed in Concord." Emerson said that Thoreau believed that "the best place for each is where he stands.... To him there was no such thing as size. The pond was a small ocean; the Atlantic, a large Walden Pond" (quoted in Baym and Gottsman, p. 1043). Harding (p. 197) quotes a friend of Thoreau's: "Henry talks about nature just as if she's been born and brought up in Concord." Concord is a small, ordinary town, but Thoreau poured all his passion into it. He became a kind of embodiment of Concord: a Concord walking, writing and talking. Nowadays, this strong sense of place of Thoreau's has become an important characteristic of contemporary American nature writing, and a large group of writers focusing on the theme of nature and writing in various geographical environments has appeared.

The new lifestyle that Thoreau advocated is not only promoted in American literature nowadays, but is also accepted by American city-dwellers. In the mid-twentieth century, a Boston couple went to live in the wilderness as Thoreau had, and wrote *At Home in the Woods: Living the Life of Thoreau Today*. The very first sentence of this book is: "We went to the wilderness because 100 years ago a man wrote a book" (quoted in Buell 1996, p. 327). Another modern woman, Anne LaBastille, built a small wooden hut by the side of Black Bear Lake in the mountains in northeastern America and wrote the *Woodswoman* series, recording her decades of life in the wilderness, her experiences, dreams and reflections there. From this we can see how strong the charm of Thoreau has been: it is not only a "late glory," but an everlasting "actual glory."

On 6 May 1862, after a weak Thoreau calmly breathed his last breath, his sister Sophia said, "I feel as if something very beautiful had happened—not death." Another friend who saw his passing said that he had "never spent an hour with more satisfaction, never saw a man dying with so much pleasure and peace" (quoted in Harding, pp. 460, 466). After attending Thoreau's funeral, American writer Louisa May Alcott wrote in a letter to a friend that

> it seemed as if nature wore her most benignant aspect to welcome her dutiful & loving son to his long sleep in her arms. as we entered the church yard birds were singing, early violets blooming between his father & his brother we left him, feeling that though his life seemed too short, it would blossom & bear fruit for us

long after he was gone, & that perhaps we should know a closer relationship now that even while he lived." (quoted in Harding, p. 468)

Thoreau said goodbye to life as gracefully as the leaves he had described in *Autumnal Tints*. However, Thoreau's "leaves" (his writings) have been enriching the world all these years (Milder, p. 328).

Matthiesen, in *American Renaissance*, mentions many times the similarity between Thoreau and Whitman. He believed that both were very close to ordinary people's lives, and that both engaged in unprecedented experiences throughout life (pp. 78, 80). However, in my view, the most marked commonality between the two is their love of the wilderness. On a winter's day in 1856, Thoreau was accompanied by his friend Amos Bronson Alcott to Whitman's home in Brooklyn, New York. Alcott remembered that when Thoreau and Whitman first met, both felt a bit awkward and acted reserved. Alcott thought they looked "like two beasts, each wondering what the other would do, whether to snap or run" and each also assessing the other, "'well, you're almost as great as I am!'...but after a while they fell into a pleasant talk." Whitman gave Thoreau a copy of the 1856 edition of *Leaves of Grass* and wrote "Walter Whitman to H. D. Thoreau." Thoreau gave Whitman a copy of his *A Week on the Concord and Merrimack Rivers*, and added a few lines that had been left out. Though they had different views sometimes, they seemed to admire each other. Whitman admitted that Thoreau "was a man you would have to like, an interesting man, simple, conclusive," and Thoreau told Alcott as they left "that is a great man." Later, talking about the relationship between Thoreau and Whitman, Emerson said that the former's fancy of the latter "grew out of his taste for wild nature, for an otter, a woodchuck, or a loon" (quoted in Harding, pp. 372–376).

Both Thoreau and Whitman were men of strong character and strong belief in the self. However, whether from their experiences or their writings, it is easy to see their affinity, that is: their love of the wilderness, their yearning for freedom and their challenge to traditions and conventions. This is why they could, in their different ways, link elegant literature with land, wilderness, rivers and mountains, birds and animals, and could become bold practitioners of Emerson's theory.

## Lecture 10. Whitman: A Singer of Nature

John Burroughs (1837–1921), Whitman's good friend and passionate follower, and also a nature writer at the turn of the century, built a small cottage in the place he called Whitman Land near the Hudson River, and there wrote and revised and completed his work on Whitman titled *Whitman: A Study* (1896). This is a place that has an affinity with Whitman: gigantic rocks covered with green moss, fields as vast as grassland, three clear and cold springs falling from grey cliffs. Occasionally there are bald eagles on the top of the cliffs. If you climb up to the cliffs, you can see the magnificent Hudson River and the farms in the distance. Early in the morning, birds from the orchard and those from the forest gather under the eaves of his house. In the evening, the singing of the nightingales and that of frogs intermingle. Burroughs believed that only in the environment closest to nature could he appreciate Whitman's feelings when he was writing *Leaves of Grass*, as Whitman was essentially a wilderness poet. This is not only because Whitman sang of nature, but because his personality had the same tone as that of nature (Burroughs 1896, pp. 1–2).

## *An Old Man's Last Pastoral Song*

Although Walt Whitman did not inhabit the Concord literary circles led by Emerson, his feelings and thoughts were in the same vein as Emerson's. Emerson declares solemnly in *The American Scholar*, which is considered to be the declaration of American intellectual independence, "Our day of dependence, our long apprenticeship to the learning of other lands, draws to a close." Whitman called on Americans to celebrate and sing themselves while Emerson called on people to "enjoy an original relation to the universe." Whitman believed that

> we have had man indoors and under artificial relations…but never before have we had man in the open air, his attitude adjusted to the seasons and as one might describe it, adjusted to the sun by the day and the stars by night. (quoted in Matthiessen 1941, p. 626)

Emerson not only gave his full approval of the publication of *Leaves of Grass*, enabling it to grow from tender leaves to flourishing grass, so to speak, but also greatly influenced Whitman's thinking and writing. As Whitman once said, "I was simmering, simmering, simmering; Emerson brought me to a boil" (quoted in Matthiessen, p. 523).

Burroughs, in *Whitman: A Study*, comments that *Leaves of Grass* was not a sacred palace, but a forest, a field, a path in the woods, and it had all the tones of the outdoor. Whitman was very much afraid of catching a "disease of beauty." To him, beauty was a spirit, not a decoration. It was said that Whitman often recited aloud his poems outdoors in the wild to mull over the most appropriate words in life and natural environment to see if they were as true and vital as nature (Burroughs 1896, p. 27). Whether in terms of themes or content, *Leaves of Grass* is a song about nature and the main theme is the song of self, except that the self is not the self in the ivory tower but the self in nature. However, it is in another of Whitman's works, a prose essay in diary form titled *Specimen Days and Collect* (1882), that we see the special significance of nature to Whitman.

In the introduction of the London edition of *Specimen Days and Collect*, Whitman explains that the book is about the personal experience of an ordinary man in the new world. It is an autobiographical essay collection consisting of three parts: the experiences of his youth; memories of his experiences between 1862 and 1865, during the American Civil War; and the nature notes he made between 1877 and 1881, and his travels in the western part of America and in Canada. People generally regard Thoreau's *Walden* as a spiritual autobiography, and are moved by the inspiring and youthful last sentence of the book "The sun is but a morning star." Unlike Thoreau's *Walden*, Whitman's *Specimen Days* is the work of his old age, when he had experienced life. Alfred Kazin, in his introduction to *Specimen Days* (1971), says that "The Whitman who had created a new self in 1855 was now, with Specimen Days, to show what a really determined genius in the nineteenth century could do by way of an autobiography." He then points out that in *Specimen Days* Whitman lost some of the fever of his *Leaves of Grass*, but gained more maturity because at this time he wanted to show that "with the aging, he can think in a relatively slow and meditative way to give some good effect" (in Whitman, *Specimen Days*, [1882] 1971, p. xix).

From the title of the book it is already clear that Whitman chose his personal experiences and natural scenery as he collected specimens of plants in the wilderness. The beauty and permanency of nature is its main theme. Opening *Specimen Days*, people will be immediately attracted by the unique beginning. An old, weather-beaten man walking in the woods and contemplating natural scenery collects specimens of his own memories of life, those unforgettable days: a happy childhood surrounded by wild nature on Long Island; scenes witnessed at the war hospitals, where young lives had been destroyed by war; and his healing and recuperating in nature, accompanied by streams, trees, the sun and the stars, after his two strokes. Now, with the sun shining, a gentle breeze under the blue sky, and happy grasses opening their tender leaves, the air of nature had never so filled Whitman's body. Whitman said that it was precisely because he was driven by this "happy hour's demand" that he thought of telling the story of himself in the fastest and most natural way, and henceforth, *Specimen Days* (Whitman 1882, p. 3).

Although this book is an autobiography, nature is at its center from beginning to end, so it can be said that the theme of this book is, again, nature. It is present not only in the many "Nature Notes"; its influence is present even in the narration of war and people: soldiers like shadowy columns moving through the night; President Lincoln riding back to the White House after a night among the soldiers; many young soldiers dying in pain, silently, like leaves of grass. The beautiful weather posed a great contrast to the cruelty of war, shocking people's hearts. When commenting on English writer Thomas Carlyle, Whitman describes him as "rugged, mountainous, volcanic, he was himself more a French revolution than any of his volumes." When he learned of the death of Carlyle, Whitman felt the death of a great man together with the starry sky on a quiet night: "With me, too, when depress'd by some specially sad event, or tearing problem, I wait till I go out under the stars for the last voiceless satisfaction" (Whitman, p. 104).

An unknown contemporary commentator of Whitman's time says in an article on *Specimen Days* that

> his prose is better than his poetry. It is clean. It is sane. It is intelligible. It is often readable. Much of it is really interesting, because either of its autobiographical effect, its reminiscential quality, or its frank fresh mirroring of out-of-door life and sensation. The book is virile. In many places it has the smell of damp loam or of new-mown grass. ("All about Walt Whitman" 1882)

Alfred Kazin considered *Specimen Days* to be an old man's last pastoral song. We see, indeed, the singing of an old man who goes back to nature, the songs from the innermost heart when it is returned to nature. As my book is about American nature writing, I will focus on the chapters on nature notes and travels in western America in *Specimen Days*.

As early as in Emerson's *Nature* in 1836, Emerson had already issued the call to use nature as the textbook for studying nature. Many years later, we see in *Specimen Days* an old man reading the book of nature with childlike curiosity and playfulness. Sitting in the woods, Whitman writes at the beginning of his nature notes, saying that "after what I have put in the preceding fifty or sixty pages—temporary episodes, thank heaven!—I restore my book to the bracing and buoyant equilibrium of concrete outdoor Nature, the only permanent reliance for sanity of book or human life" (Whitman 1882, p. 61). He then explains that "the pages now ensuing may carry rays of sun, or smell of grass or corn, or call of bird, or gleam of stars by night, or snow-flakes falling fresh and mystic…" (p. 61).

The "nature notes" were written in this unique environment, and was presented as "the textbook of nature." The titles of the essays show the range of the course Whitman studied: "New Themes Entered upon," "The Lesson of a Tree," The Oaks and I," "A July Afternoon by the Pond," "The Sky—Days and Nights—Happiness," "To the Spring and Brook," "Birds and Birds and Birds," "A Winter Day on a Sea-Beach," "Wild Flowers," "A Sun-Bath Nakedness," and "Sundown Perfume—Quail Notes—The Hermit Thrush."

Whitman's classroom had none of Emerson's abstract colors, but real wild nature. Every "nature note" of his was written in wild nature. Every page exudes fresh air, showing scenery that the Hudson River School artists sketched in nature.

> *June 19th, 4 to 6-1/2, P. M.*—Sitting alone by the creek—solitude here, but the scene bright and vivid enough…
>
> …the quiver of leaf-shadows over the paper as I write.
>
> *Sept. 1.*—I should not take either the biggest or the most picturesque tree to illustrate it. Here is one of my favorites now before me, a fine yellow poplar, quite straight, perhaps 90 feet high, and four thick at the butt. How strong, vital, enduring! how dumbly eloquent! What suggestions of imperturbability and *being*, as against the human trait of mere *seeming*.
>
> I regularly come down to the pond, and ensconce myself under the lee of the great oak; I am here now writing these lines. The dark smoke-color'd clouds roll in furious silence athwart the sky; the soft green leaves dangle all around me; the wind steadily keeps up its hoarse, soothing music over my head—Nature's mighty whisper. Seated here in solitude I have been musing over my life—connecting events, dates, as links of a chain, neither sadly nor cheerily, but somehow, to-day here under the oak, in the rain, in an unusually matter-of-fact spirit. (Whitman 1882, pp. 63, 64, 75)

If, in *Leaves of Grass*, we see a passionate Whitman who called for the self of the Americans, then in *Specimen Days* we see a Whitman who calmly and carefully observed, reflected on, studied and was always surrounded by nature, incorporating sunshine, trees, blue sky, white clouds, grasses and wild flowers, and the singing of the birds into his life and his writing. It can be said that in *Specimen Days* we witness the best student of nature, the meaning of harmonious coexistence between man and nature, and the meaning of "enjoying an original relation to the universe" (Emerson, "Nature" [1836] [1903] 1979, p. 3). Unlike Emerson, Whitman felt deeply the effect of the American Civil War and the subsequent rapid industrialization. From nature he not only learnt "reason and faith," but also inspirations of how to survive and the matter-of-fact spirit.

## *Between the Creations of Nature and the Artifices of Man*

The difference between *Specimen Days* and *Leaves of Grass* is that the former was written after the author had experienced a cruel civil war and two strokes, when both the country and he himself urgently needed healing physically and spiritually. Like Mark Twain, Whitman believed that post-war industrialization in America happened too fast, and resulted in the loss of something courageous and wise, simple and sincere. The development of science was forming a scientific hell of science, and therefore many Americans looked back at this time to the country's "childhood," yearning for the simplicity and beauty of the past. The author, after experiencing tragedy both on a national and a personal level, also looks back to the farm on Long Island where there were orchards and fields. For Whitman childhood, whether America's or his own, was connected to nature. Nature was the mother of humanity, and she could heal both physical and spiritual wounds.

If we say that the Whitman who wrote *Leaves of Grass* expressed Emerson's ideas of love of nature and pursuit of the individual self, in *Specimen Days* we see more of Thoreau. Here Whitman becomes an energetic and vital Thoreau, walking under the blue sky and in the wilderness all day long. Nature is no longer an object of admiration, but a medication that heals the wounds, a nourishment that enhances the physical and spiritual health of man. We do not see an old man troubled with illnesses and with groans and complaints, but a cheerful and happy scene of a naughty boy playing in the arms of great Mother Nature: listening to the concert of nature by the pond, studying the lessons of the oaks, sunbathing in the sunny valley, and taking a cold bath in the clear river.

Whitman's feelings for nature at this moment are like those of a child for its mother, inseparable from each other. When he recorded all this in the "nature notes," there appeared a leisure and humour not seen before in other forms of literature.

> A solitary and pleasant sundown hour at the pond, exercising arms, chest, my whole body, by a tough oak sapling thick as my wrist, 12 feet high—pulling and pushing, inspiring the good air. After I wrestle with the tree awhile, I can feel its young sap and virtue welling up out of the ground and tingling through me from crown to toe, like health's wine. (Whitman 1882, p. 69)

> *Sunday, Aug. 27.*—Another day quite free from mark'd prostration and pain. It seems indeed as if peace and nutriment from heaven subtly filter into me as I slowly hobble down these country lanes and across fields, in the good air—as I sit here in solitude with Nature—open, voiceless, mystic, far removed, yet palpable, eloquent Nature. I merge myself in the scene, in the perfect day....
>
> ...Two months (July and August) have I absorb'd them, and they begin to make a new man of me. Every day, seclusion—every day at least two or three hours of freedom, bathing, no talk, no bonds, no dress, no books, no manners. Shall I tell you, reader, to what I attribute my already much-restored health? That I have been almost two years, off and on, without drugs and medicines, and daily in the open air....
>
> ...Never before did I get so close to Nature; never before did she come so close to me....
>
> ...Nature was naked, and I was also. (pp. 71–72)

Like Thoreau, Whitman took up the cup of health, the wine of nature, and sought childish pleasure and carefree freedom. However, as he was burdened with the suffering and pain of illnesses, his need for nature seemed to be more urgent than that of Thoreau. If Thoreau considered nature "good medicine," Whitman regarded it as an "invisible physician": "Thanks, invisible physician, for thy silent delicious medicine, thy day and night, thy waters and thy airs, the banks, the grass, the trees, and e'en the weeds!" (p. 72) In nature, Whitman completely let go of his constraints. He opened all his pores to absorb the nourishment of nature, and became a Thoreau who experienced nature with all his five senses.

> Perhaps the inner never-lost rapport we hold with earth, light, air, trees, etc., is not to be realized through eyes and mind only, but through the whole corporeal body, which I will not have blinded or bandaged any more than the eyes. (p. 90)

It is precisely Whitman's real connection and communication with nature that made everything in nature his friend. Through his personal experience of nature, he told people that nature is not only the "invisible physician" that heals people's physical and spiritual wounds and keeps us healthy and rational, but also the friend we cannot afford to be separated from.

Both Thoreau and Whitman had a yearning and passion for the American West. This impulse arose from their disgust at the "disease of beauty" brought about by modern civilization, and from their yearning for the wilderness which symbolizes freedom and vitality. In the autumn of 1879, Whitman set out on a

4-month, and 10,000-mile journey across the American Midwest. To Whitman, this was a new experience. On the vast plains of the Midwest Whitman felt, for the first time, helplessness as a poet. When he was invited to give a poem at a meeting in Kansas, he made the following speech:

> My friends, your bills announce me as giving a poem; but I have no poem—have composed none for this occasion. And I can honestly say I am now glad of it. Under these skies resplendent in September beauty—amid the peculiar landscape you are used to, but which is new to me—these interminable and stately prairies—in the freedom and vigor and sane enthusiasm of this perfect western air and autumn sunshine—it seems to me a poem would be almost an impertinence. (p. 90)

After extoling the beauty of the Midwest's prairies, Whitman calls them "the home both of what I would call America's distinctive ideas and distinctive realities" (p. 90).

What attracted Whitman the most was the wildness of the Midwest, the place where Thoreau placed the future and hope. To Whitman, the whole of the American West was, in a sense, the interminable mountain ranges and the continuous flow of the rivers. For Thoreau, going westward remained imaginary, while for Whitman it was a real, personal experience. This experience resulted in two questions, one related to literary art and the other to the environment.

In *Specimen Days* Whitman writes that

> the pure breath, primitiveness, boundless prodigality and amplitude, strange mixture of delicacy and power, of continence, of real and ideal, and of all original and first-class elements, of these prairies, the Rocky mountains, and of the Mississippi and Missouri rivers—will they ever appear in, and in some sort form a standard for our poetry and art? (Whitman 1882, p. 95)

While Emerson and Cole proposed to seek literary and artistic inspiration in nature and Thoreau proposed to use a language connected with the land, in *Specimen Days* Whitman proposes to use nature as a standard to measure literary art.

As an artist and a socially responsible person, Whitman also raises the "tree question" on his journey in the Midwest.

> The tree question will soon become a grave one. Although the Atlantic slope, the Rocky mountain region, and the southern portion of the Mississippi valley, are well wooded, there are here stretches of hundreds and thousands of miles where

> either not a tree grows, or often useless destruction has prevail'd; and the matter of the cultivation and spread of forests may well be press'd upon thinkers who look to the coming generations of the prairie States. (Whitman 1882, p. 95)

Commentaries on Whitman tend to focus on the poet and neglect his other achievements as a nature conservationist and prophet who raised the "tree question" as early as the autumn of 1879. In my view, Whitman's proposal to use nature as the standard against which literature and art should be measured and his consideration of the "tree question" made him one of the major founders of American nature writing.

Whitman finished his travels in the Midwest in January 1880, and the first edition of *Specimen Days* was published two years later. Another decade on, Whitman passed away in his residence in New Jersey. In my copy of the 1971 edition of *Specimen Days*, there are photos of the tomb and the graveyard, designed by Whitman himself: located on a mountain slope covered with vegetation and surrounded by trees, grassland, blue sky and mountains, everything is natural and of nature.

Charles R. Metzger, in the introduction to his *Thoreau and Whitman: A Study of Their Esthetics* (1961), says that as contemporaries Emerson, Thoreau and Whitman all held transcendentalist views, but that, although Thoreau and Whitman put forward some very interesting ideas, they themselves insisted that they were practitioners, not theorists. Metzger suggests that, in fact, Whitman combined Emerson's theory with Thoreau's practice (Metzger, p. iv). In my view, as far as their attitude toward nature goes, this is an appropriate representation of their respective significance. In Emerson's eyes, nature was a spiritual symbol. Emerson loved nature, but in essence he was still in love with society. He believed that people existed for society, just as the eyes were for seeing and the ears for hearing (Porte, p. 12). Thoreau liked wild nature. He avoided society so he could go deep into nature, as he believed that "in society you will not find health, but in nature" (quoted in Metzger, p. 7). Whitman is the bridge between the abstraction of Emerson and the concreteness of Thoreau, the combination of culture and nature. Whitman believed that human creations and creations of nature were equally great. He wrote:

> It was a happy thought to build the Hudson river railroad right along the shore…I like both sight and sound. (Whitman 1892, p. 132)

Nature writing before the twentieth century had a very strong idealistic tone. Authors held optimistic and positive attitudes toward nature. They wished to seek freedom for the individual, the roots of the culture and the sublimation of the spirit. However, their thinking and writing were limited to the relationship between nature and the self or between nature and individual thoughts and behavior. It was only in the twentieth century, with the appearance of nature writers like John Burroughs, John Muir and Mary Austin, that nature writing took on a broader perspective.

## References

Abbey, Edward. (1968) 1971. *Desert Solitaire: A Season in the Wilderness.* New York: Ballantine Book.

Abrams, Mayer Howard, and Geoffrey Galt Harpham. 2009. *A Glossary of Literary Terms,* 9th ed. Boston: Wadsworth.

Baym, Nina, and Ronald Gottsman, eds. 1989. *The Norton Anthology of American Literature,* 3rd ed. Vol. I. New York: Norton.

Buell, Lawrence. 1973. *Literary Transcendentalism: Style and Vision in the American Renaissance.* Ithaca: Cornell University Press.

Buell, Lawrence. 1996. *The Environmental Imagination: Thoreau, Nature Writing, and the Formation of American Culture.* Second printing. Cambridge, MA: Harvard University Press.

Buell, Lawrence. 2004. *Emerson.* Cambridge, Massachusetts: Belknap Press of Harvard University Press.

Burke, Kenneth. 1966. "I, Eye, Ay—Emerson's Early Essay on "Nature": Thoughts on the Machinery of Transcendence." *The Sewanee Review.* 74.4:875–895.

Burroughs, John. 1896. *Whitman: A Study.* Boston: Houghton Mifflin. http://www.gutenberg.org/files/30342/30342-h/30342-h.htm

Cole, Thomas. (1836) 2011. "Essay on American Scenery," in Grese, Robert ed. *The Native Landscape Reader.* Amherst: University of Massachusetts Press, pp. 27–36.

Emerson, Ralph Waldo. (1903) 1979. Centenary Edition, *The Complete Works of Ralph Waldo Emerson.* Vol. I: Nature Addresses and Lectures, 2nd ed. New York: AMS Press.

Emerson, Ralph Waldo. 1979. Centenary Edition, *The Complete Works of Ralph Waldo Emerson.* Vol. IX: Poems, 2nd edition of the centenary edition. New York: AMS.

Emerson, Ralph Waldo. (1862) 1989. "Thoreau" in Baym, Nina, and Ronald Gottsman, eds. *The Norton Anthology of American Literature,* 3rd ed. Vol. I. New York: Norton, pp. 1033–1045.

Finch, Robert, and John Elder. eds. 1990. *The Norton Book of Nature Writing.* New York: Norton.

Fritzell, Peter A. 1990. *Nature Writing and America: Essays upon a Cultural Type.* Iowa City, IA: Iowa State University Press.

Frothingham, Octavius Brooks. 1972. *Transcendentalism in New England: A History.* Philadelphia: University of Pennsylvania Press.

Gifford, Terry.1996. "The Social Construction of Nature." *ISLE* 3.2 (1996): 27–35.

Harding, Walter. 1982. *The Days of Henry Thoreau: A Biography*. New York: Dover Publications.

"Hudson River School." *I Hear America Singing (IHAS)*. N.d. Web. 17 May 2013. <http://www. pbs. org/wnet/ihas/icon/hudson.html>.

Huth, Hans. 1957. *Nature and the American: Three Centuries of Changing Attitudes*. Berkeley: University of California Press.

Irving, Washington. 1821. *The Sketch Book of Geoffrey Crayon, Gent*. New ed. London: J. Murray.

Koster, Donald N. 1975. *Transcendentalism in America*. Boston: Twayne Publishers.

Lyon, Thomas J., ed. 1989. *This Incomperable Lande: A Book of American Nature Writing*. Boston: Houghton Mifflin Company.

Marshall, Ian. 1998. *Story Line: Exploring the Literature of the Appalachian Trail*. Charlottesville, VA: University Press of Virginia,

Matthiessen, F. O. 1941..*American Renaissance: Art and Expression in the Age of Emerson and Whitman*. Oxford: Oxford University Press.

Metzger, Charles R. 1961. *Thoreau and Whitman: A Study of Their Esthetics*. Seattle, WA: University of Washington Press.

Milder, Robert. 1995. *Reimagining Thoreau*. Cambridge: Cambridge University Press.

Miller, James E., Jr., ed. 1991. *Heritage of American Literature: Beginnings to the Civil War*. Vol. I. San Diego, CA: Harcourt Brace Jovanovich.

Miller, Perry. 1971. *The Transcendentalists: An Anthology*. Cambridge, MA: Harvard University Press.

Myerson, Joel. 2000. *Transcendentalism: A Reader*. Oxford: Oxford University Press.

Nash, Roderick. 1967. *Wilderness and the American Mind*. Second printing. New Haven, CT: Yale University Press.

Paul, Sherman, ed. 1962. *Thoreau: A Collection of Critical Essays*. Englewood Cliffs, NJ: Prentice-Hall.

Paul, Sherman. 1992. *For Love of the World: Essays on Nature Writers*. Iowa City, IA: University of Iowa Press.

Plato, *The Republic*, translated by Benjamin Jowett, https://www.gutenberg.org/files/1497/1497-h/1497-h.htm#link2H_4_0013, accessed on 12 December 2020.

Porte, Joel, ed. 1982. *Emerson in His Journals*. Cambridge, MA: Belknap Press of Harvard University Press.

Richardson, Robert D. Jr. 1995. *Emerson: The Mind on Fire: A Bbiography*. Berkley: University of California Press.

Ronald, Ann, ed. 1987. *Words for the Wild*. San Francisco: Sierra Club Books.

Ruland, Richard, and Malcolm Bradbury. 1991. *From Puritanism to Postmodernism: A History of American Literature*. New York: Penguin Books.

Rusk, Ralph L. 1957. *The Life of Ralph Waldo Emerson*. New York: Columbia University Press.

Sanders, Scott Russell. 1993. *Staying Put: Making a Home in a Restless World*. Boston: Beacon Press.

Scheese, Don. 1996. *Nature Writing: The Pastoral Impulse in America*. New York: Twayne Publishers.

Slovic, Scott. 1992. *Seeking Awareness in American Nature Writing*. Salt Lake City, UT: University of Utah Press.

Spiller, Robert. 1955, 1967, the 10th print. *The Cycle of American Literature: An Essay in Historical Criticism*. New York: The Macmillan Company.

St. Armand, Barton Levi. 1980. "Luminism in the Work of Henry David Thoreau: The Dark and the Light." *The Canadian Review of American Studies* 11.1, pp. 13–30

St. Armand, Barton Levi. 1989. *Emily Dickinson and Her Culture: The Soul's Society*. Cambridge: Cambridge University Press.

Thoreau, Henry David. 1975. *Essays, Journals, and Poems*. Ed. Dean Flower. Greenwich: Fawcett Publications.

Thoreau, Henry David, (1947) 1977. *The Portable Thoreau*. Ed. and with introduction by Carl Bode, New York: The Viking Press; Penguin Books.

Thoreau, Henry David. (1862) 2002. *Wild Apples and Other Natural History Essays*. Ed. William Rossi. Athens, GA: University of Georgia Press.

Thoreau, Henry David. (1854) 2004. *Walden*. Boston: Shambhala.

Turner, Frederick. 1992. *Spirit of Place: The Making of an American Literary Landscape*. Washington, D.C.: Island Press,

Whitman, Walt, 1882. *Specimen Days and Collect*. London: Academy Publishing Co.

Whitman, Walt, 1892. *Complete Prose Works*. Philadelphia: David Mckay Publisher.

Whitman, Walt. (1882) 1971. *Specimen Days*. Boston: David R. Godine Publisher, with introduction by Alfred Kazin.

"All About Walt Whitman" 1882. https://whitmanarchive.org/criticism/reviews/days/anc.00102.html. Accessed on 11 December 2020

Yaeger, Bert D. 1996. *The Hudson River School: American Landscape Artists*. New York: Smithmark Publishers.

# 4

# The Wilderness Complex Reflected in Nature Writing

## Introduction

John Burroughs, John Muir and Mary Austin played important roles in American nature writing. As writers straddling the nineteenth and twentieth centuries, they promoted connections between man and nature and, in the twentieth century, further developed the nature views of Emerson, Thoreau and others. From these three writers, we can also see the expansion of place in nature writing, as well as diversity of participants. The appearance of John Burroughs in the Catskill Mountains in the east and John Muir in the western mountains shows that nature was no longer the local product of Concord. The "song of the land" that Emerson, Thoreau and others promoted had already broken away from New England, crossed the plains of the Midwest and resounded in the mountains in the West. The appearance of Mary Austin added a female voice to the predominantly male nature writing, making American nature writing a multifaceted composition. From the friendship between John Burroughs and John Muir and the contact between Muir and Mary Austin, we also see the beginning of the unique phenomenon of writer groups driving the development of American nature writing.

Speaking in California in 1911, two years after leaving office, former president Theodore Roosevelt said that the world needed more people with scientific imagination, people who recorded scientific facts accurately, explained them clearly, and also gave them a poetic soul, and thus made them literary works. He made it clear that he was referring to people like Burroughs and Muir, both in the audience at the time and both in their seventies. In American nature writing, Burroughs and Muir are often mentioned together, not only because they had the same given names but also because of their similar adventures, their similar experiences and their equally important positions in nature writing.

Burroughs and Muir were very close in age, with the former just 1 year older than the latter. Both were very interested in nature, although the focuses of their attention and writing were different. Burroughs was referred to as "John o' Birds" while Muir was "John o' Mountains." Both were admirers of Emerson and advocators of transcendentalism. Burroughs, living in New York State in the east, attended one of Emerson's lectures and met him in person. Muir accompanied Emerson's travels around "the Yosemite" (Yosemite Valley), and established a friendship with the sage from Concord. Both Johns lived when Theodore Roosevelt was president, each accompanied him on his different trips of inspection of the American Midwest, and played an important role in Roosevelt's nature conservation measures. In 1899 both participated in the Alaska Expedition, and Muir later wrote a book titled *Travels in Alaska* (1915). Both set new standards and forms for nature essays of the time, and helped to establish the nature essay's definite position in American literature. However, they had different writing styles, with Burroughs's being gentle and subtle like Emerson's and Muir's rougher and livelier like Thoreau's. Each had a distinctive personality. More importantly, both—as writers at the turn of the century—served as a bridge between the two centuries for nature writing. Their languages and actions taught Americans how to appreciate the nature around them. Before, land was something to be experimented on or to be developed, while after them, it was admired and treasured.

American nature writer Ann H. Zwinger edited and published *Mary Austin and John Muir: Writing the Western Landscape* in 1994. In the introduction to this collection of Austin and Muir's essays Zwinger tells how the two writers met, and why she put their works together in one collection. Zwinger believed that Muir and Austin were the first "group" of nature writers in the American West. From their time on and even today, their writings, views and descriptions of the mountain ranges and deserts in the west have been influencing the nature writing of the American West (pp. vii–viii).

While both Muir and Austin inherited the tradition of transcendentalism and were deeply influenced by Emerson and Thoreau, they went beyond their predecessors on the East Coast by observing and analyzing nature in a more direct way. Emerson and Thoreau had only gone for walks in the countryside of Concord, they had never climbed the granite mountain cliffs of the West or walked through the waterless desert. Muir and Austin chose the most desolate, least welcoming natural environment to explore their spiritual pursuit on the land. Thoreau had been to the mountainous areas of Maine, but he was only there as a hurrying passerby, spending the majority of his time in the relatively gentle natural environment of Concord and expressing his love of and yearning for the West and the wilderness from there. In contrast, Muir explored the high mountains in the West and based *The Yosemite* on the six years he spent living in the mountains, while Austin broke away from the tradition of the gentlewoman by going to live in a small town in the desert for 12 years, and her work *The Land of Little Rain* was based on her experiences there. More importantly, from the experiences and works of Muir and Austin we see nature writing "get over the color green." This idea of getting over the color green came from the American historian and writer Wallace Stegner. Known for his love and descriptions of the deserts and mountains, Stegner writes in his *Thoughts in a Dry Land* (1972) "You have to get over the color green; you have to quit associating beauty with gardens and lawns; you have to get used to an inhuman scale; you have to understand geological time" (quoted in Scott Slovic's *Getting over the Color Green: Contemporary Environmental Literature of the Southwest*, 2001, p. xvii). Ann Ronald, who lived in the western state of Nevada for many years, summarizes Stegner's views and provides an explanation. She says that modern people tend to prefer green and blue to ocher and beige, and to value redwoods and oceans more than short bushes. Stegner believed this was because modern people inherently lacked appreciation of such things as sagebrush, raw earth and alkali flats, considering these things to be acquired tastes which were not easily adopted by conventional sensibilities (quoted in Slovic 2001, p. xix). While Stegner's exhortation to "get over the color green" was published in 1972 and Ronald commented on this in 1995, Austin, a woman writer, had already experienced life on the red earth and in the desert at the turn of the century, writing her "get-over-the-color-green" book *The Land of Little Rain* in 1903. This representative work of Austin's not only showed the desert as a symbol of wilderness with aesthetic value; her holistic understanding of the desert also outlined the core idea of the "land community" proposed by Leopold nearly 50 years later in *A Sand County Almanac: And Sketches Here and There*, again showing Austin's considerable foresight.

Muir and Austin described a multidimensional nature and a land co-owned by plants, animals and humans, therefore becoming spokespeople for nature. Different from their predecessors, both Muir and Austin personally participated in the fight to preserve nature and witnessed their defeat: Muir saw his beloved Hetch Hetchy valley cut by a dam, while Austin saw the green Owens Valley where she once lived turned into an alkali desert. Perhaps it was this overwhelming industrialization and unsustainable development that made Muir and Austin see the consequence, for both humanity and the world, of the shrinking of wild nature. In their works, therefore, in addition to description and appreciation of nature there is passion in its defense and rational wisdom regarding its protection. Muir and Austin broke through the limitations of place. "So, by the end of the 1800s, the contemplative spirit of the transcendentalists had married the energy of the pioneers, metaphorically speaking, and the two together were giving birth to a host of twentieth-century progeny" (Ronald, p. 42).

## Lecture 11. Burroughs: John o' Birds

On 3 April 1912, in the lounge outside the Hall of Birds of the World in the American Museum of Natural History, sat an old man with white hair. It was John Burroughs, and he turned 75 that day. Around Burroughs were more than 600 children from various countries. They learned from their teachers that this old man who looked like Santa Claus was actually a great writer who wrote about nature. Children read and recited Burroughs's well-known essays and sayings, and waited for the old man to give his speech. However, they were surprised when Burroughs told them that everything in the museum was sham, and that he felt as if he was at a funeral every time he visited a museum. He advised the children not to go to the museums to look for nature, but to ask their parents to take them to the park or the beach to see the sparrows flying around, hear the calls of the seagulls and follow the squirrel to its nest in the oak tree. Nature "is only real when you reach out and touch it with your hands" (quoted in Edward J. Renehan Jr., *John Burroughs: An American Naturalist*, 1992, pp. 7–9).

John Burroughs once said, "I am myself never so well pleased as when I can bring [my readers] a fresh bit of natural history, or give them a day in the fields and woods or along the murmuring streams" (in Ronald, p. 93). In his life and works Burroughs had one aim: to send people to Mother Nature. Unlike his transcendentalist predecessors, Burroughs had no profound new theories. Unlike his contemporary John Muir, he was not a man of action; Burroughs was a quiet man. Out of his love of nature, in his farmhouse by the Hudson River, he synthesized all sorts of nature essay forms and told nature stories one after another. If we compare American nature writing to music, John Burroughs's works would be its gentlest pieces; and yet, like *The Blue Danube*, also its most fascinating.

Burroughs wrote more than 20 books, most of them descriptions of nature, especially of birds, including the nature essay collection *Wake-Robin* (1871) and later *Winter Sunshine* (1875), *Birds and Poets* (1877) and *Locusts and Wild Honey* (1879). The scenery he describes was mostly familiar and accessible: woods, fields, birds and animals, winning many readers.

Ronald (p. 91) states that Burroughs was the most popular writer of his time, pointing out that even Muir's writing could not compare with Burroughs's 25 books and their record sale of 1.5 million copies. Another American writer, Ralph H. Lutts, in his *The Nature Fakers: Wildlife, Science & Sentiment* (1990), praises Burroughs's literary achievement, his vast natural knowledge, his prolific writing and the special position he occupied in nature writing, saying that among the nature writers at the time, Burroughs belonged to both the earlier generation and the contemporary. He also explains that in Burroughs's time many people—including President Theodore Roosevelt—grew up with his books and followed his example. "He helped to establish nature study as a popular pursuit, set many of the standards for nature writing, and gave the natural history essayist definitive form" (p. 8).

## *The Forest is Awake*

For John Burroughs (1837–1921), his love of nature and of writing about it came, to a large extent, from his childhood experiences, and later the influences of Emerson, Thoreau and especially Whitman.

Burroughs was born on a farm in the Catskill Mountains. His forebears were farmers. According to his memory, his father was a good husband and citizen, but "had no sentiment about things and was quite unconscious of the beauties of nature over which we make such an ado. The primrose by the river's brim would not have been seen by him at all." Burroughs's mother was illiterate, but she liked to take him to their farm to pick strawberries. There was grassland on the farm that had wildflowers, and from which he could see the long ranges of the Catskill Mountains. It was from his mother that Burroughs learned to love nature. The colorful songbirds in the woods of his childhood were another reason he was fascinated by nature. His biographer Clara Barrus, in *The Life and Letters of John Burroughs* (1968), told the reader that when Burroughs was about 7 or 8 years old, he found a songbird in the woods that was pale blue with white spots on the wings. This bird was the guide that led him to the kingdom of birds and inspired him to discover more beautiful birds. Also unforgettable to Burroughs was a flock of pigeons that flew into the forest of beeches on a spring morning, making the sky, the land and the woods all blue. When he grew up, his enthusiasm for birds was further enhanced by Audubon's masterpiece *The Birds of America*, and this eventually made him the spokesperson for the kingdom of birds (Paul Brooks, *Speaking for Nature: How Literary Naturalists from Henry Thoreau to Rachel Carson Have Shaped America*, 1980, pp. 5–6).

The guides that led Burroughs into the kingdom of nature writing were Emerson, Thoreau and Whitman. James D. Hart's *The Oxford Companion to American Literature* (1995) says that Burroughs "through keen observation of his native Catskill and under the influence of Emerson and Thoreau became the greatest writer of nature essays after his two Transcendentalist masters" (p. 99). From the age of 19, Burroughs wrote essays and also taught. Later, when he discovered the pale blue songbird, he also discovered Emerson and Transcendentalism. Talking of Emerson's influence on him, Burroughs says: "I read him in a sort of ecstasy. I got him in my blood, and he colored my whole intellectual outlook. He appealed to my spiritual side; his boldness and unconventionality took a deep hold upon me" (quoted in Brooks, p. 6).

According to Ralph Rusk in his biography of Emerson (*The Life of Ralph Waldo Emerson*, [1949] 1957), after attending one of Emerson's lectures for the first time at West Point in 1863, Burroughs was so excited at having seen the "sage of Concord" that he could not fall asleep. He later went with his friends, country poet Myron Benton and country writer Donald Mitchell, to visit the transcendentalist master. The three of them found Emerson by the side of the Hudson River, about to leave by ferry. They followed Emerson through the crowd toward the wharf, Emerson talking to them as they walked, and as he bought his ticket and boarded the boat; and as it moved away, Emerson was still talking to them (p. 421). Burroughs's fascination with Emerson in his youth led him to regard Emerson as his "spiritual father." His early essays were once thought to be written by Emerson. Burroughs felt this to be both an honour and a worry, as he knew that to learn the essence of Emersonian spirit was to find oneself and the characteristics of the individual self. Therefore, he looked back at his childhood and his hometown where the woods resounded with the singing of birds and wild strawberries grew in the fields. We can say that as a young writer Burroughs showed signs of transcendentalism, but when he became more mature, he found his own voice: all his writing was based on the firsthand outdoor materials and nature.

Burroughs's thinking and writing were mainly carried out in his hometown in the Catskill Mountains region in the state of New York state, so his works all have a strong flavor of the land there. This style, writing from the microcosm to understand the macrocosm, came from Thoreau. Burroughs agreed with Thoreau's view that a person "only has to stay at home and see the procession pass by" (in Ronald, p. 92), and he put this overarching theme into his essays. Burroughs's friend William Sloane Kennedy, in his *The Real John Burroughs* (1924), writes that "the idée fixe of John o's the Catskills is that there is no place like his home locality, and there are no birds, anywhere in the world equal to those of his native hills" (quoted in Lyon 1989, p. 64). This was a comment reminiscent of Thoreau: Thoreau talked about nature as if nature was a native of Concord. Similarly, if Thoreau considered Concord the microcosm of the universe, Burroughs regarded the Catskills the same way. Both held firmly onto the spirit of the land under their feet and felt and saw the outside world through it.

Like Thoreau, Burroughs was a habitual diary writer. He captured the fast-changing natural phenomena: the changes of the seasons, the arrival of birds, the blooming of flowers, and recorded them all in his diaries. In the "Spring Jottings" chapter of his book *Riverby* (1894), he writes about the benefits of writing diaries:

> I remember that Thoreau says in a letter to a friend, after his return from a climb to the top of Monadnock, that it is not till he gets home that he really goes over the mountain; that is, I suppose, sees what the climb meant to him when he comes to write an account of it to his friend. Every one's experience is probably much the same; when we try to tell what we saw and felt, even to our journals, we discover more and deeper meanings in things than we had suspected. (Burroughs, p. 156)

After describing his own journey to Maine, Burroughs expressed the same feelings: "It was not till after I got home that I really went to Maine" (Burroughs, p. 157). Burroughs's views on diary writing helps explain why so much nature writing was written in diary and journal form. It also echoes an ancient Chinese saying: "The reason we cannot see the real face of Mount Lu is that we are right in the midst of it." On the necessity for detachment sometimes to allow clearer perception, the West and the East concurred.

As for Whitman, Burroughs and he complemented each other in both thought and artistic style. They met in Washington DC in 1863, which means that Burroughs met both Emerson and Whitman in the same year. At the time

Whitman was a government administrator, looking after soldiers in Washington Hospital who had been wounded in the Civil War. They liked each other from the moment they met and became lifelong friends. They would take walks along the streets in Washington and in the countryside around the city, nature's observer and nature's poet accompanying each other. According to Rusk in his biography of Emerson, in the winter of 1870, when Emerson was lecturing in Baltimore, his two most passionate disciples, Whitman and Burroughs, traveled together from Washington to Baltimore to hear him speak (p. 449). Whitman, in his *Specimen Days* ([1882] 1971, p. 84), describes his stay with Burroughs in his country house. Friendship with Whitman evidently expanded and deepened Burroughs's understanding of nature, and he started to assume a romantic view of nature that went beyond the scenery. Burroughs, in his *Notes on Walt Whitman: as Poet and Person* (1867), wrote:

> The word Nature, now, to most readers, suggests only some flower bank, or summer cloud, or pretty scene that appeals to the sentiments. None of this is in Walt Whitman. And it is because he corrects this false, artificial Nature, and shows me the real article, that I hail his appearance as the most important literary event of our times. (quoted in Lyon, p. 65)

Burroughs regarded Whitman as a man returning to nature and *Leaves of Grass* as song of nature. He was a life-long supporter of Whitman and his works, and his *Notes on Walt Whitman* was not only his first published book but also the first book about the poet (Brooks, p. 9).

Burroughs and Whitman influenced each other in their literary creation, and we can say that Burroughs was Whitman's guide in observing nature. It is said that Whitman got the inspiration from Burroughs for the singing of the thrush in his poem "When Lilacs Last in the Dooryard Bloomed" (1865)[1] Whitman chose

1 According to Renehan's biography *John Burroughs: An American Naturalist* (1992), in 1865 Burroughs, who had just returned from his home to Washington, excitedly told Whitman that he heard the singing of the hermit thrush when he was climbing the mountains at home. Whitman made notes: "Sings oftener after sundown…is very secluded…likes shaded dark places. His song is a hymn…never sings near farm houses—never in the settlement—is the bird of the solemn primal woods & of nature pure & holy." At the time, Whitman was writing an elegy for Abraham Lincoln. He used the mournful singing of the hermit thrush as the foundation tone for his classic memorial to Lincoln.

the title for Burroughs's first nature essay collection, *Wake-Robin*. As Burroughs was usually regarded as the spokesperson for birds, people often thought the title meant the robin that is awake. However, according to Burroughs's explanation in the introduction to the first edition of the collection, wake-robin is actually a small white flower that bloomed in early spring in North America and was thought to indicate the return of migratory birds (p. vi). The title shows a poet's sense of suggestive beauty. In terms of titles, Whitman's first collection of poems *Leaves of Grass* and Burroughs's first collection of essays *Wake-Robin* are like sister works, showing the infinite connections between the poet and the nature writer. It can be said that Whitman guided Burroughs onto the literary road. Whitman believed that you needed to capture and express the force of nature and the content of beauty with a painter's eyes and a poet's ears. Under his influence, Burroughs was determined to free birds in nature from the constraints of scientists. He wanted to form a unique nature writing: in accordance with the facts of natural history but also indicative of the poetry and painterly beauty of life in the woods.

> The dandelion tells me when to look for the swallow, the dogtooth violet when to expect the wood-thrush, and when I have found the wake-robin in bloom, I know the season is fairly inaugurated. With me this flower is associated, not merely with the awakening of Robin…
>
> …but with the universal awakening and rehabilitation of nature. (Burroughs, *Wake-Robin*, 1871, pp. 3–4)

This description in *Wake-Robin* shows us the unique charm of Burroughs's nature essays.

## Guide to Nature

Burroughs was called the "saint of the American countryside," and "guide to nature." While he benefited from Emerson, he was also different from Emerson. The strongest charm of Emerson's works lies in the "self" and in leading the readers to the individual. Burroughs was different: "My books do not bring readers to me, but send them to nature" (quoted in Buell 1996, p. 190). No wonder it's said that reading Burroughs's books is like enjoying a summer holiday and taking walks in the fields, as his books are records of his life. He not only set the standard for nature writing, but also showed people a way of life that is close to nature and respectful of nature, and made this fashionable.

In 1873, Burroughs purchased a nine-acre farm to grow fruit, and also designed and built his house, naming it "Riverby." In 1895 he also built a simple cabin called "Slabsides," in the mountains about two miles from Riverby. Burroughs spent almost all of his last four decades in these two houses close to nature. He lived a farmer's and a writer's life there, expressing his wishes with both his pen on paper and his hoe on the land. Most of his writing was completed here, including *Fresh Fields* (1884), *The Light of Day* (1900), *Ways of Nature* (1905), *Bird and Bough* (1906), *Leaf and Tendril* (1908) and *Under the Maples* (1913), and they all exude a strong air of nature. Whitman, in a letter to a friend, praises Burroughs: "John has the real art—the art of succeeding by not trying to succeed: he is the farmer first—the man before he is the writer; that is the key of his success" (quoted in Renehan, p. 171). English writer Edward Carpenter, in his autobiography *My Days and Dreams, Being Autobiographical Notes* (1916), describes his visit to Burroughs's home by the Hudson River. His impression of Burroughs was that he had a "tough reserved farmer-like exterior, some old root out of the woods, one might say—obdurate to wind and weather." In his letter to Whitman he described Burroughs more vividly, as "a poet with field-glasses, a Thoreau but much more friendlier, who dresses the farmer, talks the scholar, and has studied well the book of nature" (quoted in Renehan, p. 131).

For Burroughs, his two small houses became not only the background where he studied and described nature, but also the venue where he exchanged views on nature with fellow nature lovers, and the base where the new generation of nature lovers were nurtured.

John Muir was one of the first guests invited to Slabsides. In 1896, when Burroughs heard that Muir was coming from the West to stay in the Northeast for a few weeks, he invited him to Slabsides. When told that he would stay the night in Slabsides, Muir said, "Oh, anywhere there in the woods will do; I'm at home anywhere out-of-doors" (quoted in Renehan, pp. 205–206). Burroughs, in his later diaries, described Muir as

> a poet and almost a Seer. Something ancient and far-away in the look of his eyes. He could not sit down in a corner of the landscape, as Thoreau did; he must have a continent for his playground. (quoted in Renehan, pp. 205–206)

Burroughs also received Theodore Dreiser, a journalist with *Success Magazine* at the time, at Slabsides. Dreiser, who later became a well-known writer, came to Burroughs to seek the "key of success." Others who visited Slabsides included American President Theodore Roosevelt, who called Burroughs "Oom John"

(Uncle John). Many years later, reminiscing about Roosevelt's visit, Burroughs described how Roosevelt "clenched his fists and gritted his teeth and came up like race-horse!" (quoted in Renehan, p. 254).

Of course, the most frequent visitors to Slabsides also included students of the nearby Vassar College. Burroughs always took them to the evergreen woods and the river, where there was a waterfall. While the students sat on the rocks for their picnic, he would stand on the steps of Slabsides and advised them not to "turn into closet naturalists" but to "learn to seek out nature on its own terms, alive in field and wood" (quoted in Renehan, p. 183).

Perhaps because Burroughs's life—with its closeness to nature—gave his writing a unique flavor, we tend to be in a very playful mood when we read his work. We feel that, rather than reading a book, we are traveling in the Hudson Valley in the company of Burroughs, getting acquainted with the birds, appreciating the fresh woods, and exploring the mysteries of nature. We go to the famous Hudson Valley to listen to the concert of birds; we go into to the evergreen woods that are full of fresh virginal air to watch the joy of birds making nests; we enter the wilderness, which is only two miles from the White House where Lincoln was sworn into presidency, and at the time was still a paradise of birds and wildflowers. We see a blue bird resting in Burroughs's own garden, listen to Burroughs extol the fact that the sunshine and the sky of the new continent have covered it with the shiny color of the land, making it more brilliant than its European cousin. No wonder that "the dusk and cool and quiet of the forest seem to wrap the reader of his book, and it is a sort of summer vacation to turn its pages" (Brooks, pp. 9–10). For people who live in the big cities and are busy as bees all day long, such a spiritual journey cannot but be a pleasure. From this, we can understand that nature has not only material value, but also spiritual value.

Among Burroughs's numerous works, *Wake-Robin* is generally considered to be the masterpiece. The uniqueness of this book lies in the extraordinary closeness and communication between the author and nature. It contains joy and pleasure spontaneously flowing from the heart, not the kind the ancient Chinese literati tended to have "in order to write a poem so pretending a sadness that is not felt." As Burroughs says in the introduction, "the writing of the book was only a second and finer enjoyment of my holiday in the fields or woods." He considered his observation of birds in the woods as seeing friends: "I saw many old friends and made some new acquaintances." The birds' singing was "like the voice of an old friend speaking my name" (Burroughs 1871, p. 79).

In *Wake-Robin*, Burroughs describes the most wonderful season of the year in the woods, April to August: birds arrive one after another—robins, bluebirds, golden brown orioles and colorful hummingbirds—filling the quiet forest with joy and liveliness. He believed that only when he listened to the singing of a bird did he understand it, because the singing contained the force of life. He vividly described the meaning of the songs of various birds: thrushes' singing expressed spiritual serenity, sparrows' symbolized faith, bluebirds' signified love, catbirds' was proud and robins' suggested something military. In the habits of various birds, he saw a similarity with humanity: in the world of birds, feminism prevailed. Male birds always circled around female birds and the female birds were the head of a family. He analyzed the different personalities of male and female birds: the former's life was poetic and romantic, while the latter's was more practical and responsible. He examined the different appearances of birds: the calm and dignified hawk, the elegant warbler, the royally indifferent black-billed sparrow, the chattering imitation birds, the solitary, tame and quiet cuckoo and the graceful thrush. Under Burroughs's pen, the world of the birds is filled with vivid stories, vivacious pictures, funny comedies and profound philosophies—a realm worth exploring and learning from.

Like Thoreau and Whitman, Burroughs was especially attracted to the hermit thrush. In *Wake-Robin*, he describes it thus:

> Ever since I entered the woods, even while listening to the lesser songsters, or contemplating the silent forms about me, a strain has reached my ears from out of the depths of the forest that to me is the finest sound in nature,—the song of the hermit thrush. I often hear him thus a long way off, sometimes over a quarter of a mile away, when only the stronger and more perfect parts of his music reach me; and through the general chorus of wrens and warblers I detect this sound rising pure and serene, as if a spirit from some remote height were slowly chanting a divine accompaniment. This song appeals to the sentiment of the beautiful in me, and suggests a serene religious beatitude as no other sound in nature does. (p. 51)

Huth (p. 102) points out that "When John Burroughs first began to write, there was little 'news' to make headlines for nature essays. However, he set a new standard for the nature essay, which soon occupied a definite niche in literature." Burroughs not only gave nature essays a definite form, but also created a way of studying nature in which personal experiences mingled with admiration. He compared the nature writer to a bee and the process of writing to that of making honey and believed that "nectar does not become honey until it has passed through the chemistry of the bee's body. Similarly, observations of nature must be processed within the writer's mind. The final product, though, still is true to the raw material" (quoted in Lutts, p. 9). Burroughs himself was the best practitioner of this way of studying nature. He believed that "unless you can write about nature with feeling, with love, with more or less hearty affiliation and comradeship.... It is no use" (quoted in Buell 1996, p. 190). Throughout his life he pursued this kind of affiliation and harmony between man and nature, and his moods and feelings were closely linked to the nature surrounding him. He regarded the natural scenery around him as an expression of his inner self. Trees were no longer unfeeling materials, they had been infused with feelings by the people who loved and described them, becoming a scenery of spirit.

Burroughs writes that

> one's own landscape comes in time to be a sort of outlying part of himself; he has sowed himself broadcast upon it, and it reflects his own moods and feelings; he is sensitive to the verge of the horizon; cut those trees, and he bleeds; mar those hills, he suffers. (quoted in Buell 1996, p. 178)

We are in awe of these words by Burroughs; before him no one had ever expressed the relationship between man and nature so intimately, so appropriately and so vividly. Burroughs not only gave nature essays a definite form, demonstrated an effective way of studying nature and developed techniques for describing it; he also implicitly guided people in how to respect nature and how to treasure the land under their feet. Seeking spiritual value from nature is the most precious legacy Burroughs left behind.

After Burroughs passed away, America set up the John Burroughs Association. Every year, on the anniversary of his birthday, the Association presents the John Burroughs Medal for distinguished nature writing. It also organizes various academic activities based on his life and works. Burroughs's cabin in the mountains, Slabsides, is listed as a national historic landmark and is sometimes open to visitors.

Decentralizing the self, a more intimate approach to nature and describing nature in a manner that is simpler and more accessible to the public was perhaps one of the characteristics of nature writing at the turn of the century. Burroughs's works opened a window on the kingdom of nature for thousands of young readers. He said in his old age: "Whenever I see young men walking through the country like that [i.e. with camping equipment on their backs] I sometimes flatter myself that maybe my books have had a share in sending them forth" (Quoted in Brooks 1980, p. 11). In his friend's words, Burroughs created a large group of disciples of nature. From his day on, they and their descendants have been fighting for the preservation of nature.

# Lecture 12. Muir: John o' Mountains

From Rusk's biography of Emerson we learn that Emerson had a list of people that he admired, calling them "my men" (Rusk 1957, p. 448). Thomas Carlyle was at the top of this list, and John Muir at the end. For many years Muir has been called "the last Concord disciple," "Thoreau in the west," "advocator of nature," and "patron saint of the American wilderness."

Muir left behind a rich legacy, both spiritual and material. He wrote 10 books describing—and conducting dialogues with—nature, including *The Mountains of California* (1894), *Our National Parks* (1901), *My First Summer in the Sierra* (1911), *The Yosemite* (1912), and *Travels in Alaska* (1915). He left behind him the Yosemite National Park for which he fought hard, Muir Glacier and Muir Woods which were named after him, the nature protection club he presided over for many years, and also generation after generation of spiritual seekers who admired him. As Muir had both the idealism of the nineteenth century and the scientific spirit of the twentieth, his vision was wider than those of his transcendentalist forbears, and his practice and actions also went much further than theirs. He started to reflect on how to solve the problems in the relationship between nature and man, and therefore holds a significant position in nature writing.

## *Glowing Mountains*

John Muir was born in 1838 in Scotland and spent his first 11 years on the country's rural northern coast. Many years later, in his autobiographical work *The Story of My Boyhood and Youth* (1913), Muir wrote: "When I was a boy in Scotland I was fond of everything that was wild, and all my life I have been growing fonder and fonder of wild places and wild creatures." He reminisced that his childhood friends were all "wild as myself." They liked to wander in the fields to hear the birds sing and to "watch the waves in awful storms thundering on the black headlands" (p. 3). When Muir was 11, his family migrated to the plains of Wisconsin in America and built a farm there. There Muir discovered a new world.

> Nature streaming into us, wooingly teaching her wonderful glowing lessons....
>
> ...Here without knowing it we still were at school; every wild lesson a love lesson, not whipped but charmed into us. Oh, that glorious Wisconsin wilderness!...

> …Young hearts, young leaves, flowers, animals, the winds and the streams and the sparkling lake, all wildly, gladly rejoicing together! (Muir 1913, pp. 52–53)

However, Muir's childhood was not all happiness. His father was very strictly religious; he believed that a copy of the Bible was enough for life, and sweat and hard work were the only way to paradise, and he regarded children's nature-loving behavior as almost evil. Therefore, Muir had to work hard from a very young age. Perhaps precisely because he had a father as a "severe god" at home, Muir turned to nature to find a god with motherly kindness. At 22, he went to the newly established Wisconsin University. The first botany lesson increased his love of nature, and he went to the forests and grasslands with wild passion. After graduating from Wisconsin University Muir turned to the "Wilderness University" and began his "life of wandering." He went first to Canada, the American East and South, and then to California and Alaska. In 1903 and 1904 he traveled around Europe, Japan, China, India, Australia and New Zealand, and from 1911 to 1912 he was in Africa. However, Muir's real homeland was the mountains in the American West. There lay the source of his inspiration and his real life. Very appropriately, Michael P. Cohen named his biography of Muir *The Pathless Way: John Muir and American Wilderness* (1984). We can imagine how Muir hiked and explored, alone in the wild mountains. He devoted most of his life to exploring the mountains of the American West, about which he was very passionate.

Huth says that the concept of nature held by both Burroughs and Muir "stemmed directly from Emerson and Thoreau and continued and expanded the trends which came from Concord" (p. 105), while Paul says that a transcendental thread evident in both Muir the person and his writings, stating that "Muir would not be Muir without it" (Paul, p. 243). Muir also acknowledged that, from the very beginning of his study of nature, he was attracted to and inspired by the Harvard group headed by Emerson. In the summer of 1893, the 55-year-old Muir visited Concord. He visited Emerson's and Thoreau's old homes, and laid flowers at their graves. In his letter to his wife, he wrote: "I did not imagine I would be so moved at sight of the resting places of these grand men." Muir also had dinner with Emerson's son Edward Emerson and his family. As a classmate of Thoreau, Edward Emerson's father-in-law welcomed Muir "as if it were a long lost son" (Buell 1996, pp. 318, 317).

Muir's trip to Concord was, in effect, a return visit for Emerson's to Yosemite more than 20 years earlier. At that time Muir had written to Emerson, inviting him to go to Yosemite with him "in a month's worship with Nature" and

Emerson, despite being 67 by then and advised against the trip by his fellow travelers, had accepted the invitation. In his diary entries regarding his travel to the West, Emerson recorded his impression: "In Yosemite, grandeur of these mountains perhaps unmatched in the globe" (quoted in Huth, p. 151). When he returned to Concord, Emerson added Muir to his list of people he admired, and invited Muir to visit Concord. It was not until 20 years later that Muir was able to make the visit, by which time his idols Emerson and Thoreau had both passed away—so Muir's trip was, in effect, a spiritual pilgrimage. What he was seeking was a spiritual light to guide his life and a spiritual home to provide some satisfaction, for after all Muir was, in many aspects, very similar to Emerson and Thoreau. Like Thoreau, who was still putting his *Maine Woods* in order when he passed away, Muir had his manuscript of *Travels in Alaska* beside his bed at the end of his life.

Also, like Emerson's and Thoreau's, Muir's journal entries (he wrote a total of 60 diaries) were the materials and sources of his writing. Muir's habit of diary writing can be traced back to Professor James David Butler at Wisconsin University, who not only introduced him to Emerson and Thoreau but also advised him to keep a diary like they did. Although Muir adopted Thoreau's method of jotting down notes on the spot to record the nature he saw in the mountains and glaciers, his notes were very casual while Thoreau's were more methodic. Muir's first book coming from his diaries was *My First Summer in the Sierra*. In terms of form, this book is like Whitman's *Specimen Days* in that it is arranged chronologically. Whitman believed that people "needed a 'feeding visit' in the west" (quoted in Huth, p. 156), and this book of Muir' is a record of such a visit. It vividly records his exciting and fresh experience of visiting the Yosemite for the first time. Ellery Sedgwick, the editor of *The Atlantic Monthly*, said excitedly after reading the manuscript, "I felt almost as if I had found religion!" (quoted in Brooks, p. 28). Some scholars believe it is actually possible to regard *My First Summer* as co-written by two Muirs:

> It is useful, in fact, to think of the book as being co-authored by two Muirs: the 31-year-old Muir encountering the high Sierra for the first time in the summer of 1869, and the 73-year-old Muir using his literary skills to convey the importance of that encounter in a book published 42 years later. (Branch, "John Muir's My First Summer in the Sierra" 2004, p. 142)

Muir was in love with the mountains in the West, but had no inclination to write about them until he was in his old age. This made it possible for him to represent the lively and curious experience of his first acquaintance with the Yosemite through the depth and wisdom of rich experience in his later years.

From Emerson, Muir learned to observe nature with a spiritual eye. Some commentators see Muir's love of nature as having a religious force. Paul reports that in the copy of his *Essays* that Emerson gave to Muir, Muir wrote: "in your own path there is always an angel guiding gently" (Paul, p. 238). In Emerson's view there was a universe in every heart, and he compared nature to a spiritual symbol. On the other hand, Muir believed that all things in nature were signs of the glory shown by God's thoughts. The Yosemite became his paradise, a cathedral of nature where, under the holy light, he saw a window through which he could see God. He became a disciple of the god of nature, bringing light to the unenlightened victims of modern civilization.

> No wonder the hills and groves were God's first temples....
>
> ...Yonder, to the eastward of our camp grove, stands one of Nature's cathedrals, hewn from the living rock, almost conventional in form, about two thousand feet high, nobly adorned with spires and pinnacles, thrilling under floods of sunshine as if alive like a grove-temple, and well named "Cathedral Peak." (Muir, *My First Summer*, [1911] 1997, p. 146)

When he looked at Cathedral Peak from another perspective, Muir exclaimed:

> This I may say is the first time I have been at church in California, led here at last, every door graciously opened for the poor lonely worshiper. In our best times everything turns into religion, all the world seems a church and the mountains altars. (Muir [1911] 1997, p. 250)

Although Muir's works often show Emersonian style and thoughts, as he aged Muir moved closer to Thoreau. It is said that it was due to the encouragement of his friend Mrs. Carr and also of Emerson that Muir first read Thoreau's books at the Yosemite (Paul, p. 237). Richard F. Fleck, in his book on the relationship between Muir and Thoreau *Henry Thoreau and John Muir Among the Indians* (1985), describes Thoreau as Muir's spiritual and literary mentor (p. 22). Buell thinks this is an overstatement, but that Fleck "is right that Muir carefully read Henry Thoreau cover to cover." Muir owned, marked and indexed all Thoreau's published works (Buell, p. 479), and some refer to Muir as "Thoreau in the West." This is because the two had many similarities in temperament, ideas and even expressions.

Thoreau and Muir both were well known for their love of the wild. They preferred living under the blue sky to staying indoors, and being lost in the wilderness and the mountains to living conventionally in civilized society. According to Thoreau, "all good things are wild and free (Thoreau 'Walking'" [1862] 1957, p. 618)" while in Emerson's words, Thoreau made the wilderness "home." Muir regarded wild nature as a classroom, and his life was spent attending the "University of Wilderness." However, the land Muir traveled was much vaster and he lived much closer to nature, often sleeping on a mountain slope beneath the stars.

Both Thoreau and Muir particularly liked walking. Thoreau regarded walking as a life vocation. His well-known essay "Walking," which summarizes his experience of and thoughts about walking throughout his life, is widely considered to be his best writing in his later years. Muir, on the other hand, believed that by walking in the wilderness people can "be born again and make a new beginning" (Paul, p. 245). With the blue sky and the wind as his guides, he too spent his life engaged in the vocation of walking. We can say that like Thoreau, Muir was a great walker and also a wild writer; however, Muir seems to have gone further and been freer. Apart from traveling in the mountains and glaciers in the American West, he also walked from the Midwest to Mexico Bay. His journal of this trip was edited and published posthumously under the title of *A Thousand Mile Walk to the Gulf* (1916).

The similarity between Thoreau and Muir also lies in their deep understanding of the value of the wilderness. Thoreau says in "Walking": "In wildness is the preservation of the world." It is reported that Muir greatly liked this saying, echoing it with his own: "In God's wildness lies the hope of the world." Paul comments that Muir and Thoreau "make 'companionable' reading." It can be seen they complemented each other, although Muir's writing had a more modern flavor, and with the rapid development of industry, he was more concerned with and more focused on the issue of harmonious co-existence between man and nature.

In his biography of Thoreau, *The Days of Henry Thoreau: A Biography* ([1965] 1982, p. 210), Walter Harding tells us that, having climbed to the top of Mount Katahdin in the summer of 1846, Thoreau felt uncomfortable faced with the vastness and power of nature. Harding comments that this was not the gentle and intimate nature around Walden that Thoreau was familiar with, and he therefore did not like it. Muir was different; the nature he experienced was far rougher and harsher: camping in the mountains of Nevada, encountering storms at the summit of Mount Shasta, and risking his life among the glaciers of Alaska. He approached the harsh natural environment with a positive attitude, making each experience of the wilderness a subject of his writing; in a sense, the earth was his playground and the mountains were his home.

Although Muir absorbed Emerson's idealistic concept of nature and Thoreau's views on the wilderness, and also took on elements of their writing styles, he later exceeded his two transcendentalist masters, establishing his own concept of nature—which in a certain sense had already departed from the conventions of transcendentalism—and gradually forming his own theory and practice system.

## *Guardian of the Wildness*

Emerson called on people to look at nature with new eyes and to seek in it the spirit and the self. However, in his eyes, nature was subservient to the needs of man; the relationship was actually that of master and servant. Thoreau was much closer than Emerson to nature, in terms of both feelings and actions, declaring that he wanted to be part of nature. However, in his writings we can still see a very strong self. Muir was different; he opposed the idea that nature exists to support and serve humans, which excused the unchecked use of nature in pursuit of profit. Muir criticized the materialist attitude that all things that provide benefits to humans represent culture and that nature is forever in opposition to culture, using his own experience to show that this was not so.

Muir, in his *Wild Wool* (1875), clearly explains the above view. He found in Mount Shasta that the wool of a wild sheep was much softer than that of a farm-raised animal. When he looked closer, he found that this soft wool always grew close to some rough wool. This new discovery made Muir think about the relationships between nature and man and nature and culture. He believed that this "wild wool" was designed by Mother Nature for the wild sheep, not for humans and that this showed that nature was scientific: the fine wool growing near the rough wool balanced hardness and softness, as without the support of the rough wool, the fine wool would not have survived in bad weather.

From this analysis, Muir came to two conclusions: first, that natural things can compare with man-made products in terms of both beauty and quality; and second, that human's indiscriminate eradication and taming of the rough parts of nature is against the 'rule of nature.' Just as the fine wool could not survive without the rough wool, human and culture are in a balanced relationship with nature; a relationship that is not one of master and servant, but an intricately intertwined, codependent coexistence. In *Wild Wool* Muir quotes an aphorism which one of his friends liked a great deal: "Culture is an orchard apple, nature is a crab" (Muir, p. 598), pointing out through his analysis of the wild wool that it is not so. This insightful analysis has had a positive impact, encouraging later American nature writers to seek culture in rough nature, and human in general to consciously protect primitive nature and seek to coexist with it.

Zwinger, in her introduction to her edited collection of essays by Austin and Muir, points out that even though Muir lived a long life, in his autobiography he only wrote about his life up until he entered Wisconsin University. Zwinger explains that this was because "He found it always more interesting to write about the natural world than about himself" (p. xxiv). Another scholar comments that: "Muir wrote about trees as no one else in the whole history of trees, chiefly, because he loved them as he loved men and women" (quoted in Brooks 1980, p. 24). If Burroughs's books send people on their road to nature, Muir's books start to teach people to decentralize the self and to venture deeper into nature. Muir once exclaimed in the wilderness that in face of the wonders of nature he felt insignificant. In *The Story of My Boyhood and Youth*, he called plants "plant people" and animals "my hairy brothers" (pp. 137–138). In his eyes, plants and animals were like humans and had spirit and consciousness. In fact, he saw himself as being completely on the same level as the animals and plants in the natural world.

In the same vein, Muir wrote an affectionate book about—and named after—his dog, *Stickeen* (1909). Stickeen accompanied Muir on his travels in the Alaskan glaciers; they kept each other company, endured difficult situations together, and became best friends. Like Muir, Stickeen seemed to disguise his real self so that only the keenest observer would notice the light shining in his eyes. Wherever Muir went, he followed. Only when they were facing a crack in a glacier—in other words, facing danger or possible death—did he show hesitation; however, at Muir's encouragement, he leapt across the crack. Through his relationship with Stickeen, Muir found more evidence that animals were like humans, that they had a sense of morality and had spirit. Buell calls *Stickeen* "The most popular work Muir ever wrote" (1996, p. 195). St. Armand, in a conference paper (1998) called this book Muir's "spiritual autobiography and symbolic romance," just as *Walden* is often described as Thoreau's spiritual autobiography. From the titles of these two "spiritual autobiographies," we see again Muir's closer relationship with the animals and plants of the natural world. Of particular significance is the fact that Muir also wrote two books that demonstrate his awareness of the need to protect nature: *Our National Park* (1901) and *The Yosemite* (1912), which later nurtured the ideas underpinning the environmental protection movement.

Muir was not only a man who put forward theories, but also a man of action. In protecting nature he went further than the transcendentalist masters Emerson and Thoreau. Unlike Emerson and Thoreau, Muir loved not only nature but also science, and was interested in scientific inventions when he was young. In *The Story of My Boyhood and Youth* he writes about his obsession with inventions; at the Wisconsin Agriculture Fair in 1860, a small mechanical device he had invented was a great success and attracted the attention of the audience, with one person saying: "I wish I had that fellow's head. I'd rather have it than the best farm in the State" (p. 212). When Muir was studying at Wisconsin University, he designed a bed that would rise automatically and wake him up in the morning; small devices that could turn on a light and open and close books; and a lighter whose timing could be pre-set. He studied physics, geography, mechanics, botany and other subjects, thereby gaining a solid scientific foundation from which to observe nature. Muir later became a Member of the Washington Academy of Science, and was awarded honorary doctorates by both Harvard and Yale. (Buell 1996, p. 189)

Perhaps it was because of his scientific knowledge that Muir could examine the relationship between man and nature calmly and rationally. At a time when people were unaware of the need to protect the natural environment, he stood up in its defense—not out of humanistic sentiment or impulse, but out of a love for and sense of responsibility toward the world that humans, animals and plants co-own, and a concern for the future of this world. In his travels and explorations in his beloved wilderness, Muir had discovered the damage it had suffered, and he realized the importance and urgency of the protection of the natural, wild environment. In 1889 he published two essays in *Century* magazine describing the wonderful scenery of the Yosemite and the prospect of its becoming a national nature reserve, and he went on to work with the editor of *Century*, Robert Johnson, towards the establishment of Yosemite National Park in 1890. In 1892 Muir helped to establish the Sierra Club, the aim of which was to protect the natural environment of the mountains in Nevada, and was elected its first president. The club is still thriving today, largely because of its publication of unique books and magazines, and its influence now extends to the whole country. Muir was also a member of the American National Forests Committee, the members of which traveled all over the mountain areas in the American West in order to provide advice to Congress on the regulation of forestry management. In 1897 they submitted a report to Congress which included the proposal to establish two new national parks, one surrounding the Grand Canyon and another around Mt. Rainier.

According to Huth's biography of Muir, he accompanied the then US President Theodore Roosevelt when he traveled around the Yosemite.

> Roosevelt and Muir disappeared into the woods of the Yosemite Valley for a weekend. No report has ever been published about this meeting; but from what we know of later happenings, a good deal of Roosevelt's policy concerning conservation was probably devised then, based on the counsel John Muir gave the President, who was not "building this country of ours for a day," but "to last through the ages." (p. 179)

In 1908 Muir, who was already in his old age, started to make calls and write articles against the building of the Hetch Hetchy dam. According to Roderick Nash, even though Muir did not succeed in this endeavour, he'd done his very best, and felt that "public opinion is not yet awakened" (quoted in Nash, *Wilderness and the American Mind*, 1967, p. 164). Although frustrated, Muir could see that his participation in the action had begun to awaken a sense in his fellow Americans that there was a need to protect nature. This became a significant issue, and remains a difficult mission for later generations.

Commenting on the literary history of American nature writing, Lyon points out that

> Muir's contribution to the developing American nature essay was threefold: he enlarged it to include wild-country adventures, while essentially continuing Thoreau's philosophy of wildness; deepened its evolutionary and ecological content with his emphasis on flow and harmony, and added a strong new note of militancy, in response to the dramatic environmental conditions of his time. (p. 60)

Indeed, Muir appreciated the value of the wilderness as much as Thoreau did, and his understanding and experience was broader and deeper. He was not only a defender of the wilderness, but also a spokesperson for it, and took action, while his numerous adventures in the mountains and glaciers and his tireless work broadened his literary vision and expanded the content of nature writing.

## Pulsing Nature

Zwinger, in the introduction to *Mary Austin and John Muir*, comments that "Muir, who never planned to be a writer, turned out to be an uncommonly fine one...a nature writer's nature writer" (p. xxii). Her comment is not an overstatement: it is true that Muir did not set out to be a writer, and he first started to write at the urge of his friends and the editors of some literary journals he felt his writing was as slow as the formation of glaciers. When he apologized to Walter Hines Page, the then editor of *The Atlantic Monthly*, the latter replied:

> I thank God that you do not write in glib, acrobatic fashion: anybody can do that. Half the people in the world are doing it all the time, to my infinite regret and confusion.… The two books on the parks and on Alaska will not need any special season's sales, not other accidental circumstances: they'll be Literature! (quoted in Brooks, p. 25)

What Page referred to as "Literature" is the "pulsing nature" that Muir presented to readers, resulting from the merging of his heart and the wilderness.

Vividness and a sense of movement are characteristics of Muir's writing. The mountains were not only his homeland, but also the source of his inspiration. The higher he climbed, the greater the vision in his writing. When he put his experience into words, they were infused with the vitality and the spirit of the mountain. In *My First Summer*, for example, Muir describes his experience of climbing to the top of a mountain through the wilderness:

> From garden to garden, ridge to ridge, I drifted enchanted, now on my knees gazing into the face of a daisy, now climbing again and again among the purple and azure flowers of the hemlocks, now down into the treasuries of the snow, or gazing afar over domes and peaks, lakes and woods, and the billowy glaciated fields of the upper Tuolumne, and trying to sketch them. In the midst of such beauty, pierced with its rays, one's body is all one tingling palate. Who wouldn't be a mountaineer! Up here all the world's prizes seem nothing. (p. 153)

Like Emerson, Muir sought the beauty and totality of an interconnected and harmonious nature. Natural scenery, under his pen, consisted of bushes dotted with small flowers that grew out of the moist earth in early spring; grasses hanging over streams, swaying in the flowing water; and glaciers nourished by small hills and sunshine. His natural scenery was, therefore, often a very vivid composition.

> Every morning, arising from the death of sleep, the happy plants and all our fellow animal creatures great and small, even the rocks, seemed to be shouting, "Awake, awake, rejoice, rejoice, come love us and join in our song. Come! Come!" (Muir [1911] 1997, p. 67)

This cheerful and harmonious scene takes us to a fairyland far away from modern society. In the following narration from *My First Summer* we can feel Muir merged into the lively natural environment:

> We are now in the mountains and they are in us, kindling enthusiasm, making every nerve quiver, filling every pore and cell of us. Our flesh-and-bone tabernacle seems transparent as glass to the beauty about us, as if truly an inseparable part of it, thrilling with the air and trees, streams and rocks, in the waves of the sun,—a part of all nature, neither old nor young, sick nor well, but immortal. (p. 16)

The rivers, to Muir, were also different.

> Out of the grassy meadow and down over this ice-planed granite runs the glad young Tamarack Creek, rejoicing, exulting, chanting, dancing in white, glowing, irises falls and cascades on its way to the Merced Canon, a few miles below Yosemite, falling more than 3,000 feet in a distance of about 2 miles....
>
> ...Sing on, brave Tamarack Creek, fresh from your snowy fountains, plash and swirl and dance to your fate in the sea; bathing, cheering every living thing along your way. (Muir [1911] 1997, pp. 102–103)

The wind was "the very breath of Nature, whispering peace to every living thing," flowers were "wild children of light," and stars were in the sky "like one vast lily meadow in bloom." Nature, to Muir, was a combination of hardness and gentleness. "Nature in the midst of her beauty-loving tenderness! Painting lilies, watering them, caressing them with gentle hand, going from flower to flower like a gardener while building rock mountains and cloud mountains full of lightning and rain" (Muir [1911] 1997, p. 133).

It is precisely because he captured the movement of nature that Muir felt electrified, and he conveyed this feeling to his reader through his moving prose. When he camped in the mountains for the first time, he exclaimed: "Never while anything is left of me shall this first camp be forgotten. It has fairly grown into me, not merely as memory pictures, but as part and parcel of mind and body alike" (p. 83). This kind of body-and-mind communication with nature naturally infused Muir's works with an awareness of the need for ecological protection and an urge to fight humanity's ravaging of nature through blind development.

Apart from vividly describing nature, Muir also studied the connection between nature and the mind and put forward new views in this regard. He believed that the mind and the body absorbed two different kinds of nourishment. Insipid life in the city would make us sick and unable to know ourselves, and as our benumbed minds and limbs were interconnected, the life of a mountaineer seemed to him particularly beneficial to both body and mind (Cohen, pp. 44–45). This gives another perspective on why Muir loved climbing mountains.

For him, "going out, I found, was really going in" (quoted in Zwinger, p. xvii). Having climbed to the top of a mountain, he exclaimed:

> Wonderful how completely everything in wild nature fits into us, as if truly part and parent of us. The sun shines not on us but in us. The rivers flow not past, but through us, thrilling, tingling, vibrating every fiber and cell of the substance of our bodies, making them glide and sing. (quoted in Buell 1996, p. 192)

This description not only shows us the pulsing of nature but also makes us realize why the heart needs the nourishment of the wild nature, why "going out was really going in," why people today follow the footsteps of Muir into the wilderness, deep into the mountains to seek spiritual thrill and connect with nature's glory.

Ronald praises Muir as "the sage who foretold the entire American conservation movement," and believes that "his words and his actions taught Americans a twentieth-century way of viewing their natural surroundings." Many people have wanted to be like him, to love nature and describe it as he does. Many have been engaged in the nature conservation movement and advocated for the natural environment. However, few have successfully imitated him in his almost primal passion and his wholeheartedness. As Ronald says, "John Muir, I think, stands alone" (pp. 99–100).

## Lecture 13. Austin: Walking into the Depths of the Desert

In her autobiography *Earth Horizon* (1932), describing meeting John Muir and others at a friend's place in Carmel, Californian in 1904, Mary Austin says: "Of all these l recall John Muir the most distinctly" (Zwinger, p. vii). Austin was most impressed with John Muir, because the two of them were kindred spirits. As Esther Lanigan quotes in her biography *Mary Austin: Song of a Maverick* ([1989] 1997, p. 31), literary historian and critic Franklin Walker called Austin a woman with "a keen ability at perception and recording impressions, and an appreciation of transcendental thought as expressed by Emerson, Thoreau, and John Muir."

However, Austin was different from her male predecessors. While Emerson reflected on nature in the woods under the stars in the small town of Concord, Thoreau experienced nature by the side of Walden Pond, and Muir appreciated it in the high mountains in the West, Austin cast her eyes to the harsh, desolate desert, into which even men would not want to go; and she, with her soft, gentle tone, gave the dry and abstract landscape charm and a captivating force.

Before Austin, nature writing had seemed the monopoly of men, but with the publication of her *The Land of Little Rain* in 1903, we are able to see nature under a female pen, adding another dimension to American nature writing and making it richer and fuller. As Austin's life straddled two centuries, both she herself and her writings have a cross-century color and an influence that connects the earlier with the later nature writing in America. The decentralizing of the self and the foregrounding of ecological groups in *The Land of Little Rain* was far ahead of her time, and opened up a new path for twentieth-century nature writers. What makes Austin stand out, too, is her unique writing style and her "personal dictionary" of desert descriptions developed during her 12 years of living in a small desert town. It is not surprising that Buell describes Austin as "the first creative writer to make a sustained attempt to map the western desert as an eco-cultural region" (1996, pp. 78–79).

## *Song of a Maverick*

Mary Austin (1868–1934) was born in Carlinville, a small town in Illinois. According to Thomas. M. Pearce's biography *Mary Hunter Austin* (1965), her literary inclination came from her father George Hunter, who went to the Civil War and later became a lawyer. Her father was interested in British and American literature and had a large collection of books. Mary's childhood home was located on a slope which led down to a small creek, faced a flat plain, and had a big walnut tree in front of the door. For every child, childhood includes life both indoors and outdoors. For Mary, however, outdoor life was the most important, and it became the driving force behind her writing, for it was outdoors that Mary had her unforgettable experience. According to her memory, when she was five she was walking near a tall walnut tree, alone; when suddenly, after a moment of peace and silence, the sky, earth, trees and grasses, and the child in between, all came alive in a pulsing light. Thirty years later, Austin wrote: "To this day I can recall the swift inclusive awareness of each for the whole—I in them and they in me and all of us enclosed in a warm, lucent bubble of livingness" (quoted in Pearce, pp. 20–21).

In the eighteenth century, Jonathan Edwards had had an experience of communing with God on his father's farm. In comparison, Austin's childhood experience was a communion with everything in the world. It was not only a spiritually overwhelming experience, but also gave her power and energy. It urged the adult

Austin to be in touch with and write about the desert in the West with extraordinary courage, and to view the world from the perspective that humans and nature are an inseparable whole. Due to the power and energy she drew from this perspective, Austin gave up her pursuit of the English major to which those around her believed she was suited, choosing instead to study natural science, believing that she could study English on her own, while natural science needed teachers and labs. Many years later, she wrote that "a new type of writer will have to be evolved, writers whose approach is purely literary but who are capable of immersing themselves in the data of science to the point of saturation" (quoted in Lanigan, p. 24).

Austin, in *Earth Horizon*, also talked of Emerson's influence on her. "Oddly, the only writer out of those days who affected her style was Emerson…don't know why, the predilection showed her early in her college life" (Zwinger, p. ix). Austin's autobiography differed from most in that it was told in the first and the third person alternately. After graduating from university, as her father had passed away, Austin moved to southern California with her mother and brothers; and it was there that her fascination with the desert began, and that it became the object and source of inspiration for her writing. Life in California evoked mixed feelings in Austin that were difficult to express: it was there that she found love and lost it, had a daughter and lost her. After marrying, she followed her husband Stafford Austin to the Owens Valley, where she gave birth to a daughter, Ruth. Ruth was born with a mental disabilities and died of illness in childhood. Later Austin and her husband divorced. Although her life in the Owens Valley did not turn out to be happy, it was there that Austin created a new life for herself, a life connected with the land and specifically the desert; the time she spent in the desert later crystalized and provided the material for her desert classic, *The Land of Little Rain*.

In 1906, three years after the publication of *The Land of Little Rain*, Austin left—with reluctance—the small desert town in the Owens Valley where she had spent more than 10 years. She had taken part in the effort to stop the Los Angeles Water Committee from building a dam in the Owens Valley, but it had not been successful. Like Muir she could not stand seeing her beloved land being ravaged, so she moved away to live in Carmel-by-the-Sea, a town with Spanish flavor.

Austin chose Carmel because the environment around it was still wild, plus a group of writers and artists had gathered there; as early as 1904, Carmel had become an artists' colony, and meeting poet George Sterling, writer Jack London, journalist and novelist James Hopper, John Muir and others helped Austin's writing career. According to Lanigan in her biography of Austin, the writer's workshop was a platform built in the branches of a large oak tree, and Austin would climb on a "rickety ladder" to the platform to work. Sometimes she wore a Greek robe, at other times she dressed as an "Indian princess"; she was quite exotic, and quite well known in the Carmel literary circle (Lanigan, pp. 93–95).

In 1908, Austin learnt that she had breast cancer. She went to Europe seeking treatment and she traveled there for two years, and Brooks reports that she found that she was quite famous in London. She met with George Bernard Shaw, visited Joseph Conrad and listened to H.G. Wells talking about his personal problems, the latter calling her "the most intelligent woman in the United States" (Brooks, p. 191). Austin returned to America in 1910—in order to help people on the East Coast understand the American West better, and also because it was more convenient for her book to be published there—Austin settled in New York and lived there for more than a decade. During those years, she made frequent trips between the West and the East Coast, but the object of her writing continued to be humans and nature in the West. In the end, Austin found the East did not suit her and she disliked the environment in New York, where people were too eager to make quick profits, and missed the fresh air and sunshine of the West, which she knew was the base of her life and her writing. In 1924, Austin built a Spanish-style mudbrick house in Santa Fe and spent her last 10 years there, passing away in August 1934.

Apart from the ten plus years in New York, Austin spent most of her life in California and New Mexico, and all her work had a strong flavor of the American West. Lanigan calls Austin's voice "one of the most unusual, gifted, eccentric, exasperating, tragic, elitist, and idiosyncratic in American literature" (p. 1). Austin was a prolific writer with broad interests: she wrote 32 books, including novels; more than 200 essays; and three plays, as well as poems, news reports and many letters. However, another Austin biographer, Pearce, tells us that of all of Austin's works, the books that describe the harmony between nature and man—such as *The Land of Little Rain* (1903), *The Flock* (1906), *Lost Borders* (1909), and *The Land of Journeys' Ending* (1924)—are her best literary achievement, saying that after reading them, the reader will feel "he has walked in sunlight, gazed at far horizons, breathed clear air, and met unforgettable people" (Pearce, p. 135). There is no doubt that this is the real charm and achievement of Mary Austin.

## *The Land of Little Rain*

In his introduction to the fourth edition of *The Land of Little Rain*, contemporary American writer Terry T. Williams points out that Austin's eyes were focused on the desert landscape of the American West and that "in many ways, I view Austin as a sister, soulmate, and a literary mentor, a woman who inspires us toward direct engagement with the land, in life as well as on the page" (in Austin [1903] 1997, p. x). Austin's best-known book is a "desert classic."

*The Land of Little Rain* was written in prose and contains 14 short stories and essays. It shows the charm of the desert, where Austin lived for 12 years, from various perspectives. This "land of little rain" is located to the south of the Yosemite Valley and the Sierra Nevada, extending across Death Valley to the Mojave Desert. Talking about writing the book, Austin said, "I was only a month writing *The Land of Little Rain* but I spent twelve years peeking and prying before I began it." She explained further:

> The real heart and core of country are not to be come at in a month's vacation. One must summer and winter with the land and wait its occasions. Pine woods that take two and three seasons to the ripening of cones, roots that lie by in the sand seven years awaiting a growing rain, firs that grow fifty years before flowering—these do not scape acquaintance. (quoted in Brooks, pp. 185, 187)

Both Austin's personal experience and this book clearly demonstrate this opinion of hers: only someone who knew the land deeply could have described the strange and remote desert so vividly: its wild beauty, its plants and animals, its rare passersby, its mountain slopes white under the sun, the tear-like streams in the mountains, the wonder of the cactus, the persistence of the animals seeking water, the patience of the vultures who waited for animals to die so that they could feast on the dead bodies, the deep love of the Indians for the land, and the oasis in the desert where the cheerful and leisurely Spanish town was. Under Austin's pen, the naked wilderness has a captivating magic and nourishment for the human spirit. No wonder Williams says that through *The Land of Little Rain*, Austin makes her beloved desert and her spirit stay with us (in Austin [1903] 1997, p. xv).

When talking about readers' changing taste of literary works, Lanigan quoted a critic,

> A century, even a quarter of a century, dies around a book; and then the book lies there, a shaming thing because it shows us how much worse we once were to have liked it; and something else too; it demonstrates how the world shakes off what it does not need, old books, old notions of aesthetics, old mind-forms, our own included.… The world to the eager eye is a tree constantly pruning itself, and writers are the first to be loped off. (Lanigan, p. 78)

If this view stands, then Austin's *The Land of Little Rain* is an exception. When it was first published by Houghton Mifflin in 1903, it immediately attracted large groups of readers and was received very well by the critics, and its charm has continued to grow. In 1950, Mifflin published a second edition; in 1974, the University of New Mexico Press published a third; and in 1988, Penguin published a fourth, since which *The Land of Little Rain* has also been published in paperback, hardback and large character editions, the most recent being a 2011 paperback edition. It is worth mentioning that the book's 1988 and 1997 Penguin editions have introductions by two contemporary American nature writers, Edward Abbey and Terry T. Williams respectively. The latter said that almost 100 years after *The Land of Little Rain* was first published, it remains a desert classic and the admiration for it continues to grow. (Austin [1903] 1997, p. x)

Why is the charm of this book so long-lasting while Austin's other books have faded in people's memories? Lanigan points out that "while many Americans turned to the old cathedrals and traditions of Europe, or even to the Far East, to alleviate anxieties emanating from an increasingly mechanized and technologized world, Austin found a solution in America itself." She goes on to explain that Austin turned her eyes to the undeveloped American West, and in *The Land of Little Rain* she showed readers a free and carefree pre-industrial world. This, no doubt, sounded a clarion call for starting over again to thousands of Americans who yearned for a simpler and more spiritual life based on primal myths. Lanigan believes that in this action of Austin's there was an "anti-modern" and "pro-primitive" yearning, and the kind of harmony between people and land that Austin describes was beyond the Americans who lived in the cities at the time and beyond most Americans today (pp. 78–79).

Buell (1996, p. 80) says that *The Land of Little Rain* "is a mark of the maturity of her environmental vision." This is because, throughout the book, Austin put the story of the natural environment above the story of the people. Buell explains further that

> her protagonist is the land, more particularly the geography of its watercourses and the patterns of life created by water scarcity: the habits of life, the routes of travel, the visible marks on the landscape all being shown as shaped by the need of both animals and humans to adapt themselves to this basic necessity. (1996, p. 80)

In fact, in *The Land of Little Rain* Austin presents neither a simple nature nor complex humans, but a community of animals, plants and humans who are connected by the same needs and the need to adapt themselves. We can say that through this community, we can see the whole world, and even a microcosm of the world in the future. On this point Austin exceeded her predecessors.

Unlike her predecessors, Austin seldom mentions "I" in her works; what she describes is a social community in the American West. Her interest in the environment is greater than her interest in people, and her main theme is the land and the habit of life. The book begins with a description of this borderless land with a scarcity of water. On the map it is called "desert." but the American Indians gave it a more appropriate name, "the Country of Lost Borders." "Not the law, but the land sets the limit." Desert often implies that the land does not nourish people, but Austin finds this a very inaccurate description because "void of life it never is, however dry the air and villainous the soil" (p. 1). Under her pen, therefore, we see on this "land of little rain" living a community: the mountains that are wind scraped, sun baked into yellow or brown; the valleys covered in blue mists and filled with hot sun; the water that flows through sparse grass like a thin white ribbon; the desert plants that have to root deeply into the land due to the harsh natural environment, but show a tenacious life force; the animals who chase each other like squabbling children on the streets, or look at the circling vultures with fear and resignation and wait for death in desperation; and of course also the indigenous Shoshone people who live far apart from each other just like the desert trees do.

We see that Austin considered all living things in the desert to be equally important and every living thing to be an integral part of the whole. This makes us feel as if it's not Austin who is describing the desert, but the living things telling their own stories of their lives. The human voice, the usual center of literary works, is gradually fading away and is replaced by multiple voices. This approach of Austin's is appropriate, for she understood the desert deeply, saying that "of all its inhabitants it has the least concern for man" (p. 25). In the desert, therefore, humans, who have always acted as if they are superior, have to seek survival on an equal footing with all the plants and animals. The people who survive in the desert are the people rooted in the land.

The above views are particularly evident when Austin talks about the Shoshone; she writes: "Once a Shoshon always a Shoshon," thus demonstrating the connection between this tribe of people and the land. "The Shoshones live like their trees, with great spaces between, and in pairs and in family groups they set up wattled huts by the infrequent springs." She goes on to explain that it seems that only in the Southwest could the Shoshone build a home, because it was not the building but the surrounding environment that gave meaning to home.

When discussing another American Indian tribe, the Paiute, she said: "Not the wattled hut is his home, but the land, the winds, the hill front, the stream" (pp. 31, 33, 65). Through this book, Austin sends us a message: modern people should give up their anthropocentric concepts and instead approach, experience and merge into nature on an equal footing.

The charm of Austin also lies in the fact that she showed—through both her own experience in the desert and her writing—that women have a special affinity with nature. Traditionally, women stayed in a natural environment that had been tamed by humans, in comfortable houses and gardens, but Austin was different: she went deep into the wilderness where, until then, only men went. In my view, Austin's actions were not challenges to conventions, but the result of female intuition. As women have a special affinity with nature, they do not have a strong desire to conquer and control it, but think more of how to care for it, beautify it so as to make it their home. As a result, the nature in women's writing is different from the nature in men's writing. As mentioned before, when Emerson was writing his essay "Nature," even though he was in pretty Concord, what he heard was the roar of the tall willow tree on a stormy night, "Hoarse concert: they roar like the rigging of a ship in a tempest" (Emerson, in 1964, p. 384). When Austin started to write *The Land of Little Rain* she was in a comparatively rougher natural environment, living in a small desert town, yet her writing shows a completely different attitude:

> There was a weeping willow whose long branches moved back and forth across my window like blowing hair, like my memory of my mother's long and beautiful hair. I think it was this which gave the reminiscent touch to my mood. For though I was in the midst of it, I began to write of the land of little rain as of something much loved, now removed. (quoted in Lanigan, pp. 76–77)

Both were writing about a willow tree in nature, but Emerson's writing shows stormy passions while Austin's has soft, gentle feelings for the land—precisely because of which nature had, under Austin's pen, a charm different from that of male writers.

> If one is inclined to wonder at first how so many dwellers came to be in the loneliest land that ever came out of god's hands, what they do there and why stay, one does not wonder so much after having lived there, none other than this long brown land lays such a hold on the affections. The rainbow hills, the tender bluish mists, the luminous radiance of the spring, have the lotus charm. (Austin, [1903] 1997. p. 6)

When writing about the water source, Austin also showed soft love.

> By the end of the dry season the water trails of the Ceriso are worn to a white ribbon in the leaning grass, spread out faint and fanwise toward the homes of gopher and ground rat and squirrel…to the little people the water trails are as country roads, with scents as signboards. (Austin, p. 9)

Austin's descriptions are like a series of small landscape paintings. She depicts how rain was formed:

> Out over meadow or lake region begins a little darkling of the sky,—no cloud, no wind, just a smokiness such as spirits materialize from in witch stories…
>
> …a wind comes up and drives the formless thing across a meadow, or a dull lake pitted by the glancing drops, dissolving as it drives. Such rains relieve like tears. (Austin, p. 95)

The snow in the mountains had a crystal beauty. "It increases, is wet and clogging, and makes a white night of midday." From her contact with the basket-weaving American Indian women, she felt that "every Indian woman is an artist, sees, feels, creates, but does not philosophize about her processes" (Austin, pp. 97, 63). In a word, Austin, with her unique female sensitivity and feelings, offered a more direct and more sensitive interpretation of nature, giving her writing a moving charm. However, it would be wrong to understand Austin simply through the traditional stereotype of "man thinking and woman weeping." Austin's writing has a distinct female charm, but its influence on American nature writing goes far beyond that.

Williams believes that the talent of Austin as a writer is evidenced in her essays, and that her works went beyond the constraints of her time (Austin [1903] 1997, pp. xi, xiv). The endurance of Austin's writings lie in the fact that she experienced the hardship in the desert for many years, and developed a style that combines magnificence and beauty to describe it; and also in the personal "desert description" dictionary she accumulated through her personal experiences. Under Austin's pen the desert was both beautiful and mysterious, and also terrifying and therefore exudes a strange enchantment: "East away from the sierras, south from Panamint and Amargosa, east and south many an uncounted mile, is the country of lost borders." There are mountains of all sizes and shapes, which have been tinted yellow or red, and it is wind not water that had molded these mountains into their shapes and sizes. It is a "land of lost rivers, with little in it to love; yet a land that once visited must come back to inevitably" (Austin, p. 2).

The small landscape paintings Austin presents to the reader include: the mountain slopes colored by the rainbows, the trees nourished by the cloud mists, the desert baked under the hot sun, a parasol-like shade over your head provided by the wings of an eagle, the dead and dried up trees standing in the desert like skeletons under the moonlight. With magnificent and terrifying beauty, they presented a unique and enchanting scenery. People may complain about this land, yet still feel an inevitable pull towards it.

In Austin's personal dictionary, the adjective that is most used to describe the desert is "white": the water traces were "white ribbons," all the paths of the wild animals were marked by "white," the sands showed "silvery white," and the direction marks of the American natives were in the shape of "white hearts." Williams points out that only someone who had been through the trial of the desert would describe it as being "white," the shade associated with bones, dry skin, a thirsty mouth, and the illusion of water that represents the desperate hope of tired travelers. White is the pulsing heart of the desert, luring you to go deeper and lose yourself there (Williams in Austin [1903] 1997, p. xii).

Austin's *The Land of Little Rain* changed people's perception of the desert. The traditional image of the desert was an ugly, barren, unpopulated land of little use, but under her pen it became a charming landscape full of life and of lives, just like the Walden Pond in New England and the Yosemite in California. Unlike people before her who described the rich land, dense forests and wet riverside areas, she wrote about the arid, barren and vast desert, creating a "desert aesthetics" that no one in America had considered before her.

Scheese reports that the period between 1880 and 1920 has been called "the age of the 'back to nature cult'" (Scheese 1996, p. 28), and in my view John Burroughs, John Muir and Mary Austin were spokespeople of this age. As the "guide of nature," Burroughs—with his non-didactic and beautiful essays—has sent many readers onto the journey of observing and appreciating nature. Muir combined his interest in science, faith in the wilderness and active protection of nature, and merged natural history with nature writing. Austin showed people an inclusive nature that contained ecology, genders and races. She also, through her praise of the harmonious relationship between the American Indians and their natural surroundings, provided a good example for the industrial society in how to improve the tense relationship between civilization and nature. Despite their respective uniqueness, these three nature writers have one thing in common: all of them found magnificent beauty and consolation in natural surroundings.

## References

Austin, Mary. (1903) 1997. *The Land of Little Rain*. 4th edition. New York: Penguin Books.

Branch, Michael P., and Scott Slovic, eds. 2003. *The ISLE Reader: Ecocriticism, 1993–2003*. Athens, GA: University of Georgia Press.

Branch, Michael. 2004. "John Muir's *My First Summer in the Sierra*." *ISLE* 11.1 (2004): 139–152.

Brooks, Paul. 1980. *Speaking for Nature: How Literary Naturalists from Henry Thoreau to Rachel Carson Have Shaped America*. Boston: Houghton Mifflin.

Buell, Lawrence. 1996. *The Environmental Imagination: Thoreau, Nature Writing, and the Formation of American Culture*. Second printing, Cambridge, MA: Harvard University Press.

Burroughs, John. 1871. *Wake-Robin*. New York: Hurd and Houghton.

Burroughs, John. 1894. *Riverby*. Boston and New York: Houghton, Mifflin and Company, The Riverside Press, Cambridge.

Burroughs, John. 1896. *Whitman: A Study*. Boston: Houghton Mifflin.

Burroughs, John. 1904. *Wake-Robin*, *Works of John Burroughs*. Vol. 1. Boston: Houghton Mifflin Company.

Burroughs, John. 1904. *Indoor Studies*, *Works of John Burroughs*. Vol. 8. Boston: Houghton Mifflin.

Burroughs, John. 1904. *Literary Values and Other Papers*, *Works of John Burroughs*. Vol. 12. Boston: Houghton Mifflin.

Cohen, Michael P. 1984. *The Pathless Way: John Muir and American Wilderness*. Madison, WI: University of Wisconsin Press.

Cole, Thomas. 1965. "Essay on American Scenery." In *American Art, 1700–1960: Sources and Documents*, edited by John W. McCoubrey. Englewood Cliffs, NJ: Prentice-Hall.

Emerson, Ralph Waldo. 1964. *The Journals and Miscellaneous Notebooks of Ralph Waldo Emerson*. Vol. 4. Cambridge, MA: Harvard University Press. Ed. Alfred R. Ferguson.

Fleck, Richard F. 1985. *Henry Thoreau and John Muir Among the Indians*. Archon Books.

Harding, Walter. (1965) 1982. *The Days of Henry Thoreau: A Biography*. New York: Dover Publications.

Hart, James D. 1995. *The Oxford Companion to American Literature*. 6th edition. New York: Oxford University Press.

Huth, Hans. 1957. *Nature and the American: Three Centuries of Changing Attitudes*. Berkeley: University of California Press.

Lanigan, Esther F. (1989) 1997. *Mary Austin: Song of a Maverick*. Tucson, AZ: University of Arizona Press.

Lutts, Ralph H. 1990. *The Nature Fakers: Wildlife, Science & Sentiment*. Golden, Colorado: Fulcrum Publishing.

Lyon, Thomas J., ed. 1989. *This Incomperable Lande: A Book of American Nature Writing*. Boston: Houghton Mifflin Company.

Muir, John. (1875) 1997. *Nature Writings: Wild Wool*. New York: Literary Classics of the United States.

Muir, John. (1911) 1997. *My First Summer in the Sierra*. New York: Penguin Books.

Muir, John. (1913) 1965. *The Story of My Boyhood and Youth*. Madison, WI: University of Wisconsin Press.

Nash, Roderick. (1965) 1967. *Wilderness and the American Mind*. Second print, New Haven, CT: Yale University Press.

Nash, Roderick. 1989. *The Rights of Nature: A History of Environmental Ethics*. Madison, WI: University of Wisconsin Press.

Paul, Sherman, ed. 1962. *Thoreau: A Collection of Critical Essays*. Englewood Cliffs, NJ: Prentice-Hall.

Paul, Sherman. 1992. *For Love of the World: Essays on Nature Writers*. Iowa City, IA: University of Iowa Press.

Pearce, Thomas Matthew. 1965. *Mary Hunter Austin*. New York: Twayne Publishers.

Renehan, Edward J., Jr. 1992. *John Burroughs: An American Naturalist*. Post Mills, VT: Chelsea Green.

Ronald, Ann, ed. 1987. *Words for the Wild*. San Francisco: Sierra Club Books.

Rusk, Ralph L. (1949) 1957. *The Life of Ralph Waldo Emerson*. New York: Columbia University Press.

Scheese, Don. 1996. *Nature Writing: The Pastoral Impulse in America*. New York: Twayne Publishers.

Scheese, Don. 1996. "Desert Solitaire: Counter-friction to the Machine in the Garden." In *The Ecocriticism Reader*, edited by Cheryll Glotfelty and Harold Fromm. Athens, GA: University of Georgia Press.

Slovic, Scott, ed. 2001. *Getting over the Color Green: Contemporary Environmental Literature of the Southwest*. Tucson, AZ: University of Arizona Press.

Whitman, Walt. (1882) 1971. *Specimen Days*. Boston: David R. Godine Publisher.

Zwinger, Ann H., ed. 1994. *Mary Austin and John Muir: Writing the Western Landscape*. Boston: Beacon Press.

# 5

# The English Cultural Heritage in Nature Writing

## Introduction

In comparison with colonial America, England has a long cultural history of having a close relationship with the land. W. J. Keith, a professor of English at Toronto University, Canada quotes the English writer Harold J. Massingham as having said: "In our depths we are a country, not an urban, people" (in *The Rural Tradition: A Study of the Non-fiction Prose Writers of the English Countryside*, 1974, p. v). In the first lecture, titled "Origins of American Nature Writing," I have already mentioned the influence of English Romanticism on nature writing. One of the features of Romanticism is the worship of nature, and the natural environment and its landscapes, including animals, were the focus of its writers and poets. In America, the period between 1830 and 1865 represented by Emerson, Thoreau and other writers has sometimes been called "the American Romantic period" (Abrams, pp. 357, 351). In this chapter, we will explore the English cultural tradition of nature writing through the works of five English writers.

As mentioned above, English culture has deep roots in the land, and this tie coincides with the concepts in American nature writing. In *The Country and the City* (1973), Raymond Williams describes and analyzes the images of country and city in English literature which constructed the notion that a simple and natural country life was the symbol of a golden age. Keith focuses his research on prose writing about the English countryside and comments on prose writers since the seventeenth century, rural literature, and the rural tradition reflected in their representative works. Buell frequently mentions the influence of the English rural tradition—and especially that of one of its representatives, Gilbert White—on America nature writing. Buell regarded Thoreau as "an American Isaac Walton or Gilbert White" and John Burroughs as "the same breed as Gilbert White" (Buell 1996, pp. 560, 347).

In *The Norton Book of Nature Writing*, the editors not only present an excerpt from White's *The Natural History and Antiquities of Selborne* (1789) as the first item in the collection, but also discuss the influence of the English rural tradition—as represented by White—on American nature writing at length in the introduction, quoting American historian Donald Worster's comment (from his *Nature's Economy: The Roots of Ecology*, 1977): "The rise of the natural history essay in the latter half of the nineteenth century was an essential legacy of the Selborne Cult." In Worster's view, writers like John Burroughs, John Muir, William Henry Hudson and Richard Jefferies all wrote in a style that blended human life into a beloved landscape, as did White (in Finch and Elder, pp. 19–20). Rachel Carson, on winning the John Burroughs Medal Award in 1952, said in her speech at the award ceremony that Burroughs's tradition had its beginnings in even earlier writings that had flowered on both sides of the Atlantic. She cited Thoreau in America and Hudson and Jefferies on the other side of the Atlantic as having followed the same tradition (p. 94).

The experiences and works of the five English writers discussed in this chapter demonstrate different aspects of the rural flavor of traditional English culture. Of these five writers, three—Gilbert White, Mary Mitford and Richard Jefferies—were born and grew up in small country towns, while Edward Thomas, was born and grew up in a big city and William Henry Hudson was born and grew up overseas. Despite their diverse backgrounds and life experiences, these five writers shared a love of the countryside and the themes of their writing. However, in describing their beloved country life, their focuses and writing styles differ. As Mary Mitford says, "every separate work would bear the stamp of the author's mind, of his peculiar train of thought, and habits of observation" (quoted in Keith, p. 23).

Gilbert White lived in the small English village of Selborne, Hampshire, and devoted his whole life to the observation and study of local plants and animals. His book does not use a dramatic language, but merely tells the story of the simple and leisurely life of an English gentleman. Although this lifestyle has disappeared from busy, restless modern society, through reading White's book the reader can experience the leisure and ease of such a country life, and also learn of a healthy lifestyle and an enlightening philosophy that enrich the body and the soul. White was engrossed in English country life, remaining single and staying very contentedly on the land his whole life. Unhurriedly observing the birds and animals, flowers and insects, he described a leisurely rural life and produced a very interesting book of observations on life in nature. Selborne was not only where he lived, but the place to which he devoted his whole life, studying the local natural history.

Mary Mitford was born in the English county of Hampshire, as was one of the country's most famous female writers, Jane Austen. Although not as famous in English literature as Austen, whose work she liked, Mitford's influence in nature writing cannot be ignored. In Keith's *The Rural Tradition: A Study of the Non-fiction Prose Writers of the English Countryside*, Mitford is the only female writer among the 11 studied. Buell twice mentions Mitford's significant influence on "the local landscape and culture" in American nature writing. He says that female nature writers, and also Thoreau, displayed similarities with the rural sketches of Washington Irving and Mary Mitford, adopting their minute, precise portraitures of village life (Buell 1996, pp. 26, 405). Unlike Austen, Mitford was not a novelist; she was content to use prose essays to describe life in the English countryside. Her work *Our Village* draws on her own experiences in the English countryside; with uniquely female insight and gentleness she describes the natural scenery and cultural life, charming her contemporary and modern day readers alike. This book holds a lasting charm; there have been many editions and reprints, partly because the scenery described is, as Mitford says, "so peaceful, so cheerful, so varied and so thoroughly English" (Mitford [1824] 1893, p. 250).

To Jefferies, writing more than a century after White, life in the countryside mirrored his inner life. Unlike the other four English country writers featured here, Jefferies was a farmer's son, and lived in the countryside in Wiltshire, in England's West Country, until he was almost 30. However, he based most of his nature essays on memory after he had left his home county and was living in the suburbs of London. Even though the city was already spreading into the countryside, he saw clearly that

> the country has also penetrated the mind of the town. No sooner has a man made a little money in the city, than he rushes away to the fields and rivers, and nothing would so deeply hurt the pride of the nouveau riche as to insinuate that he was

> not quite fully imbued with the spirit and the knowledge of the country. (quoted in Keith, p. 9)

Jefferies's words reveal the significance of rural life in the hearts of the English. The countryside, under his pen, is in fact a spiritual land, a land where the divine and the human unite and blend into one.

Hudson was born and grew up in the Pampeana region in Argentina in South America, and was a contemporary of Jefferies. Like Jefferies, Hudson grew up in the countryside, and lived there until he was 30, when he moved to London. To Hudson, nature was first of all a medication that could cure illness and make the body strong. With the help of Mother Nature, Hudson, who was told by doctors that he "had a permanently bad heart and might drop down at any moment" when he was 15 years old (Hudson, [1918] 1939, p. 263), miraculously lived to 81. To Hudson, contact with nature was a necessity; he says in his autobiography: "I who feel, when I am out of sight of living, growing grass, and out of sound of birds' voices and all rural sounds, that I am not properly alive!" (Hudson, p. 2) Hudson grew up on the South American prairie among herdsmen, but a copy of White's *Natural History of Selborne* connected him with the cultural traditions of England in his teens. The wetlands and forests became "the Selborne Parish" in Hudson's heart, and the countryside of England coexisted in harmony with the wide plains of South America in his mind. This unusual juxtaposition gives Hudson's works a strong sense of place, where the wide, wild prairies of South America—with their birds and flowers—mingle with the English countryside across time and space, exuding an amazing charm.

Of the five writers this chapter will discuss, Thomas is the only one born and bred in metropolitan London. He embodied the conflict, confusion and anxiety of modern city dwellers. Nature became a sanctuary for his heart and a home to return to. He grew up in a busy city, but fell in love with the homeland of Jefferies and took frequent walks in the English countryside. His wife said that during his walks "his soul was revived when it was faint with despair, and comforted as some are by religion or music" (quoted in R. G. Thomas 1985, p. 159). He loved the landscape outdoors. He did not enter the stable career path of a public servant that his father had envisaged, but lived a financially precarious life, and died in war. Unlike Jefferies, who described the gentle rolling hills of southern England, or Hudson, who reflected on South American rural life, Thomas—with no roots in the countryside—seemed to be looking for a spiritual homeland throughout his life.

Discussing the rural tradition in English literature, Raymond Williams says: "Country life has been seen, repeatedly, as the life of the past, of the writer's childhood, or man's childhood life, in Eden and the Golden Age" (quoted in Keith, p. 13). Although Williams believed this kind of nostalgia in the rural tradition of English literature would inevitably lead to the phenomenon of "romanticizing the countryside," it cannot be denied that for English people, deeply rooted in their cultural tradition, the countryside and country life have a special meaning. The countryside is more or less synonymous with fields and gardens, with the flowers in May, the golden elms in autumn, the birds singing and flowers blossoming in the woods, the rolling hills and the fields, and the myths and symbols of ancient England. The land is not only the conduit for inspiring a love for nature, but also a spiritual vehicle that carries the ancient cultural traditions and allows unlimited scope for the imagination. In American nature writing, we can also often see this deep connection with the land. It is true that the picturesque English scenery described by the above writers differs from the wild and grand scenery presented in American literature, yet the care for and love of the land is the same. Their writing styles, in the sense that they blend the external and inner landscapes, are also similar. In their description of the relationship between humanity and nature, they complement each other and serve the same purpose.

## Lecture 14. White: Selborne's Idyllic Country Life

As my research area is nature writing, I have long been interested in the eighteenth-century English naturalist and writer Gilbert White (1720–1793); and in 2004 I had my most direct experience of him. That summer, after attending a conference on "Ecological Criticism" in England, I paid a visit to White's home in Selborne. The village is small and inconspicuous, and even though we carefully checked the signs along the way, we still managed to miss it. However, the extra distance we traveled was not a waste, as it took us through unforgettable scenery: along the sides of the road, against green trees and golden wheatfields grew masses of lavender, like purple clouds of fragrance in the fine drizzle. We could feel the atmosphere of Selborne before we even arrived. White's former residence, The Wakes, is located beside the main street of Selborne. It was in this very old house, quiet and unpretentious, that White wrote his famous book more than 200 years ago, and its manuscript still lies on the desk in his old bedroom. From the back window we could see a wide garden and green lawns, with a gentle slope extending to the green woods. White once said that he was a gentleman and had only ever wanted to be a gentleman; and it was in this small place, Selborne, that he led a quiet gentleman's life, observing birds, animals, flowers and insects in a leisurely manner and contentedly growing flowers, beans and fruit.

White sent his life's collection of firsthand observations of Selborne and its surroundings to two well-known scholars: Thomas Pennant, a famous Welsh zoologist, and Daines Barrington, an English lawyer and judge; and later, at the urging of these two correspondents, he edited his letters and published them as *The Natural History and Antiquities of Selborne* (1789). English nature writer Grant Allen, in his introduction to the 1996 edition of the book, which has the shortened title *The Natural History of Selborne*, calls White a typical gentleman of the eighteenth century, and says:

> Time was not then money; it was opportunity for enjoyment, for self-development, for culture. And as such White used it with a consciousness of dignity and a sense of worthiness in life which have almost faded out of our hurried modern existence. (in White [1789] 1996, p. xv)

Fortunately, more than 200 years after White's passing, Selborne still remains intact: simple, peaceful, and away from the madding crowd, an uninterrupted world that has an almost primitive calm and peace.

## *White and Selborne*

Gilbert White was born in 1720 in Selborne—a small, quiet village located about 50 miles to the southwest of London, among the hills and fields of Hampshire—and lived there his whole life. In my copy of *Selections from White's Natural History of Selborne*,[1] a book with a faded red cover, there is a black-and-white photograph of Selborne seen from the woods. Below the photograph is White's description of his village, with which the White family had several generations' connection, as his grandfather and father both worked in the parish. White, the eldest of seven children, was the only one who remained living in Selborne throughout his life. Little is known of his early years, but his correspondence with friends shows that he was interested in natural history from childhood, when he used to spend days observing swallows. White went to Oriel College, Oxford University when he

1 I bought this beautiful little book in the souvenir shop at White's residence, The Wakes. *Selections from White's Natural History of Selborne* was first published by Macmillan and Co. Limited in 1909, with subsequent editions in 1913, 1914, 1919 and 1924. My copy, the 1924 edition, contains 40 images.

was 19, obtained his Bachelor of Arts after four years, and later spent some time as the college's Dean, before returning to Selborne.

Some sources say that White was once vicar of the parish of Selbourne, and others that he did not hold an official position, but occasionally presided over church services. However, one thing is certain: White loved Selborne. He declined several job invitations from universities, preferring to enjoy the leisurely life of a country gentleman. He never married and devoted all his energy to observation and study of local plants and animals. His life, as his nephew said, "passed tranquil and serene, with scarcely any other vicissitudes than those of the seasons" (White [1789] 1996, p. x), but it was not an isolated life, as he kept up his correspondence with Pennant and Barrington. The correspondence with Pennant, a member of the Royal Society and the author of *British Zoology* (1766), was made up largely of lively discussions about bird and animal habits. White's firsthand information about the habits of animals in the south filled a gap in the first edition of Pennant's *British Zoology*; when the second edition was published Pennant added the information White had provided, and acknowledged White's contribution at the end of the book. White's other main correspondent, Barrington, came from a prestigious family and, like Pennant, was a member of the Royal Society. Although Barrington was a lawyer and later a judge by profession, he was passionate about natural history, science and art. It was at the urging of these two correspondents that White, late in his life, sorted out his letters to the two for publication as a book. Four years after the publication of *The Natural History and Antiquities of Selborne*. White passed away, leaving behind a scholarly monument. In the cemetery of the small church in Selborne, there stands a very simple gravestone, inscribed only with his initials "G. W." and the date of his passing. White's grave is as his life—quiet, modest and simple, without any ornament.

However, it would have surprised White, who was never interested in fame, that his book became extremely popular, and that as a result scientists, naturalists and writers all visit Selborne to pay their respects. The nineteenth-century English naturalist and natural historian Charles Darwin visited Selborne in the 1850s. James Russell Lowell, a nineteenth-century American writer, visited The Wakes twice, and on his first visit he called White's book "the Journal of Adam in Paradise." The late-nineteenth century American nature writer John Burroughs also visited White's home and wrote articles in praise of his works. A twentieth-century American nature writer, Edwin Way Teale, stayed at Selborne for three days in the spring of 1966, and later published a beautiful travel essay titled "The Selborne Nightingale" (1970).

The worship of White and Selborne in modern society shows a nostalgia for a world that has gone, and a yearning for an elegant and peaceful life that no longer exists. The connection between White and Selborne is similar to that between the nineteenth-century American nature writer Thoreau and his Walden Pond. The connection is so close that it is difficult to mention one without thinking of the other. When commenting on or introducing *The Natural History of Selborne*, more than one writer or critic tells people who want to understand and appreciate the letters to visit Selborne and experience personally the places White used to frequent: to climb the wooded hills, experience the peace and quiet of the valley, and take in the light and color of the vast scenery. The attraction of Selborne lies in the fact that it has remained untainted since the eighteenth century. The beech forest behind the village is still green and luxuriant; the 1000-year-old yew tree in the church's graveyard still hangs its green branches and leaves over White's gravestone; the winding path still gives you a feeling that this is a place closed and almost cut off from the world. This lively scenery reduces the distance between us and White, enabling us to see a real person in a real world. As the English writer Hudson said after visiting Selborne, it brings you "the thought of one whose memory was interwoven with living forms and sounds" (Hudson [1901] 1930, p. 243).

Selborne also shows the world that White was not only the author of a masterpiece, but also an active and creative man whose lifestyle has become a model for later generations. He helped with building the road to the woods; he and his brother built two huts on the slope of a wooded hill, and he held many tea parties there amid the beautiful surrounding scenery. He was a passionate gardener, once planting in one day more than 10 kinds of flowers, including primroses and bluebells. In his garden there were not only traditional flowers and plants, but also wild flowers he had transplanted from the fields. At the height of his activities he managed front and back gardens, a forest, an orchard, an outer garden of about six acres and lawns, and also two fields. It is said that he grew more than 40 kinds of vegetables, 6 varieties of plums and 17 of peaches. In the small world that he managed, birds sang, flowers bloomed, vines and roses climbed walls, and a grassy path led to the orchard. On the edge of the vegetable gardens bloomed tulips and wallflowers, and at the edge of the outer garden was a pond. White's achievements in natural history were also remarkable. He was the first to distinguish three kinds of English warblers and to describe the European mouse and the English brown bat. He was also the first to declare that earthworms were hermaphroditic (both male and female in one body) and to describe the relationship between birds' selection of habitat and their numbers. He also discovered that swallows, like many insects, copulate and also eat, drink and bathe in flight. These discoveries belonged to the science of animal behavior and zoology, and at the suggestion of Barrington, White submitted and presented many essays on swallows—including the swift, the sand martin and the barn swallow—at the Royal Society all of which were published in the Royal Society's journal *The Philosophical Transactions.* Apart from *The Natural History of Selborne*, White also wrote *Calendar of Flora and the Garden* (1765) and *The Naturalist's Journal,* which starts in 1768 and ends in 1793. No wonder people say that White was a naturalist and also a horticulturalist, an observer and also a poet. He infused scientific observation and hard work with a romantic passion.

## White and The Natural History of Selborne

The first edition of White's best-known book was published in 1789 under its full title, *The Natural History and Antiquities of Selborne.* In the later editions, the words "and Antiquities" were dropped. The main part of the book consists of 110 of White's letters, 44 to Pennant and 66 to Barrington, about wild animals, plants, seasons and antiquities in the parish of Selborne. The book has been so popular that there have been more than 100 editions to date. Apart from the letters, each edition includes different additional content. For example, the copy I have is the 1996 Wordsworth edition with an introduction and notes provided by Grant Allen. In addition to the 110 letters, this edition contains White's manuscripts "Observations on Various Parts of Nature," and "The Naturalist's Calendars," while the 1813 edition contains six of White's poems. Allen's edition, generally considered the most comprehensive, gives us an opportunity to see White as both a naturalist and a literary writer.

*The Natural History of Selborne* is not a sensational book. When John Burroughs first read it, it did not particularly interest him, but many years later he discovered its charm and value. As he says in his essay "Gilbert White's Book,"

> Indeed, the work of the Selborne naturalist belongs to the class of books that one must discover for himself: their quality is not patent; he that runs may not read them. Like certain fruits, they leave a lingering flavor in the mouth that is much better than the first taste promised. (Burroughs, *Indoor Studies*, 1904. p. 177)

Burroughs then says in his essay "Gilbert White Again," "One of the few books which I can return to and reread every six or seven years is Gilbert White's *Selborne.* It has a perennial charm. It is much like country things themselves" (Burroughs, *Literary Values*, 1904, p. 185). James Russell Lowell comments that

White's book "neither stimulated people's thoughts nor stirred any passionate emotions, nor could even be called popular because it only attracted a limited number of readers, but that number has been kept full by new recruits in every generation" (quoted in Keith, p. 39). In his study, Teale[2] has 16 different editions of *The Natural History of Selborne*, adding weight to Lowell's evaluation of White's book.

The most noticeable feature of *The Natural History of Selborne* is that it is in the form of letters, some of which serve to introduce and conclude to the book. The first nine letters describe and introduce the geography, scenery and history of Selborne and its surrounds. They are not dated, and do not have the names of the addressees or sender, or the usual salutations and valedictions of letters, but are more like short descriptive essays, with the last six letters serving as a conclusion. They focus on the variety of weather conditions in Selborne and their impact on local plants and animals, showing a microcosm to provide insight to the great wide world.

Now let's open *The Natural History of Selborne* and follow its pages to experience White's village. The first letter introduces the geographical location:

> The parish of Selborne lies in the extreme eastern corner of the county of Hampshire, bordering on the county of Sussex, and not far from the county of Surrey; is about fifty miles south-west of London, in latitude fifty-one, and near midway between the towns of Alton and Petersfield. (White [1789] 1996, p. 3)

Then we see the unique "hanging wood," a steep slope of about 300 feet covered with trees. People call it "the Hanger" because the branches of the trees hang over the slope. From the hanging wood, you can see a panorama of the small village lying at the foot of the mountain, not far from the wood. The only road leading into the village winds among the small green trees that fill the valleys. The village lies from southeast to northwest, with a spring at each end. The northwestern spring is often dry, while the southeastern one bubbles all year round. However,

2 Edwin W. Teale (1899–1980) wrote more than 30 books on nature, the best known being his four seasons series about American scenery: *North with the Spring* (1951), *Journey into Summer* (1960), *Autumn Across America* (1956), *Wandering Through Winter* (1965). These books are not armchair discussions but rather the results of field studies that Teale and his wife conducted when they drove tens of thousands of miles across and around America. Teale was awarded the Pulitzer Prize for *Wandering Through Winter* and the John Burroughs Prize for his achievements in literary nature writing.

in this small parish there exists a great variety of scenery, soils and flora. On the small hill there is also a paddock for sheep. Standing on its edge, you see an amazing picture before you: hills, dales, woodlands, water and heath-covered wild land. There is clay soil here that suits wheat, white soil for hops, and sandy loam soil for timber. There are rich black and white sandy soils, and also lean and barren sandy soil. The water system is also unique. The spring at the southeast end of the village forms two streams, one going south to the estuary of the English Channel and the other going north into the estuary of the North Sea. In the eight letters following the introduction, White uses the same simple, gentle tone to describe farms, woodlands, lakes, wild plains, cottages on slopes, sunken rocky paths, animals in the woods, waterbirds and plants beside the lakes. Every letter is intimate and charming, exuding the freshness of the fields of the eighteenth century. A picturesque English village is presented to us like an elegant Chinese mountain and water painting.

White's actual correspondence with his friends starts from his 10th letter, as this letter has a date and it concludes thus: "The parish I live in is a very abrupt, uneven country, full of hills and woods, and therefore full of birds" (White [1789] 1996, p. 42). From then on, White leads us into a world of birds, insects, plants and animals, and into the colorful natural environment where they live. In these letters, we listen to him chatting about how many kinds of English warblers there are, why male and female chaffinches migrate separately to different places in winter, why the old tortoise is afraid of rainwater, and how to entice a field cricket out of its hole with a grass stalk. From the letters we learn that nature is a great economist, because it transforms one animal's leisure into another's means of survival. For example, when a cow cools itself in a pond, its manure provides food for insects, and in turn insects provide food for fish. Earthworms are food for many kinds of birds, birds in turn become food for animals, and animals for humans. White teaches his readers how to observe and sets a good example. If a new bird arrives on his land he would be the first to notice, and if a swallow is a few days earlier or later to arrive than usual, this would not escape his attention either. What his letters convey is not only a naturalist's field experience and discoveries, but also his joy and contentment. To borrow Hudson's quoted and revised line, "His image is stamped on every grass" ([1901] 1930, p. 242).

A literary atmosphere naturally permeates White's correspondence with his two friends. There are vivid descriptions, such as "This morning I saw the golden-crowned wren, whose crown glitters like burnished gold. It often hangs like a titmouse, with its back downwards" (White [1789] 1996, p. 62) and "I heard many birds of several species sing last year after Midsummer; enough to prove that the summer solstice is not the period that puts a stop to the music of the woods" (p. 150). There are also often quotations from famous poets, such as the ancient Greek Oppian, the ancient Romans Horace and Virgil, the English Milton and Scottish Thomson. It is worth mentioning that at the end of his 24th letter to Pennant, White adds a poem of his own, "The Naturalist's Evening Walk," considered by many to best express White's understanding of Selborne. In it, he sets out on a walk at dusk.

> When day declining sheds a milder gleam, / To see the swallow sweep the dark'ning plain, / What time the timorous hare limps forth to feed, / To hear the clamorous curlew call his mate. / Each rural sight, each sound, each smell combine; / The tinkling sheep-bell, or the breath of kine; / The new-mown hay that scents the swelling breeze, / Or cottage-chimney smoking through the trees. (White [1789] 1996, pp. 87–88)

White's letters combine science with literature and nature with art. No wonder *The Natural History of Selborne* has attracted a wide range of readers. To Coleridge, a poet, it was "this sweet delightful book," while Darwin, who read the book in his youth, wondered "why every gentleman did not become an ornithologist" (quoted in Keith, p. 56). Talking about White and his book, Burroughs comments: "As a person, he had flaws, but as the author of this book, he was close to perfection" (1904, p. 187), and perhaps best summarizes the secret of the book's everlasting charm:

> When one reads the writers of our own day upon rural England and the wild life there, he finds that they have not the charm of the Selborne naturalist; mainly, I think, because they go out with deliberate intent to write up nature. They choose their theme; the theme does not choose them. (Burroughs 1904, p. 187)

In other words, White did not love nature because he wanted to write about it, he just loved it. *The Natural History of Selborne* expresses his life-long love of his home surroundings, where the birds and animals, plants and flowers all have features and atmosphere that are unique to Selborne. Everything is presented in White's every day, chatty style and intimate tone which people who deliberately set out to write about nature cannot hope to achieve.

When we re-read *The Natural History of Selborne* more than 200 years after White's passing, we feel not only a nostalgic yearning for a pastoral time, but also admiration for his foresight. In White's time, field naturalists were rare. The way he conducted his field observations of animals and plants was unique, and he even ran the risk of being laughed at. When other country gentlemen went out, it was to enjoy outdoor activities that involved guns, hunting dogs or fishing rods, but *The Natural History of Selborne* opened their eyes to an amazing world out there. White's book and his lifestyle introduced society not only to the mystery and energy of natural history, but also to a new way of observing nature and to a full and meaningful way of life, as people who love this kind of research experience not only the joy of discovery, but also sensual pleasure from the natural scenery. The eighteenth-century White walked into nature, observed it, described it and lived in harmony with it—behavior that became a tradition, and one that nature writing still respects and promotes today. The pastoral-flavored rural literature and the cross-disciplinary features of his works have become the keynote of nature writing, and even of today's ecological criticism. Thoreau, Burroughs and Muir in the nineteenth century, as well as many nature writers today, have all followed White's path after reading his book. The very first piece in *The Norton Anthology of Nature Writing*, published in 1990, is an extract from White's *The Natural History of Selborne*: and White and his book have also been discussed in the opening chapters of a number of books on the history of ecological thinking. This is because White's book goes beyond everyday observation and entertainment. White regarded Selborne as a complete and diverse ecological whole, which is why his book about the village is one of the greatest early contributions to the field of ecology in British science. It is, naturally, also a significant means by which people living in a high-tech era can learn how to find a balance between human and nature, and also between love of nature and enjoyment of city life.

## Lecture 15. Mitford: Our Village

James T. Fields, a nineteenth-century American writer and publisher, wrote a book titled *Yesterdays with Authors* (first published in 1883) in memory of his friendships with six (by then deceased) authors. Mary R. Mitford was the only female author included, alongside the eminent writers and poets Thackeray, Hawthorne, Dickens, Wordsworth and Alexander Pope. Fields had a unique way of prompting his reminiscences of his friends: he would write down his stories from memory while looking at their portraits. The chapter about Mitford begins: "That portrait hanging near Wordsworth's is next to seeing Mary Russell Mitford herself as I first saw her, twenty-three years ago, in her geranium-planted cottage at Three Mile Cross." Then Fields offers a description: "It will be observed that the dress of Miss Mitford in the picture before us is quaint and somewhat antiquated even for the time when it was painted, but a pleasant face is never out of fashion." Fields tells readers that one day, more than 20 years earlier, he began "an acquaintance with the lady of *Our Village*" in Three Mile Cross in the south of England, and they became good friends (Fields, pp. 263–265).

What impressed Fields most about Mitford was that "[h]er fondness for rural life, one may well imagine, was almost unparalleled." He went on to explain that as a result of loving nature ever since childhood:

> [s]he knew all the literature of rural life, and her memory was stored with delightful eulogies of forests and meadows. When she repeated or read aloud the poetry she loved, her accents were like flowers' voices, if they could but speak. (Fields, p. 272)

Fields described Mitford as "always cheerful," while another friend of hers, the English writer and critic John Ruskin, called her "merry Miss Mitford" (quoted in Keith, p. 83). However, merry Miss Mitford did not live a carefree life. In her time, she was one of the few writers who lived on their writing, in other words, her pen supported her family. Mitford, born in Alresford, Hampshire, was an only child. Her father came from a prestigious family and graduated from Edinburgh University with a degree in medicine, while her mother came from a well-to-do minister's family. Mitford's childhood was carefree; she could read by the age of 3 and could recite traditional ballads from an even younger age. However, the good life did not last; her father was extravagant, and also a gambler. He squandered not only his own fortune, but also the 20,000 pounds Mitford won in a lottery when she was 10, and he later had to depend on his daughter's writing. The family moved several times, into ever smaller houses, before eventually settling in Three Mile Cross—the small village that Mitford's *Our Village* refers to—and living there for more than 30 years. Before moving to Three Mile Cross, Mitford had already published several collections of poems, but—as some critics have speculated—may only have become an ordinary poet if her father had not squandered all the family fortune; Mitford herself said: "Our family losses made me an authoress" (quoted in Keith, p. 84).

Mitford's life at Three Mile Cross were her materials for *Our Village*. Surprisingly, despite writing under the pressure of earning a living, she presented readers with a bright rural landscape and cheerful country life. She added beauty to her life with nature, decorated the dull rooms with violets picked from the fields, and drove worries from her life with the singing of the nightingale outside her window; as she said herself: "To live in the country is, to my mind, to bring the poetry of nature home to the eyes and heart" (quoted in Keith, p. 88). Fields, who she met in her old age, said that "[a]lthough she had been prostrated by the hard work and increasing anxieties of forty years of authorship, when I saw her she was as fresh and independent as a skylark" (Fields 1883, p. 267).

*Our Village* was first published (1824) as a series of rural sketches in a women's magazine, then as a collection of five volumes, one of which came out every two years from 1824 to 1832. When the first book was published, it immediately became popular among readers and was also received well by critics. It was reprinted in England and America, and between 1824 and 1835, *Our Village* went through 14 editions. As a result, Mitford became the most popular writer of her day, and was also well paid. Admiring visitors flocked to her cottage; artists came to sketch the scenery she had described; passing coachmen and the postman pointed the scenery in the book out to people; and children were named after those in her book. Some readers, feeling that Mitford reproduced in language the landscapes idealized by the seventeenth-century French landscape painter Claude Lorrain, call her "the Claude of English village life" (Feldman 1997, p. 453). Others believed that her writing style was very much like a Dutch oil painting—intimate, realistic, brightly colored and interesting. The English writer William Howitt visited Three Mile Cross in 1835 and later published "A Visit to Our Village" in the then famous literary journal *The Athenaeum*.[3] Of course, what is more important is that *Our Village* started the tradition of rural prose writing and laid the foundation for a brand new branch of literature.

## *Authentic English Country Flavor*

The nineteenth-century American writer Thoreau called nature his "bride," and like him Mitford never married. Fields said that "Her dogs and her geraniums were her great glories" (Fields, p. 271). Mitford's glories were, of course, not only these but also her authentic work about English rural life, *Our Village*. This series of sketches is based on Mitford's diary entries, in which she described her

3 It should be pointed out that William Howitt himself was also a nature poet and writer. Buell's *The Environmental Imagination* (1996) has a chapter about Howitt, describing in detail his influence on American nature writing and saying that the first book of nature Thoreau read seriously was Howitt's *The Book of the Season, or the Calendar of Nature* (1831), which started the literary tradition of using the passing of the four seasons as an organizational structure. Howitt also wrote *The Rural Life of England* (1838) and *The Year-Book of the Country* (1850).

experiences and feelings when taking walks in and around the village in the four seasons. These writings are associated with the land; the words exude the flavor of the land and the fragrance of the flowers and plants. Take the first volume of *Our Village* as an example: Mitford starts with the first snow and frost and ends with the first autumn leaf falling onto the ground. The titles of the sketches include "Walks in the Country," "Violeting," "The Cowslip Ball," "Nutting," "The Wood," and "The Dell." There is movement in the stillness of the exquisite scenery, showing a unique English country charm.

Mitford said:

> I have always had a preference for close, shut-in scenes, both in landscape and in nature, and prefer the end of a woody lane, with a porch, and a vine, and clustered chimneys peeping out among trees, to any prospect I ever saw in my life." (quoted in Keith, p. 89)

This is exactly what *Our Village* describes, with an undeniably unique local color. In the first sketch, titled "Country Pictures," the author introduces "our village" in a gracious tone:

> Of all situations for a constant residence, that which appears to me most delightful is a little village far in the country; a small neighborhood, not of fine mansions finely peopled, but of cottages and cottage-like houses…with inhabitants whose faces are as familiar to us as the flowers in our garden; a little world of our own, close-packed and insulated like ants in an ant-hill, or bees in a hive, or sheep in a fold, or nuns in a convent, or sailors in a ship; where we know every one, are known to every one, interested in every one, and authorized to hope that every one feels an interest in us. (Mitford, [1824] 1893, p. 3)

The author then explains that even in books she prefers "a confined locality," and likes staying in the small villages featured in Jane Austen's novels or visiting naturalist Gilbert White to wander around the fields together. What she did not like would be to dash through places all over Europe: go to sleep in Vienna and then wake up in Madrid; for her this would produce "a real fatigue, a weariness of spirit." With her writing, therefore, she wanted to introduce the reader to a small village surrounded by green mountains and guide them along a pleasant journey. She calls us gently: "Will you walk with me through our village, courteous reader? The journey is not long. We will begin at the lower end, and proceed up the hill" (Mitford [1824] 1893, pp. 3–5).

We follow her past small cottages in the shade, admiring the green lawns and gardens around cottages and listening to her stories about their owners: the retired publican who is proud of a life without care or worry; the hardworking shoemaker who is so busy from morning till night, "[a]n earthquake would hardly stir him" (p. 7); the blacksmith with eight children who is the king of his little kingdom; the well-known and highly skilled carpenter who has a lovely daughter only 3 years old but with the body and brain of a child of six. At the end of the village lived the shortest man and his tall, beautiful wife. The short man had a giant's voice, as loud as a trumpet when he opened his mouth. When we climb up the hill with the author, this is the picture we see in front of us: The little path that goes down the hill makes a turn half way; and at the place where it turns, there is a red cottage covered with vines, a symbol of comfort and content. Further down, there is a white cottage; then a large patch of woods, a rope workshop and the street of the small village. Through the green trees, we can see various kinds of rooftops and chimneys and partial views of walls. After enjoying the overview of the village, we can see tender green lawns on one side of the valley: soft moss, purple and white orchids, bright blue hyacinthine flowers and waving willows in the wind, decorating the hill slope, making it like in a picture and in a poem. Walking to the other side of the valley,

> now we seem hemmed in by those green cliffs, shut out from all the world, with nothing visible but those verdant mounts and the deep blue sky: now by some sudden turn we get a peep at an adjoining meadow, where the sheep are lying, dappling its sloping surface like the small clouds on the summer heaven. (Mitford [1824] 1893, p. 112)

Howitt called *Our Village* "sunshiny pictures of an English village" (quoted in Keith, p. 92). Mitford did indeed add romanticism to rural England, showing the sunny side of the Berkshire countryside. In the chapter "Violeting" she found it difficult to suppress the sudden rush of joy she felt at seeing violets.

> Through this little gate, along the green south bank of the green wheat field, and they burst upon me, the lovely violets, in tenfold loveliness! The ground is covered with them, white and purple, enameling the short dewy grass, looking but the more vividly colored under the dull, leaden sky. (Mitford [1824] 1893, p. 67)

Green trees, wildflowers scattered all over hills, and bees and butterflies dancing and darting among the flowers fascinated her. She would exclaim:

> What happiness to sit on this tufty knoll, and fill my basket with blossoms! What a renewal of heart and mind! To inhabit such a scene of peace and sweetness is again to be fearless, gay, and gentle as a child. Then it is that thought becomes poetry, and feeling religion. (Mitford [1824] 1893, p. 68)

Deidre Lynch, a professor of English at Toronto University, thinks that in *Our Village* Mitford wanted to demonstrate a sense of contentment that could be achieved when living in a particular and limited place. She used her own village as a background for literary scenes, and presented a "village geography" that has, according to Lynch, "*déjà vu, déjà lu*" qualities for the reader (Lynch 2000, p. 1104). Keith points out that Mitford had no desire to describe rugged or sublime natural scenery; her scenes always contained people, she rarely wrote about deserted nature (Keith, pp. 89–90). Lynch and Keith both note two common features that are characteristic of both Mitford's works and American nature writing: a "sense of place," and humans living in harmony with nature. Let's look at another example that illustrates this. The chapter titled "The Queen of the Meadow" in the third volume of *Our Village* begins:

> In a winding unfrequented road, on the south side of our Village, close to a low, two-arched bridge, thrown across a stream of more beauty than consequence, stood the small irregular dwelling and the picturesque buildings of Hatherford Mill. It was a pretty scene on a summer afternoon, was that old mill, with its strong lights and shadows, its low-browed cottage covered with the clustering Pyracantha, and the clear brook.…
>
> …Many a traveler has stayed his step to admire the old building of Hatherford Mill, backed by its accompanying figures, the jolly miller sitting before the door, pipe in mouth, and jug in hand, like one of Teniers' boors,[4] the mealy miller's man with his white sack over his shoulders, carefully descending the out-of-door steps, and the miller's daughter, flitting about amongst her poultry, gave life and motion to the picture. (Mitford [1826] 1870, pp. 76–77)

This description from Mitford's pen helps us to appreciate deeply the uniqueness of the place with its idealized country life and pastoral scenes, offering an unusually vivid, visual picture of English country life.

4 David Teniers the Younger (1610–1690) was a painter of the Flemish baroque period, well known for his paintings of peasant life.

## *Telling a Story of the Land in a Unique Style*

The 1893 edition of *Our Village* has a lengthy 27-page introduction written by Anne Thackeray-Ritchie, daughter of the nineteenth-century English writer William Makepeace Thackeray and a writer herself. Ritchie writes about not only Mitford's life and works, but also her close connections with other eminent English and American writers of her time, including Ruskin, Fields, and the English poet Elizabeth Barrett Browning. She reports that Mitford regarded her friendship with Mrs. Browning as one of the happiest events of her life; that in her old age Ruskin's friendship cheered her, and that she was supported by every book he gave her and every care he showed her; and that Fields crossed the ocean and visited her in her home more than once, becoming a very good friend. (in Mitford 1893, pp. xxxi–liii) Mitford kept in contact with friends mainly through correspondence. *The Letters of Elizabeth Browning to Mary Russell Mitford 1836–1854*, published in three volumes in 1983, includes about 500 letters between the two; Fields, in his chapter about Mitford, includes about 50 letters Mitford wrote to him; and *The Friendships of Mary Russell Mitford* (Mitford 1882), also records her literary friendships through letters. No wonder Keith comments with admiration, "Miss Mitford is, indeed, a connoisseur of the art of letter-writing." He then quotes from a letter Mitford wrote to a friend:

> I never could understand what people found to like in my letters, unless it be that they sprang direct from the soil, that they have a root to them—the sort of quality that makes one sometimes prefer a wild plant alive and growing in its woodland nook to a fine cut flower in a rich vase. (quoted in Keith, p. 102)

This unique writing style is also demonstrated in *Our Village*, giving Mitford's work an outstanding quality and making it well loved.

In the chapter "Cowslip Ball," Mitford writes as if talking intimately to a friend in a letter: "There are moments in life when, without any visible or immediate cause, the spirits sink and fail, as it were, under the mere pressure of existence." Then, after listing various kinds of bad moods that are difficult to get rid of, Mitford turns to her magical treatment:

> I will go out into the air this cool, pleasant afternoon, and try what that will do.… I will go to the meadows, the beautiful meadows! And I will have my materials of happiness, Lizzy and May,[5] and a basket for flowers, and we will make a

5 Lizzy is the neighbor's little girl, and May is Lizzy's little dog.

> cowslip-ball. (quoted in Thackeray-Ritchie, introduction to Mitford [1824] 1893, p. xxxvi)

Happiness is usually thought of as a spiritual state, but here Mitford wants to have her "materials of happiness": the meadows, flowers, young companions and communication with nature. Thackeray-Ritchie reports: "She declares that it is only by getting out and afield that she can bear the strain and the constant alternation of enforced work and anxiety. Nature was, indeed, a second nature to her" (introduction to Mitford [1824] 1893, p. xxxvi). In Mitford's writing flows this second nature of hers and a natural rhythm. "And on we go fast, fast! Down the road, across the lea, past the workhouse, along the great pond, till we slide into the deep narrow lane, whose hedges seem to meet over the water." She lets the reader share her views and her happiness along the way, and also "a more beautiful solitude" than described in *Robinson Crusoe* when she arrives at the vast and isolated meadows: one or two miles long and gently sloping downwards on two sides, with several fenced rich pastures. A clear and lively stream winds its way through the meadows. The little stream seems to have the free and lively personality of the northern rivers and is far different from the limpid water of the southern streams. Now it leaps over bright pebbles, splashing and glistening in the sun, making sounds like that of birds in the woods, loud and pleasant; and now it hides in the bushes of golden marigolds. When it reaches alder wood, hawthorn bushes and wild roses, it appears extremely calm as if it is sleeping. However, in the silent stream sometimes trout suddenly leap, and this reminds Mitford of the seventeenth-century English angler Izaak Walton.[6] "Izaak Walton would have loved our brook and our quiet meadows; they breathe the very spirit of his own peacefulness, a soothing quietude that sinks into the soul." After enjoying the scenery, picking flowers and weaving a cowslip ball, the unhappiness in Mitford's heart has disappeared completely. She sighs: "How melancholy I was all morning! How cheerful I am now!" At the end of this sketch, she advises kindly, "Try it, my dear readers, if ever ye be nervous—I will answer for its success" (Mitford [1824] 1893, pp. 126–130, 139–140). Keith summarizes the magic of Mitford's style, describing her as "a writer for whom the individual reader could react with particular affection and warmth, since her sketches read not as literary essays but

6 Izaak Walton was an English poet and writer (1593–1683). His idyllic *The Compleat Angler* (1653), which describes the joy, art and skill of fishing, has had more editions published than most books in the history of English literature.

as personal letters. No subsequent rural writer was to enjoy so trustworthy and appreciative a public" (Keith, p. 86).

Ritchie says that Mitford "found her true romance in friendship, not in love" (introduction to Mitford [1824] 1893, p. xxxiv). This certainly had a direct influence on her writing style, as one of the main characteristics of Mitford's essays is her use of intimate, direct spoken language, as if chatting with friends or telling them a story. The writing is filled with fine feelings and vivid energy. In the chapter titled "The Bird Catcher" in the third volume of *Our Village*, she discusses two kinds of fog, that of London and that of her village. "A London fog is a sad thing, as every inhabitant of London knows full well: dingy, dusky, dirty, damp; an atmosphere black as smoke and wet as steam, that wraps round you like a blanket" (Mitford 1828, p. 254). However, the fog in the country village in Berkshire in the southern part of England was different.

> In Berkshire: days beginning in a soft and vapoury mistiness, enveloping the whole country in a veil, snowy, fleecy, and light, as the smoke which one often sees circling in the distance from some cottage chimney, or as the still whiter clouds which float around the moon; and finishing in sunsets of a surprising richness and beauty, when the mist is lifted up from the earth, and turned into a canopy of unrivalled gorgeousness, purple, rosy, and golden, disclosing the splendid autumn landscape, with its shining rivulets, its varied and mellow woodland tints, and its deep emerald pasture lands, every blade and leaf covered with a thousand little drops, as pure as crystal, glittering and sparkling in the sunbeams like the dew on a summer morning, or the still more brilliant scintillations of frost. (Mitford 1828, p. 256)

In Mitford's sketches, even the sunshine in the countryside is "the dewy sunshine," vivid and lively in contrast to London fog (Mitford 1928, p. 259). When she writes about seasons, it is as if she is talking about a person's feelings. At the beginning of a chapter titled "The Fall of the Leaf" in Volume 1 she writes:

> The weather is as peaceful to-day, as calm, and as mild, as in early April; and, perhaps, an autumn afternoon and a spring morning do resemble each other more in feeling, and even in appearance, than any two periods of the year. (Mitford [1824] 1893, p. 249)

Country life certainly brings Mitford mental peace and calm; however, it is not uneventful. In *Our Village* there is a strong sense of movement: harvesting songs from farmer Brook's fields, as ripe peas are picked; under the tree Mrs.

Allen is feeding the chicken, surrounded by three chirping granddaughters; and Mitford herself is also always busy: taking walks, going on outings, making hay, picking nuts. She conveys a sense of herself and of the movement of nature to the reader. Sentences like this pepper the book: "And on we go, fast, fast! Down the road, across the lea.... Now we go round the corner and cross the bridge, where the common, with its clear steam winding between clumps of elms, assumes so park-like an appearance" (Mitford, [1824] 1870, p. 28). This conversational prose style is intimate and rhythmic, and conveys a strong sense of participation to the reader, making them feel as if they are living in the world described by the author. This writing style of Mitford's entices people from different times and different places into the "now" she created, compelling them to follow what she is experiencing, and to hear what her heart is saying. One cannot help but admire this unique writing style.

*Our Village* doesn't only present the sunny side of life; there are some dark clouds in the blue sky. In a chapter titled "The Wood," the author leads us through a small, lush wood filled with flowers and singing birds, but there is suddenly a scene of tragedy: "Above twenty of the finest trees lay stretched on the velvet turf. There they lay in every shape and form of devastation...majestic corpses, the slain of to-day!" (Mitford [1824] 1893, p. 102) Later she records the experience of witnessing an oak tree being cut down:

> Ah! here we are at the very scene of murder, the very tree that they are felling...see how the branches tremble! Hark how the trunk begins to crack! Another stroke of the huge hammer on the wedge, and the tree quivers, as with a mortal agony, shakes, reels, and falls. (Mitford [1824] 1893, p. 102)

Unlike in other chapters, the ending of this chapter is also rather sad.

> Even the heavens seem to sympathize with the devastation. The clouds have gathered into one thick low canopy, dark and vapoury as the smoke which overhangs London...the deep flush fades away, and the rain begins to descend; and we hurry homeward rapidly, yet sadly, forgetful alike of the flowers, the hedgehog, and the wetting, thinking and talking only of the fallen tree. (Mitford [1824] 1893, p. 103)

Mitford may not have had the ecological awareness of today, or have considered nature to have rights. However, she describes the felling of trees as "murder" and "devastation" and the felled trees as "slain" and "corpses" with a heart-wrenching sadness. In this way, showing her sincere affection for the natural world around her, she taught her readers a lesson on treasuring nature.

Re-reading this work that has withstood the test of time of nearly two centuries after it was first published, we realize all the more deeply Mitford's talent for writing. Her focus is not desolate, unknown and wild nature, but familiar scenery that people seldom notice: the calm, green English scenery, which cannot be called grand; devoid of wildness and seemingly simple. *Our Village* gives the reader a feeling, as Barrett Browning writes, that

> If read by snatches, it comes on the mind as the summer air and the sweet hum of rural sounds would float upon the senses through an open window in the country, and leaves you for the whole day a tradition of fragrance and dew. (quoted in Keith, p. 20)

Ruskin was also deeply moved by Mitford's search for value and beauty in ordinary country life, saying that she found "history enough in the life of the butcher's boy, and romance enough in the story of the miller's daughter" (quoted in Keith, p. 91). Even though in her writing Mitford has a tendency to idealize English country life, we cannot help but feel her sense of beauty and optimism; as she said, "despair is not in my nature" (quoted in Keith, p. 98). Seeking energy from nature and magic cures for human worries and anxiety is also a characteristic of nature writing, as is viewing and describing the world around us optimistically, tolerantly and open-mindedly. In Mitford's works there is no sarcasm and no malice; she shows the reader the most beautiful side of nature and of the human heart, using a unique epistolary style to tell the story of the land in a conversational tone, as if chatting with friends. This generous, optimistic attitude to life and lively, carefree style of writing have become a distinguishing feature that has influenced later generations of nature writers.

## Lecture 16. Jefferies: The Story of My Heart

In 1883, English writer and naturalist Richard Jefferies (1848–1887) caused a huge uproar with the publication of his autobiography *The Story of My Heart*. In the book he describes his yearning to be in actual harmonious co-existence with "the tangible universe" and his "experience of having the sound of the earth going through my body and heart." He says he felt that trees, grass and stars "seemed like exterior nerves and veins for the conveyance of feeling to me," that he believed there was a spiritual life superior to any god, and that the orthodox Christian god did not exist. Even though Jefferies's atheist view caused outrage and criticism throughout society, the book soon came to be regarded as an outstanding and unusual autobiography, and many more editions have been published. *The Story of My Heart* reproduces Jefferies's fascinating experience among the undulating hills, meadows and woods of the countryside in southern England. He prays

> that I might have the inner meaning of the sun, the light, the earth, the trees and grass, translated into some growth of excellence in myself, both of body and of mind; greater perfection of physique, greater perfection of mind and soul; that I might be higher in myself. (Jefferies [1883] 1922, p. 80)

Jefferies differed from the other English country writers, as he came from a different background. A farmer's son, he lived in the Wiltshire countryside until he was almost 30 years old and never went to university. However, lack of institutional education did not affect Jefferies's development. He had many talents, and was a naturalist, novelist, essayist, poet, agricultural commentator, rural sociologist and historian as well as being a prolific writer. In a short life of 39 years, he wrote more than 10 collections about rural life, 8 novels, 1 autobiography and 1 autobiographical novel. Of all Jefferies's writing, the truly enduring works are his nature sketches. English writer Edward Thomas (1878–1917)—the subject of Lecture 18 of this series—comments on these sketches in his *Richard Jefferies: His Life and Work* ([1909] 1972).

> For almost the first time in English literature a pure countryman who is nothing but a countryman reveals his life and neighbourhood.… Thus at his best, he writes as if his hand had in it part of that spirit which builds the hills and lights the stars over them; in his veins are the saps of oak and ash and elm, the blood of things that run and fly and creep. (Edward Thomas [1909] 1972, p. 146)

Jefferies devoted his short life to experiencing and describing nature; and even when he was gravely ill and knew he would not live long, it was not the coming of his own death that concerned him. Observing the birds and flowers outside his window, he worried: "without me to keep records for them what are the birds and flowers going to do?" (quoted in Edward Thomas [1909] 1972, p. 301) Jefferies may not have imagined that in the mountains of New England in America another person lay on a sickbed, describing birds and flowers. Bedridden, the American poet Clara Robinson described how "the posies on her inner windowsill" and "the birds on her outer windowsill" would give wing to her thoughts and imagination, and let them fly to the natural world outside the window. This coincidence perhaps shows that the English tradition has a natural influence on, and invisible affinity with, American nature writing. American poet Robert Frost later visited Robinson's former residence with friends, and wrote the poem "Spring, Bottle, Donkey Ear and Books," included in his poetry collection *New Hampshire* (1923), about the visit.[7] Toward the end, Jefferies was still dictating the introduction he had been invited to write for a new edition of White's *The Natural History of Selborne.* He once said that "[w]ords last longer than steel" (quoted in Edward Thomas [1909] 1972, p. 54), and the writing he left behind has proved that true. His home, the countryside under his pen, is in fact a spiritual place, a realm where humans and heaven merge, and will last as long as the water and mountains in nature.

## *Jefferies's Homeland*

Jefferies was born at Coate Farm in Wiltshire on 6 November 1848, and spent nearly 30 years in this rural area. As E. M. Forster (1879–1970) once said, "the fibres of England unite in Wiltshire" (quoted in Keith, p. 128). The county of Wiltshire has a long history: its ancient war trenches, castles and Roman roads all speak of ancient times. However, its essence lies in its vast undulating green hills which roll up and down like a piece of music and are infinitely fascinating. In summer they resemble clusters of green clouds; in winter, snow transforms them into floating white clouds. Jefferies's birthplace, Coate, is located below Burderop Wood, near the most typical rolling hills. As a child, Jefferies had a great time playing in these isolated and solemnly beautiful hills; while later their silence, together with the fluid beauty of their movement and entirety, became an image of "sea" in his mind which stayed there forever and was the source of inspiration for all his writings. He once wrote, "In the hills, there is always hope" (quoted in Edward Thomas [1909] 1972, p. 174). It should be mentioned that American nature writer John Muir, a contemporary of Jefferies, expressed similar sentiments: "In God's wildness lies the hope of the world" (Muir, [1938] 1979, p. 317). Jefferies's home was not, of course, only the hills; the forest of elms, oaks and firs, beech and ash trees, and the vast meadows were the classrooms where Jefferies learnt about flowers and birds. There were a great number and variety of plants and birds for him to study: bluebells, clover, sweet basil. He liked geraniums best, because their purple flowers seemed to embody rich passion and calm thoughts at the same time, showing a character like his. A reservoir called Coate Water

7 This collection of poems won the Pulitzer Prize for Poetry in 1923.

stretched out southward from Coate Farm, shaped like a big fish with a large tail to the northeast. There were clusters of whirlpools, and occasionally a couple of small boats. Between the undulating green meadows and the clear water, in thatched houses, lived Jefferies's people—hardworking, taciturn people who had different blood from those who lived in the cities. Like Thoreau, Jefferies regarded his home as a miniature version of the wider world. No matter where he later went, his home was like a bank deposit from which he could draw sustenance for his heart whenever he felt unwell.

Jefferies's father and grandfather were both farmers, but some of his relatives on his mother's side were in printing, and therefore his uncles and aunts often gave him books. Even though he had no formal education after leaving school at 15, he read and also learned common sense from the fields and woods, and became very knowledgeable. According to his memoir, he liked to climb trees, catch birds, go fishing, swing between trees and play all sorts of naughty children's games before he was 9. He was familiar with the routes of birds and animals, claiming that he could draw a map of where they would appear. In his teens, the sunrise became his passion. Every morning he would go and watch the sun rise in the eastern sky through the elm trees. He felt his thoughts and his subconsciousness would sublimate as the day brightened, writing: "I drank the beauty of the morning; I was exalted" (Jefferies, *The Story of My Heart*, [1883] 1922, p. 76). He would walk through the dewy meadow to the green rolling hills and meditate in the morning mist, letting his thoughts and feelings merge into the beauty of nature to rid himself of an inner loneliness. In the forests around his home, the rolling hills and the valleys, he had several favourite "solitary corners." In the library of Coate Farm, he read books on astronomy, geography, agriculture and medicine, and also liked to read Shakespeare, Scott and Byron, and Greek and Roman classics. His experience in the wild combined with knowledge from books made Jefferies more perceptive and far-sighted than many other people and laid the foundation for his writing. Quiet and thoughtful, feeling lonely in a crowd but utterly at home in the natural world—all this predestined his later life.

At the beginning of 1866 Jefferies, who had just turned 17, started work as a reporter for a local newspaper called the *North Wiltshire Herald*. His essays about local life and customs published in the paper have been republished in a collection titled *Jefferies' Land*. However, In 1872 Jefferies's three long letters about the lives of farmhands in Wiltshire were published in *The Times*. These letters were clearly reasoned, his sentences were short and forceful, and they showed both his deep understanding and knowledge of country life, and his concern for farmhands.

From 1874 to 1880, he published four novels. These early novels, however, did not receive much attention, and they disappeared from the scene very quickly.

In early 1877 Jefferies, by now nearly 30 years old, moved with his wife and children to Surbiton, a suburb of London, where he worked as a writer to support his family. The 10 years from 1877 until he passed away in 1887 were his most productive time. In the last five years of his life he was very poor, and also gravely ill, and he had to dictate some of his last works for his wife to record them. During that period, in addition to novels and his autobiography, Jefferies completed 10 volumes of nature writings which are considered the best of his works. It has been noted that Jefferies's early nature writing was by a walking person, while his later nature writing was by a person sitting and lying down. However, the scenery that he described from his window while on his sickbed was no less vivid, and there was an added tragic beauty when compared with what he had observed while walking through it. In an essay in *Field and Hedgerow* (1889), a collection published after he passed away, he wrote: "Through the bars of my prison I can see the catkins thick and sallow-grey on the willows across the field" (quoted in Keith, p. 143). Due to nearly 30 years' experience of living in the countryside, Jefferies was already well versed in the book of nature, and he possessed the treasures of country life. Whether physically able to walk in the natural scenery or not, he was a spokesperson for nature.

## *Jefferies's Scenery*

Jefferies was a recluse by nature. In his *The Life of the Field* (1884) he writes: "When I look in the glass I see that every line in my face means pessimism; but in spite of my face—that is my experience—I remain an optimist." Jefferies's optimism came from nature; he says, "Let us not look at ourselves but onwards, and take strength from the leaf and the signs of the field" (quoted in Edward Thomas [1909] 1972, p. 218). Seeking happiness and a healthy lifestyle in nature is a noticeable characteristic of Jefferies's early nature writing.

Jefferies produced four early books of nature writing: *The Gamekeeper at Home* (1878), *Wild Life in a Southern County* (1879), *The Amateur Poacher* (1879), *Round About a Great Estate* (1880). These early nature writings illustrated his youth and his memories of the land. The rolling hills, lowlands with flowing streams, Burderop Wood, Coate Water, Liddington Castle, Coate Farm, the lives of the farmhands who lived on the land and the personalities of various animals and birds there—together these presented a picture of simple and rugged beauty. The cheerfulness of life is the enduring attraction of these books. In *Wild Life* he writes: "The joy in life of these animals—indeed, of almost all animals and birds in freedom—is very great. You may see it in every motion." He then cites the agile leaping of wild rabbits, the songs skylarks must sing, the love tunes of doves in the hawthorn bush, and the way blackbirds stroke their feathers on the fence to explain that the real sense of living is particularly intensely felt with the animals, and this itself is an exquisite joy. He says, "Watch the birds in the spring; the pairs dance from bough to bough, and know not how to express their wild happiness" (quoted in Edward Thomas [1909] 1972, p. 130). From the happiness of the animals in the fields, Jefferies saw the disadvantage of living indoors. In *The Gamekeeper* he advises that "[i]t's indoors, sir, as kills half the people; being indoors three parts of the day, and next to that taking too much drink and vitals." To Jefferies, nothing benefited the health more than fresh air and the smell of the forest. There was also the fresh smell of newly ploughed earth, the hedges and the meadows after rain. He even asserts that "Anything with a green leaf is the thing; depend upon it, if you want to live healthy" (quoted in Edward Thomas [1909] 1972, pp. 125–126). No wonder Thomas praises Jefferies's early nature sketches, saying that

> no one English writer before had had such a wide knowledge of laborers, farmers, gamekeepers, poachers, of the fields, and woods, and waters, and the sky above them, by day and night; of their inhabitants that run and fly and creep, that are still and fragrant and many-colored. (Edward Thomas [1909] 1972, p. 320)

Another Jefferies scholar, Keith, believed that in the tradition of nature writing Jefferies was the first to consciously guide or even coax the reader into the countryside (Keith, p. 134). In this respect, Jefferies is very much like his contemporary, American nature writer John Burroughs; both can be called "guides to nature."

The publication of *Nature near London* in 1883 marked the beginning of Jefferies's later nature writing. From 1877, when he took his family to live in Surbiton, to when he passed away in Goring in southern England, his life circumstances changed and his health deteriorated. Over this time his writings also changed noticeably. While in the early writings we see the young Jefferies busy in the hills of Wiltshire, in the later writings we see a Jefferies who focuses more on a microscopic view of nature, on colors and shapes, and whose physical strength is declining while his spiritual strength is soaring.

As long as Jefferies could walk, he always did so with a notebook in his hand. Every piece in *Nature near London* resulted from his daily walks. His natural and realistic descriptions of nature near London amazed locals. Under his pen, summer is the season with the most beautiful air, "ripe and rich, full of the emanations, the perfume, from corn and flower and the leafy tree." Autumn is totally dominated by the color brown: "The sunshine of the autumn afternoons is faintly tawny, and the long grass by the wayside takes from it a tawny undertone." He describes "brown partridges and pheasants, brown thrush and brown clover hay-ricks," while even the "pale yellow thatch on the wheatricks becomes a deeper yellow.... What can you call it but tawniness?—the earth sunburned once more at harvest time." In Jefferies's eyes, "[t]he Autumn itself is nutty, brown, hard, frosty and sweet" (pp. 49–51). He then turns his eyes from Coate Water to a small river in the county of Surrey near London, and focuses on a trout hidden under the water: "There were ways and means by which he could be withdrawn without any noise or publicity." Jefferies went to watch the trout every day for three summers, worried that some angler would take it. His curiosity and concern about the trout, the beauty of its movement and its stillness in the water, together with the lively river where it lived and the scenery on the banks, were all written into a chapter called "The little river in Surrey County" (pp. 10–11).

Jefferies's later nature writing collections also include *Red Deer* (1884), *The Life of the Fields* (1884), *The Open Air* (1885), and *Field and Hedgerow* (1889). He dictated *Field and Hedgerow* before his death, and it was published posthumously. These works were the peak of Jefferies's writing career, and the best of his prose writing.

What we notice first in Jefferies's later nature writing is the dreamily appealing style. "A fir-tree is not a flower, and yet it is associated in my mind with primroses" he says in the chapter "Wild Flowers" in *The Open Air* (Jefferies 1893, p. 30). The author then leads us through his memories to the country road lined with blooming primroses that he used to visit every day in early spring, winding through firs in the rolling hills. There were also flickertails—their color was like that of the wildflowers and the blue sky, and they never faded in the author's memory—and larks which sank into the green grass and bathed in the sea of flowers. "Give me the old road, the same flowers—they were only stitchwort—the old succession of days and garland, ever weaving into it fresh flowers from far and near...there has been a place in the heart waiting for them" (Jefferies 1893, p. 49). In *The Life of the Fields*, we read these words:

> The lost leaves measure our years; they are gone as the days are gone, and the bare branches silently speak of a new year, slowly advancing to its buds, its foliage, and fruit. Deciduous trees associate with human life as this yew never can. Clothed

> in its yellowish-green needles, its tarnished green, it knows no hope or sorrow; it is indifferent to winter, and does not look forward to summer. (Jefferies 1884, p. 106)

From his comparison of deciduous trees and evergreens, we see into the author's inner world. In the same book, we also see a small river. In summer, the grass of the meadow seems to crowd toward the side of the river, hawthorn bushes obstruct sight and the river disappears into the meadow; only the glistening of the flowers can be seen. At dusk, an oriole sings in the hawthorn bush, and in the evening, a star sometimes twinkles in the water. The author exclaims: "The brook has forgotten me, but I have not forgotten the brook" (Jefferies 1884, p. 36). In his last work, *Field and Hedgerow*, Jefferies leaves this advice to later generations: "Nature is like a child, who will sing and shout.... You do not know what you may find each day; perhaps you may only pick up a fallen feather, but it is beautiful, every filament" (quoted in Edward Thomas [1909] 1972, p. 304).

## Jefferies's Heart

When Jefferies was gravely ill, a friend sent him a copy of American poet Walt Whitman's prose collection of autobiographical experiences, *Specimen Days*. Whitman had written it while recuperating in the New Jersey countryside after having had two strokes, and it is regarded as "the last pastoral song of an old man," Whitman's swansong of returning to nature. Even though Jefferies's *The Story of My Heart*, published in 1883, was not his swansong, he started to write it at the age of 34, only five years before his death, and like *Specimen Days* it talks about the author's yearning to return to nature. This autobiography is a truthful description of how Jefferies sought spiritual sublimation in nature; he is no longer shy about revealing his past life and inner thoughts to the public. He picks up his memories of his experiences in the countryside when he was young, saying: "The story of my heart commences seventeen years ago" (Jefferies [1883] 1922, p. 1). We then see Jefferies wandering in the fields and the hills, giving all his heart to the universe:

> I was utterly alone with the sun and the earth. Lying down on the grass, I spoke in my soul to the earth, the sun, the air, and the distant sea far beyond sight. I thought of the earth's firmness—I felt it bear me up: through the grassy couch there came an influence as if I could feel the great earth speaking to me. (Jefferies [1883] 1922, p. 4)

As green grass brushes his face and shadows of the trees tremble over it, Jefferies experiences the existence of his body and the "existence of soul," completely lost in and merged into the "existence of the universe." From the grass and leaves, from the birds' singing and their feathers, and from listening to the songs of the insects and watching the dancing of the butterflies, he absorbs the essence, grandeur and beauty of nature. He admits that he "prayed with the glowing clouds of sun-set and the soft light of the first star coming through the violet sky" to seek a higher and deeper inner life (in Begiebing & Grumbling 1990, pp. 465–469). *The Story of My Heart* is not only an expressive autobiography of the heart; in it Jefferies also shows the transition of nature from a companion in childhood to a spiritual support in adulthood. The individual's merging into the whole and seeking spiritual sublimation in nature is the lesson Jefferies left in this work for later generations; he saw that all life forms constitute a whole and a whole beauty.

Nature embodies beauty. In his later works, Jefferies seems to think that all beautiful and cheerful things are right, and that it would be wrong for them to be otherwise. For him, the land is always beautiful and "the heart, from the moment of its first beat, instinctively longs for the beautiful." He wonders if it is possible "to build up a fresh system of color language by means of natural objects" (quoted in Edward Thomas [1909] 1972, pp. 303, 299). Even in art, what Jefferies saw was natural beauty, and he tried to analyze and appreciate the beauty of art from the perspective of nature. In the essay "Nature in the Louvre" he describes how he went to see the statue of Venus three times. This work of art aroused his interest because it aroused his memory of nature.

> Here there came back to me this old thought born in the midst of flowers, and wind-rustled leaves, and I saw that with it the statue before me was in accord. The living original of this work was the human impersonation of the secret influence which had beckoned me on in the forest and by the running streams. She expressed in liveliness of form the colour and light of sunny days…Though I cannot name the ideal good, it seems to me that it will be in some way closely associated with the ideal beauty of Nature. (quoted in Edward Thomas [1909] 1972, pp. 300–301)

Appreciation and experience of art from the perspective of nature is another characteristic of Jefferies's later works. Therefore, when the adult Jefferies gave up childhood's hunting and opposed the killing of birds, and when he raised the issue of protecting the land, trees, plants and flowers, maybe to a certain extent it was out of a wish to protect the beauty of nature. In *Field and Hedgerow* he writes that

> [e]very blade of grass was mine, as though I had planted it separately. They were all my pets, as the roses the lover of his garden tends so faithfully. All the grasses of the meadow were my pets, I loved them all; and perhaps that was why I never had a "pet," never cultivated a flower, never kept a caged bird, or any creature. Why keep pets when every wild free hawk that passed overhead in the air was mine? I joyed in his swift, careless flight, in the throw of his pinions, in his rush over the elms and miles of woodland; it was happiness to see his unchecked life. What more beautiful than the sweep and curve of his going through the azure sky? These were my pets, and all the grass. (quoted in Edward Thomas [1909] 1972, pp. 301–302)

Jefferies was poor all his life, and his works therefore showed his special love of sunshine. One of the themes of his works is to build a human world that is harmonious and full of sunshine. In his works, he wanted not only natural sunshine, but also sunshine for all people, especially in his later works. Caring about nature did not make him ignore cruelty in real life. Jefferies still wrote about nature, but at the same time also showed his sympathy and care for the working people. The birds that flew past in sunshine would make him think that birds had been eating food, singing and living very happily ever since ancient times. In contrast, there were "the sullen poor who stand scornful and desperate at the street corners" (quoted in Edward Thomas [1909] 1972, p. 293). As a result, he wanted to appeal for better treatment of poor people. Seeing the waving of wheat in the fields and finding it hard to describe its beauty in words, he found his sympathy for the hardworking harvesters.

> This is human life, real human life—no rest, no calm enjoyment of the scenes, no generous gift of food and wine lavishly offered by the Gods, the hard fist of necessity forever battering man to a shapeless and hopeless fall. (quoted in Edward Thomas [1909] 1972, p. 293)

He sighs with feeling:

> An endless succession of labour, under the brightness of summer, under the gloom of winter; to my thought it is a sadness even in the colour and light and glow of this hour of sun, this ceaseless labour, repeating the furrow, reiterating the blow, the same furrow, the same stroke—shall we never know to lighten it, how to live with the flowers, the swallows, the sweet, delicious shade, and the murmur of the stream? (quoted in Edward Thomas [1909] 1972, p. 222)

Jefferies wanted the hardworking poor people to also have leisure time and be able to enjoy the beauty of nature. At that time, Jefferies himself was also struggling in poverty. Jefferies is considered a genius, a dreamer, a person of foresight. This is not surprising, as he linked the description and protection of nature with human care more than a century ago. In his eyes, neither human society that excluded nature nor the natural environment that excluded humans could be entertained. In his ideal world, nature and humanity are merged and there is harmony between them. As he explains in *The Open Air*, "My heart is fixed firm and stable in the belief that ultimately the sunshine and the summer, the flowers and the azure sky, shall become, as it were, interwoven into man's existence" (quoted in Keith, p. 128).

Keith compares Jefferies to a Janus in English rural writing, one side looking backward to the English countryside of the nineteenth century and the other forward to the future, the twentieth century, which is why his works are significant (Keith, p. 129). We can also say that Jefferies's early works focus on tradition, namely the life of the English countryside in Victorian times. These works are simple and natural, and present a vivid picture of the English woods and the fine beauty of the countryside. Jefferies's later works, on the other hand, have a broader significance. He sought a spiritual sublimation in nature; he linked nature with innate human love of beauty. Believing that all things in nature are pets of humans, he promoted the protection of nature. He linked the sunshine in nature with human care, and called for the establishment of a sunny and harmonious world. His writings originated in the countryside of Wiltshire, but his vision went beyond Wiltshire. He reminded people that nature was right beside us, at home, on the roof and in a place everyone could reach. In his later works nature and spirit had merged into one, and nature had assumed a certain religious color. However, most importantly, as his biographer, the poet Edward Thomas says, "It is he who, above all other writers, has produced the largest, the most abundant, and the most truthful pictures of Southern English country, both wild and cultivated" ([1909] 1972, p. 326).

## Lecture 17. Hudson: The Wilderness of South America and the Countryside of England

On a sunny day in May 1925, the then British Prime Minister Stanley Baldwin revealed a new memorial to Hudson in Hyde Park, London. This memorial, made by the famous sculptor Jacob Epstein, was placed in the Hudson Memorial Bird

Sanctuary, facing a lawn and a pond of water. The sculptor originally planned to make a sculpture of Hudson lying on the grass watching birds, but as no sculpture of people was allowed in the royal park, Epstein carved an image of the bird-woman Rima, the protagonist in Hudson's novel *Green Mansions* (1904), which has the forests of South America as the background. Rima is a symbol of wild nature threatened by humans, and perhaps also a symbol of Hudson himself, as he was different from the crowd—he loved primitive nature and has been referred to as a product of nature. To be the subject of a public memorial in London is a great honor. Why did the Government of the United Kingdom bestow this great honor on Hudson?

W. H. Hudson (1841–1922) was an English writer and naturalist whose works were mainly about nature and the relationship between nature and man. Unlike other European and American nature writers, Hudson had grown up on the South American pampas, and only went to settle in England when he was 30. He was born on the pampas, and grew up there in a wild natural environment, in the company of sheep, cows, wild birds and herdsmen. In his childhood he watched birds and played with snakes, wandering on the vast green plains. In his youth, he galloped on horseback in the wilderness of Brazil, Uruguay and Patagonia. He collected specimens of birds, wrote essays on the habits of birds, and won the admiration of well-known English and American ornithologists and naturalists. Two South American birds are named after him.

Hudson once said that his life ended when he left the pampas. However, from another perspective, he never left; the birds, flowers, hills and waters of England made him think all the more of the rugged and free natural environment of South America. The dual cultural influences of Britain and South America gave his writings unique insights. He wrote elegantly about the bleak and desolate wilderness, describing its bold, unconstrained nature in an elaborate style. English pastoral scenes complemented the wilderness of the South American pampas, combining into uncommon natural scenery under Hudson's pen. The English writer John Galsworthy (1867–1933), in his introduction to *Green Mansions*, says:

> Writers are to their readers little new worlds to be explored; and each traveller in the realms of *literature* must need have a favourite *hunting ground*, which, in his goodwill—or perhaps merely in his egoism—he would wish others to share with him. (in Hudson [1904] 2007, p. 1)

What Hudson presented is not an ordinary, known world, but a hunting ground that is mysterious and attractive.

## *Son of the South American Wilderness*

Hudson, in his foreword to Edward Thomas's essay collection *Cloud Castle and Other Papers* (1922), explains his own background, saying: "We were poles apart in the circumstances of our lives. He, an Oxford graduate, and a literary man by profession; I, unschooled and unclassed, born and bred in a semi-barbarous district among the horsemen of the pampas" (quoted in Keith, p. 191). Hudson's parents were both children of migrants to the new world of North America. His father moved from cold New England to Argentina, where the weather was more pleasant and better for his lungs, purchasing a large farm on the pampas where they settled down. Their house was called 25 Ombu Trees because 25 ombu trees over 100 years old grew there. Ombu trees, native to South America, have a huge trunk and dark green leaves like bay leaves. Hudson was born at home at 25 Ombu Trees in 1841. He recalled that on the flat pampas, 25 Ombu Trees was a gigantic landmark that people could see more than 10 miles away. These trees not only provided shade for people and animals, but were also a playground for birds and children. The Hudsons' house was located at the top of a slope, with the vast prairies in front and a deep, wide stream at the back. Hudson's childhood memories were filled with the vast pampas, the stream, lakes, wild bulls, horses, strange flowers and birds, cardoon thistles and ombu trees. There were also the unforgettable smells: of the dust of cattle and sheep coming home, the water of the stream behind the house, and the fragrance of primroses at dusk. Sitting under the trees, Hudson could see the golden-crowned woodpeckers moving in the branches overhead and ovenbirds making nests on the roof. The liveliness of nature filled young Hudson with wonder and amazement. Peach blossoms dropped like pink rain when orioles shook the branches. When the blossoms finished, Hudson saw tender leaves come out of the branches and grow lush. He said that he could hear the growing of poplars, a sound "like the wash of the sea on a wide shore." He watched strange and beautiful birds and butterflies fly in the morning light, and saw migratory birds leave under the stars: swans, wild geese, ibis, wild ducks, red herons, curlews, rails and crag martins. When he climbed to the top of a rose willow, birds flew past and over him and he wanted very much to have wings so that he could fly with them to the glistening river and to the end of the pampas. Hudson did not receive any formal education, only lessons from itinerant teachers; his real teachers were life and nature. He could climb trees when he was five and ride horses when he was six, and he learned to use a gun when he was 10. His childhood was like that of a bird in the forest, free and carefree. As a nature writer, he had a most direct education.[8]

However, from the age of 15 Hudson suffered from various illnesses, including first typhoid and then rheumatic fever, and was told that he could die of a heart attack at any moment. Although often sick and pathologically afraid of death, he sought a rich, colorful and creative life. Luckily, his home had a very good library, and reading and studying nature became the goal in his life. Around this time, a friend noticed that he was interested in nature, especially in birds, and so gave him a copy of Gilbert White's *The Natural History of Selborne.* Hudson read White's book hungrily and felt that he had much in common with this English man who had passed away many years earlier. He was amazed that there was actually a person like White who shared his passion for nature and had the same obsession with birds and the same curiosity about the outdoors. White's book deeply influenced Hudson's later life and became a guide for Hudson in becoming a naturalist. From then on, the wetlands, woodlands and thistle bushes became his private "Selborne Parish" in his heart (Tomalin, pp. 22–23), and he started to regard England as his homeland and a land he was yearning for. Perhaps as early as his teens, the English rural scenery coexisted with the wild and vast pampas of South America in his mind.

To a certain extent, Hudson's illness was also the reason that he was able to fulfil his life's wish. Had he been healthy, he might have engaged in some common profession or job like most other people, rather than becoming a great naturalist and writer. For many years after his illness, he traveled for health reasons, conducting field observations at the same time and writing beautiful reports of his observations. He became an ornithologist and was recognized for his work in this area. His travel started from his home on the pampas, and later extended to other parts of Argentina, then to Brazil and Uruguay. He would often stay out in the wild country for months at a time, hunting for food with his gun and cooking on fires he built. At night, he would wrap himself in a poncho and sleep with cicadas and fireflies around him, with the wind in the grass, the shriek of owls and the dark sound of the wetland birds in his ears. No wonder people call him a "field naturalist" and a "bird-man" (Keith, pp. 178, 182). In all the works he later wrote, there is a permeating atmosphere of South America.

8 The above is based on Hudson's autobiography *Far Away and Long Ago: A History of My Early Life* (1918), and on Ruth Tomalin's *W. H. Hudson: A Biography* (1982).

When he was 25, Hudson started to work for the Smithsonian Institute of the United States, collecting South American bird specimens, two of which were named after him (Hudson's tyrant-bird and Hudson's canastero). The bird specimens Hudson sent to the Smithsonian Institute were later selected and sent to the Zoological Society in London, and this opened a new page in his life, as he started to write reports and essays about birds for the Proceedings of the Society. These essays had a similar flavor to those by White: although they were essays they had the style of letters, and showed the author's love of and joy when talking about birds and flowers. People commented that when he was talking about birds his words had music in them (Tomalin, p. 86). He was later made an associate member of the Royal Zoological Society.

After Hudson's father passed away in 1868, his brothers and sisters left and the house at 25 Ombu Trees was abandoned. In April 1874, Hudson boarded a ship for England. He married and settled there, becoming a British citizen in 1900. From early 1874 until 1901, when he was given an annuity of £150 per year, Hudson lived off his writing, a poor life. He was used to the vast pampas of South America, and had no interest in urban London. However, the English countryside attracted him and aroused his impulse to wander as he used to in South America. He would go on walks for as long as possible, stopping at huts in the Savernake forest and wandering around in the woods with red deer and little deer. He went to the south of England to observe the heath and the wading birds on the banks of the River Exe. He also visited Shakespeare's birthplace at Stratford-upon-Avon and, having listened to owls of the genus *Strix* singing on autumn evenings there, corrected Shakespeare's description of the singing of these birds. Hudson went to Selborne and, finding White's tombstone hidden in the grass under the tall hemlock trees, smoothed aside the grass to have a clear look at his guide's name.[9] In Richard Jefferies's home in Wiltshire, the rolling hills greeted him and became the materials of his later work *A Shepherd's Life* (1910). In a cold winter wind, Hudson and his wife walked to Three Mile Cross—Mary Mitford's home and the subject of her well-known work *Our Village*—where he met a local who had known Mitford as a child and had beautiful memories of her. Hudson enjoyed these new discoveries and unexpected pleasures. He liked to explore alone or listen to stories directly from the farmers in the countryside, which is why he preferred to stay with farmers when he was traveling, and even

9 Hudson wrote about his visit to Selborne in his essay "Selborne," which was included in his prose collection *Birds and Man*, first published in 1901.

when his financial situation improved, he still avoided hotels. He always stayed in small village inns, as the normal life of birds was not disrupted there, and he could watch them flying around and listen to the music of the fields. From Hudson's autobiography we can understand the author's life-long love for nature, "Never had I felt confinement less—I who feel, when I am out of sight of living, growing grass, and out of sound of birds' voices and all rural sounds, that I am not properly alive!" (Hudson [1918] 1939, p. 2)

Perhaps it was precisely due to this life close to nature that Hudson, despite having been diagnosed at 15 with a heart condition that could kill him at any time, lived to be 81. Hudson had no children, and died alone in his London house in the autumn of 1922. Two days before he died he met with Helen Thomas, widow of the poet Edward Thomas, and the foreword that he wrote—at Helen's invitation—for Thomas's prose collection was his last piece of writing. Hudson was buried in Worthing Broadwater Cemetery. There lie not only his wife but also Richard Jefferies, a fellow lover of nature who had died young. At the entrance to the cemetery stands a monument erected by Sussex Wildlife Trust and the Royal Society for the Protection of Birds to commemorate Hudson and Jefferies, two naturalists and nature writers who contributed greatly to the protection of nature.

## English Country Writer

Although the raw material for Hudson's writing came mostly from his life on the pampas of South America, it was in England that his writing career really started. He produced more than 20 books, including prose essays, novels and an autobiography. From the titles alone of his novels such as *The Purple Land* (1885), *El Ombu* (1902) and *Green Mansions: A Romance of the Tropical Forest* (1904), one can feel their exotic flavors. Among them, *Green Mansions*—a story of romance between a Venezuelan poet and naturalist in political exile and the bird-woman Rima—is the most successful. In this novel, Hudson linked nineteenth-century Romanticism with twentieth-century ecological concepts, the pain of loss of the wild, the dream of return to nature, and the conflict between modern and primitive humanity. Although the background to this novel is South America, the unique atmosphere and the depth and intensity of feelings are considered to be comparable to those of Emily Bronte's *Wuthering Heights*. *The Oxford Companion to Twentieth-Century Literature in English* (1996, p. 321) describes Hudson's *Far Away Long Ago* as "highly regarded" in autobiographical writing. However, Hudson's prose essays about the vast outdoors of South America and the English countryside are the most important of his works and have left the deepest impression. As Edward Thomas says in his *A Literary Pilgrim in England* ([1917] 1937), "generation after generation of school boys know his *British Birds* and everyone remembers *Idle Days in Patagonia*" (p. 184).

Hudson's important prose essay collections include *The Naturalist in La Plata* (1892), *Idle Days in Patagonia* (1893), *Birds in London* (1898), *Nature in Downland* (1900), *Birds and Man* (1901), *Hampshire Days* (1903), *Afoot in England* (1909), *A Shepherd's Life* (1910) and *Adventures among Birds* (1913). These works show Hudson's unique style. Richard E. Haymaker, a scholar of Hudson, summarizes both Hudson's life journey and the range of his writing in the title of his book, *From Pampas to Hedgerows and Downs: A Study of W. H. Hudson* (1954). Hudson's unusual life experiences give his works a rich color of locale, from the wind in the vast and wild South American prairies to the flowers and birds in the hedgerows of the English countryside, across time and space with layer upon layer of scenery, projecting an amazing charm.

Hudson fused the South American scenery with the English countryside and harmonized nature and culture into one. In his memory, his childhood crystallized into two vivid scenes which he described in *Far Away and Long Ago*. One is dynamic and vibrant: on the quiet green plain extending away to the horizon, with sunset coloring the western sky, a large herd of cattle come homewards with loud noise, raising a great cloud of dust, while the herdsmen galloping behind them with wild cries. The other is gentle and quiet. Hudson describes

> my mother at the close of the day, when we children, after our supper of bread and milk, join in a last grand frolic on the green before the house. I see her sitting out of doors watching our sport with a smile, her book lying in her lap, and the last rays of the setting sun shining on her face. (Hudson 1918, p. 10)

In a few simple words, these two images show his rich English cultural background and the reason why he could describe the wild and rough life on the South American pampas in fine detail.

Hudson's works set against the natural world in South America as their background not only contain vast, bare, bleak and desolate wilderness, but also picturesque beauty and poetic tenderness. In *Idle Days in Patagonia* we see Hudson alone with a dog, riding through wilderness and thorns. It is a world without man or sound, except the hooves of the horse and the breathing of the dog. There is deadly silence and Hudson finds it impossible to think, as thinking is like starting a noisy motor in his brain. In that desolate wilderness, Hudson feels that if he were to fall from the horse and die, birds and animals ate his body, and it became a pile of bones under the sun and wind, nobody would notice. However, he would go out there day after day as if going to a festival, as he enjoyed the free idleness and appreciated the uninterrupted peace. It was a connection with the ancient and the primitive, a simple and natural way of survival and a spiritual return to the wilderness (Begiebing and Grumbling, pp. 508–510). On the other hand, in *Far Away and Long Ago*, what we see is a pleasant picture of birds and flowers in springtime when peach trees are in blossom. Huge peach trees stand on the green lawns and peach blossoms are like pink clouds under the blue sky, and this is only half of the beautiful scene; the other half is birds singing in the trees, yellowish-green birds hidden in the pink sea of blossoms, singing wondrous songs (Hudson 1918, pp. 55–57).

Hudson grew up with shepherds on the pampas and had the instincts of a shepherd. Even in the downlands and hedgerows of the English countryside, in a place of green lawns, he could also find wild and ancient charm, beautiful and not so modern times and the home of his heart. In *A Shepherd's Life* he uses Jefferies's home in Wiltshire as the background. The vast and rolling hills make him feel that he has returned home. He writes:

> The final effect of this wide, green space with signs of human life and labor on it, and sights of animals—sheep and cattle—at various distances, is that we are not aliens here, intruders or invaders on the earth, living in it but apart, perhaps hating and spoiling it, but with the other animals are children of Nature, like them living and seeking our subsistence under her sky, familiar with her sun and wind and rain. (quoted in Keith, pp. 182–183)

In *Afoot in England*, Hudson claims that the sound of cattle is the loveliest natural sound he has ever heard, because it brings up his childhood memories. *A Shepherd's Life* and *Nature in the Downland* are prose essay collections based on Wiltshire in the southwest of England and the Sussex downlands in the south. They are records of ancient village life and history, but also descriptions of nature and natural scenery. Hudson writes of an old shepherd whose dreams were all about his sheep, farm huts covered with vines, churches and cemeteries hidden behind the trees, foxes, hedgehogs, wildflowers and birds. Hudson especially liked the perennials in the farmers' gardens. From their fragrance, he found "an atmosphere, a sense or suggestion of something immeasurably remote and very beautiful—an event, a place, a dream perhaps" (Tomalin, p. 210). *Hampshire Days* is about forests. It is a world of grass, heather, wildflowers and hedgerows, and also a woodland of birds, insects and spiders, as well as a record of Hudson watching squirrels dancing in the tops of trees and listening to the singing of orioles.

Hudson's works show an intimate relationship with nature and have a freedom and ease that is difficult to describe. They do not have the artificiality that suggests someone laboring at night under an oil lamp. On the contrary, in *Hampshire Days* he writes:

> The blue sky, the brown soil beneath, the grass, the trees, the animals, the wind, and rain, and sun, and stars are never strange to me; for I am in and of and am one with them; and my flesh and the soil are one, and the heat in my blood and in the sunshine are one, and the winds and the tempests and my passions are one. (Hudson 1923, p. 47)

Hudson was closer to nature than anybody was, but he was also cultured. In the chapter "Following a River" in *Afoot in England*, he describes the feelings of a traveler who has returned to his own home:

> For I was following not the Exe only, but a dream as well, and a memory.... My forefathers had dwelt for generations beside it, listening all their lives long to its music, and when they left it they still loved it in exile, and died at last with its music in their ears. (quoted in Tomalin, p. 101)

It is no wonder that Galsworthy comments that

> [h]is work is a vision of natural beauty and of human life as it might be, quickened and sweetened by the sun and the wind and the rain, and by fellowship with all the other forms of life—the truest vision now being given to us, who are more in want of it than any generation has ever been. A very great writer; and—to my thinking—the most valuable our age possesses. (in Hudson, [1918] 1939, p. xi)

Epstein's decision to sculpt the bird-woman Rima to commemorate Hudson was a very appropriate choice. Epstein worked on the sculpture for seven months in a workshop in the forest, with only trees and birds, fog, mist and silence for company; therefore this piece of work can be said to be part of nature. The bird-woman who is feeding birds and the eagle beside her all show Hudson's life-long love and pursuit. In nature, birds were Hudson's favorites. One of the reasons he fell in love with his wife was because she had a pleasant voice like that of a bird. Throughout his life Hudson was seeking a life of freedom, like that of a bird. This bird never seemed to fly into a house, but up in the blue sky.

## Lecture 18. Thomas: Walking in the English Scenery

In the Soldiers' Cemetery in northern France, located at No. 43, Row C, there is an ordinary-looking white tombstone which often has fresh flowers or wreaths placed on it. Lying here is not only a soldier who died in the First World War, but also a writer and poet. His home is far away in England and his name is Edward Thomas. More than half a century after his death on the battlefield, his name was carved on a stone in the "Poets' Corner" in Westminster Abbey.

Like Richard Jefferies, a writer he loved, Edward Thomas lived a short life—39 years—but he died on the battlefield, killed instantly when German bombs landed on the outpost where he was. Two years earlier Thomas, who had already published nearly 30 collections of essays, left his wife and children and joined the army, and two months before he died he had requested to be sent to the front in France. He was buried in a foreign land.

In war, numerous soldiers die on the battlefields. However, very few writers like Thomas would volunteer to go to war. When asked what he was fighting for, Thomas bent down and scooped up a handful of earth and said, "Literally, for this" (quoted by R. G. Thomas in his biography *Edward Thomas: A Portrait*, 1985, p. 249). To Thomas, England was not an idea, nor even a country, but a land of special meaning, a place a poet's body and soul would call home. We can say that Thomas put down his pen and took up arms to fight not for his country, but for the land he loved, praised and regarded as the resting place of his soul.

Unlike Jefferies and Hudson, Thomas was born and grew up in metropolitan London. However, his temperament did not agree with city life. He exhibited a modern sense of conflict between yearning for spiritual calm on one hand and facing reality on the other. He never felt at home in big cities; the land was the only place where his heart could rest, and so his eyes turned to the countryside and to nature. Like Thoreau in the nineteenth century, walking in the countryside and in the woods became his career. His footprints were left on the countryside of Kent, Surrey, Hampshire and Wiltshire in southern England. He told his friends in a letter, "Still, I usually walk twelve miles a day" (quoted in R. G. Thomas, p. 105). In his walks, he discovered the essence of the English countryside and the fleeting beauty of nature. When describing the beautiful scenery of the English countryside, he wrote in his nature writing collection *The Heart of England* (1906), "We are so rich that we do not count our treasure" (Edward Thomas, p. 171). Thomas decided to undertake the job of "counting our treasure." Although people regard him as a poet, he left behind many collections of prose nature writing. The American writer Ralph Waldo Emerson once said of Thoreau, "The length of his *walk* uniformly made the length of his *writing*" ("Thoreau," [1862] 1989 in Baym and Gottsman, p. 1037). This comment also suits Thomas as well, as his nature writing was also the result of his walking. He walked in the English countryside and offered sketches of nature to the reader as treasures. In the end, he had to say goodbye to his wife and children and died on the battlefield to protect this land.

## Away from the Madding Crowd of the City

Thomas was born in London in the early spring of 1878 to Welsh parents. His father Philip Thomas had excelled in his civil service examinations and was a public servant in the Ministry of Commerce. Philip wanted his son Edward to become outstanding in the civil service as well, urging Thomas to take the civil exams. He sent Thomas to Oxford because he believed that being with the elites there would ensure he had a good career. However, he did not know that Thomas had harbored a different aspiration ever since he was young and would take a career path vastly different from what his father expected. This, to a certain extent, was due to the influence of Jefferies. Ever since he was a child, Thomas was fascinated by the countryside described by Jefferies. Fortunately, his grandmother's home was in Jefferies's home county of Wiltshire, and from the age of 10 he would often leave London to spend his holidays there. Coate Farm—where Jefferies had lived—and the surrounding countryside stayed in Thomas's heart and accompanied him all his life. It was there that he met an old farmer named David Uzzell. Uzzell was like a character from Jefferies's books, and Thomas regarded him a fatherly person in his life, calling him "Father Uzzell." This old man led Thomas onto a different path than that designed for him by his real father. He took the young Thomas fishing, catching birds, and looking at various birds and plants. Father Uzzell could imitate the singing of birds and knew almost all plants and flowers. He changed walks in the fields into interesting nature lessons, and it was then that Thomas developed his lifelong love of nature. From the age of 14 or 15, he started to write down interesting activities in the countryside, and these writings were later referred to as "outdoor notes" or "nature notes," providing material for many of his nature writing collections. Thomas kept this habit of taking "nature notes" throughout his life, even on the battlefields of the First World War. His wife Helen Thomas remembered that "[t]he years which he had spent in the Jefferies country in Wiltshire were the most treasured memories of his boyhood" (quoted in Keith, p. 192), and when they got married, they spent their honeymoon in Father Uzzell's cottage.

The Jefferies countryside brought sunshine and happiness to Thomas's early life and made him hate the career path his father had chosen for him. After he graduated from high school and was preparing for the civil servant examinations, he wrote his first collection of essays *The Woodland Life* (1897). His father thought studying at Oxford might give Thomas the opportunity to fully develop his talents and obtain a reliable job. However, after graduating from Oxford, he did not find a stable job as his father expected. He could not stand a boring and monotonous life and was forever searching for the world Jefferies described, the carefree life outdoors. As he married early, he was burdened with the responsibility of supporting a family. He worked as a journalist, a critic for a literary journal, and became a writer and poet later. To pursue his dream he led a life of poverty. Once, when he was unable to pay his debts, he was nearly taken to court. To support him in his pursuit of his dream, his wife Helen showed great patience and love. Within a few years of their marriage the family moved from the suburbs of London, first to the Kent countryside, then to Hampshire. Helen quietly undertook the job of looking after their four children so that her husband could travel or walk at any time he wished or seek a quiet place to write. She never complained, even if he did not come home for days on end; she understood him deeply, and it was she who recommended Jefferies's *The Story of My Heart* to Thomas before they were married.

Although Thomas had the same life goals as White and Thoreau, he was not carefree as they had been. Due to financial pressure he had to edit books, write reviews and even write occasional pieces to order. On the other hand, this also made him a prolific writer. His works include *Oxford* (1903) and *Beautiful Wales* (1905) for the A & C Black Colour Book series, *Windsor Castle* (1910) and *The Isle of Wight* (1911) for the Beautiful England book series, and poetry criticisms *Feminine Influence on the Poets* (1910) and *Keats* (1916), as well as his biography of Jefferies, *Richard Jefferies: His Life and Work* (1909). His work *A Literary Pilgrim in England* (1917) explored the development of writers and poets from the perspective of the influence of their environment on their life and writing, which has an affinity with the emphasis on a "sense of place" in nature writing today. He also wrote many collections of essays on nature, such as *The Heart of England* (1906), *The South Country* (1909), *The Light and Twilight* (1911), and *In Pursuit of Spring* (1914).

At the beginning of 1913, Thomas met the American poet Robert Frost in England. At the time Frost was not well known and had not yet published a collection of poems, while Thomas had already published nearly 30 collections of essays. They discovered that they shared a common love for nature and poetry and found in each other a kindred spirit. Their friendship started there and then; they often visited each other, took walks in the countryside and discussed poetry and literature. Thomas became the main critic and introducer of Frost's poetry in England, while Frost discovered poetry in Thomas's prose essays and encouraged him to shift his attention to writing poetry. However, even when writing poems, Thomas's main concern was still nature and human moods. For example, among the poems he sent to Frost are "After Rain," "March" and "Birds' Nests." After he joined the army, Thomas worked first in training young officers in England, then he asked to be assigned to the artillery on the battlefront in France. While he was there, he learned that three of his poems had been accepted for publication by an English poetry magazine. In February, the army unit he was in was preparing for the spring offensive and they were faced with a life-or-death situation. He wrote to Frost, "I should like to be a poet, just as I should like to live, but I know as much about my chance in either case, and I don't really trouble about either" (R. G. Thomas, p. 28). Thomas died in action on Easter Monday morning in April 1917, having written most of his poems in the two years between his enlistment and his death on the battlefield—some of them on the train home from his post and some in huts on the battlefields. Apart from a few published under the pen name of Edward Eastaway between 1915 and 1917, the majority of his poems were published posthumously, and the most complete collection is *Collected Poems of Edward Thomas* (1978), edited by his biographer R. G. Thomas.

Even though Thomas only started to write poetry three years before he died, he was first and foremost a poet, whether in people's hearts or in most English literature anthologies and dictionaries, and his poetic temperament derived mainly from the nurture of nature and from his childhood love of the outdoors, which developed in Jefferies's countryside. This love and fascination made him turn his back on the city life mapped out for him by his father to take up a life of wandering, thereby becoming a failure in his father's eyes. It made him write romantically of swallows coming home and of plants in spring even when at war, as the diary he kept at that time reveals. Thomas's poetic temperament also made his nature sketches, which predominated among his writings, poetic.

## *Exploring the Heart of England*

In Thomas's nature writings, we read of the inner dialogue of the author, the anxiety of urban residents, and a persistent yearning for the countryside and the outdoors, to find roots in the land and to protect a piece of pure and clean place.

The industrial revolution brought an anxiety and a sense of crisis and restlessness. The search for a sense of spiritual belonging and a home is one of the themes of Thomas's nature writings, which we can see in R. G. Thomas's biography. Between 1900 and 1906 the Thomas family moved house five times, twice in the London suburbs and three times in the Kent countryside (R. G. Thomas 1985, p. 94). Frequent moving and his love of taking walks in the countryside not only provided material for his writing, but also to a certain degree reflected the restlessness in his heart. The industrial revolution and modernization seemed to have changed the world into a kaleidoscope; everything seemed to have become fleeting and elusive. Knowing this, it is not difficult for us to understand why Thomas gave the first essay in his *The Heart of England* the title of "Leaving Town." Even though Thomas was born and grew up in London, in this essay he looks at the city from an outsider's perspective: uniform houses, desolate backyards, and on the side of a small piece of green, wizened trees, scaffolds and bricks. People lived an aimless life in the caves and deserts of the city. These were strange streets and a prison-like city, and living in a city like this, "[t]he end of the day was like what is called a natural death in bed; an ill-laid fire dies thus" (Thomas, [1906] 1909a, p. 5). In the very next piece, Thomas describes the joyful scene of a farmer with a white horse and two chestnut-colored horses pulling a plough in the vast fields in the morning light. He writes:

> Hour after hour he (the farmer) crosses and recrosses the field, up to the ridge, whence he can see miles of hills and wood; down to the Woodside where the rabbits hardly trouble to hide as he appears, or to the thick hedge with marigolds below and nearly all day the songs of nightingales. (Thomas, [1906] 1909a, p. 22)

In Thomas's eyes this is a truly meaningful life, because looking at the farmer ploughing his land "is to see one who is in league with sun and wind and rain to make odours fume richly from the ancient altar, to keep the earth going in beauty and fruitfulness for still more years" (Thomas [1906] 1909a, p. 25).

In his book *The Country* (1913) Thomas expressed his opinion that "apart from nature we have nothing else to rely on" because

> [w]e cannot make harmony out of cities.... We may feel the painful splendor of our humanity in the town, but it is in the country more often that we become aware, in a sort of majestic quiet, of the destiny which binds us to infinity and eternity. (quoted in R. G. Thomas, pp. 94–95)

Thomas regarded nature as a holy church, the home of the heart. In the thick forest he saw "their leaves were fresh and wet as when beauty and joy shed a few tears.... The shadows were as solemn and imperturbable as to a child a cathedral." He exclaims, "And for ourselves—we seemed to be home from a long exile" (Thomas [1906] 1909a, p. 46). In addition, Thomas gave an insightful comment in another book of his *The South Country*: literature sends us to nature mainly to satisfy our sensual pleasure; science sends us to nature mainly for us to obtain knowledge; and the industrial society and the big cities send us to nature to meet our physical and psychological needs, or in other words to help us escape from the former (Thomas [1906] 1909, p. 163). In this respect, Thomas and his American friend Robert Frost were kindred spirits. Frost lived for many years at White Mountain in New Hampshire and in the Green Mountains of Vermont. He wrote many poems based on his experiences there, and these were included in his collection titled *New Hampshire*. American professor of English Ian Marshall says in his book *Story Line: Exploring the Literature of the Appalachian Trail* (1998) that most of the poems in *New Hampshire* give off "a sense of the wilderness," as Frost wished the mountain he lived on higher and "wilder." He makes special mention of the first poem in the collection, which has a line that can be considered what is called in Chinese culture "the eye of the poem," that is, its essence: "The line where man leaves off and nature starts." Marshall comments that "New Hampshire is the place where he's most willing to venture across the line into the realm of nonhuman nature" (p. 183).[10] Marshall then uses the narrative poems in the section of "Notes" of *New Hampshire* to show how they "stress our ability to achieve some kind of integration with nature, telling stories about how nature enters our lives, is part of our reason for being, suggesting that the lines of separation between us and nature are easily crossed" (Marshall, p. 199). We can see that even though Thomas and Frost lived in different countries, they had a common yearning for nature and desire to describe it. The natures that appeared under their respective pens were, naturally, different: Thomas's was the rural scenery of the old English countryside, while Frost's was the wild "new" land of America.

10 Like Edward Abbey, who floated down Green River in the American West with a copy of Thoreau's *Walden*, Marshall took a copy of *New Hampshire* with him when he went to White Mountain in America's New England to visit the places Frost describes in his poems. Marshall recorded his thoughts and feelings in his book *Story Line.*

Even in his native land, Thomas was also different from Jefferies, discussed earlier in this lecture series. While Jefferies's works came mostly from his memories of his childhood in the countryside, Thomas—born and bred in London—went the opposite way, leaving the city and going to the countryside to make up for his lack of country roots. In a letter to a friend he says: "Walking you will perhaps see suits me. Really I am never so well as when I am rid of the postman and all company walking 20 or 30 miles a day" (quoted in R. G. Thomas, p. 162). Thomas's wife Helen said that during his walks "his soul was revived when it was faint with despair, and comforted as some are by religion or music" (quoted in R. G. Thomas, p. 159). Walking was a need of his heart and also the source of inspiration for his writing. His nature sketches were mostly the result of his rambles in the countryside. This method of writing and walking makes his works appear more relaxed structurally, but also gives them a distinctive style. In *The Heart of England* ([1906] 1909), Thomas presents the oldest and most picturesque scenes of the English countryside to the reader: small villages in valleys, cottages surrounded by hedgerows, simple old inns, lone travelers, green grassland that is like a green stream, sheep and cattle grazing in a leisurely manner. We follow him to an ancient woodland and see a beautiful, melancholy scene: blossoms of the chestnut trees fall silently onto a small stony path crisscrossed with the shade of branches; only a lone swallow flying past disrupts the falling of the flower rain (p. 49). In autumn we follow him to a farmhouse in the valley. We see how the colors of the flowers and fruit of the hills meet harmoniously on the walls and roofs of the farmhouse, making the building "the royal flower of autumn" (p. 93). We also follow him to a country churchyard, but find it difficult to link it with death because even though the caretaker is busy, he is busy planting flowers and cutting grass. The variety of epitaphs or inscriptions on the tombstones shows the changes over time. It is interesting and quite poetic (p. 112). Thomas's description of the scenery "uses language to explain paintings," as the Chinese would say.

> The road curves gently but goes almost east and west; on the north side is a bank surmounted by dense thorns; on the south is a low hedge, over which can be seen two broad meadows, with oaks here and there, running up in shining bays among tall woods, and above and beyond those woods several wooded ridges, hunched and blue, on the last them a windmill like a stag's antlered head that somehow hints at the sea beyond. (Thomas [1906] 1909, p. 77)

Thomas was an Oxford graduate. He was not only a person who loved the countryside, but also a complicated and temperamental poet, which added literary color and personal feelings to his observation of nature. It has been commented that, in comparison with other nature writers, Thomas seemed to be "too literary" (quoted in Keith, p. 200). However, that emotions change with the changes of the natural scenery is, in fact, another of the distinctive features of Thomas's writing. In his eyes, all scenery contained imaginative feelings. Although he was describing natural scenery, he was actually revealing his own inner nature. Take *The Heart of England* as an example: we follow him to a pond on a moonlit, hazy evening when the cherry blossoms are blooming,

> Walking slowly thus, with a bowed head, you find an image of yourself and the universe in a shallow pool among the trees. The pool is your own mind. The flowers at its edge—hyacinth, primrose, marsh-marigold, and all the trees and their foliage are the intimate and permanent companions of your life, and they are clearly mirrored in the water. An oak draws a long, snaky shadow from side to side, with the head and neck of a sitting dove among the leaves. And there, too, are the stars and the moon, brought down to that homely company of trees and flowers by the shining water, and preserved there in strangely woven patterns; just as in your mind you mingle visions of the world, the past, the imaginary, with your own domestic surroundings and acts, and a changing mood, like a puff of wind across the pool, erases them. (Thomas, [1906] 1909, p. 148)

This description illustrates, to a certain extent, the idea that life is a dream. On an autumn day, with fine rain kissing the land, Thomas comes to a walnut tree beside a haystack which has died due to a lack of water. The old wagon against the tree trunk, the ladder in the branches, the shaft and wheels of the old, rusted plough and chain harrow in the grass make him sad, "In the soft, straight rain these things are a buried world, the skeletons of a fair-seeming old life mingled with a sort of pleasant tranquility as on the calm dim floor of a perilous main" (Thomas, [1906] 1909, p. 98). This description reminds us of his poetic line that is often quoted, "The past is the only dead thing that smells sweet" ("Early One Morning" in Thomas 1917, p. 27).

One of Thomas's essays is about a willow tree in front of his house. He writes:

> There is a willow before my house. It is covered in ivy by day and by night in mystery and when I lie in bed I see men and women through it passing by. I see it fantastically in the pages of books I read and a true sense I see life through the leaves of willow tree. (quoted in R. G. Thomas, p. 19)

From this we can see that to Thomas natural scenes and humanity, the human mind and human culture all blended into one. However, he was very clear about the position of humans in relation to nature; as he says in *In Pursuit of Spring*, "The earth does not belong to man, but man to the earth" (quoted in R. G. Thomas, p. 222). This may explain why Thomas devoted the best time of his life to the land and to walking in the English countryside.

When we talk about Thomas and his works, we cannot help but think of two other nature writers at the turn of the last century, Jefferies and Hudson, and their relationship with Thomas. Even though Thomas was born 30 years later than Jefferies, there are similarities between the two. Both started off as journalists, and then turned to nature writing. Both died young, at the age of 39. Jefferies was Thomas's idol when Thomas was in his teens, and greatly influenced the latter's life. In addition to writing a biography of Jefferies, Thomas edited a collection of his writing, *The Hills and the Vale* (1909). We can call the relationship between Thomas and Jefferies that of a spiritual friendship (since they never met), while Thomas not only met Hudson, but was friends with him for more than 10 years; the correspondence between the two was published after Thomas died. Two days before his own death, Hudson met with Thomas's widow Helen. The foreword for Thomas's essay collection *Cloud Castle and Other Papers* (1922) was the last piece Hudson wrote, and in it he calls Thomas "my friend and one of the most lovable beings I have ever known," although he also humorously points out the differences between them, Thomas being an Oxford graduate and a professional writer while Hudson was not formally educated and had grown up on the wild South American pampas. Perhaps the two factors that distinguished Thomas from Jefferies and Hudson the most were that Thomas received an elite education at Oxford and that, having grown up in the city, he lacked country roots.

Thomas embodied the conflict, confusion and anxiety of modern urban society. Although he was born and grew up in a metropolis, he was fascinated by the outdoors and the countryside. All his life he chased spring days, wishing to experience a rebirth like that of nature; he had the ideals and imagination of a poet, but could not rid himself of the shackles of reality; there were more disappointments than successes in his life; he wanted to escape from the prison of the city, but poverty and an unstable career drove him to the edge of despair and even to the idea of suicide. When he finally went to the trenches of the First World War, he wanted to protect England, and even sacrificed his own life for this land that he loved. He also had a desire to turn his life to dust, as he could not escape from conflict and contradictions. On that April morning in 1917 when Thomas left this world, he abandoned his pursuit of spring and his confusion about life and the question of how, in the process of modernization, human beings could live harmoniously with nature.

## References

Abrams, Mayer Howard, and Geoffrey Galt Harpham. 2009. *A Glossary of Literary Terms*. 9th edition. Boston: Wadsworth.

Austin, Mary R. 1882. *The Friendships of Mary Russell Mitford: As Recorded in Letters from Her Literary Correspondents*, Edited by The Rev. A. G. L'strange. New York: Harper & Brothers, Franklin Square.

Begiebing, Robert J., and Owen Grumbling, eds. 1990. *The Literature of Nature: The British and American Traditions*. Medford, OR: Plexus Publishing.

Buell, Lawrence. (1995) 1996. *The Environmental Imagination: Thoreau, Nature Writing, and the Formation of American Culture*. Second printing, Cambridge, MA: Harvard University Press.

Burroughs, John. (1889) 1895. *The Writings of John Burroughs*, Vol. 8: Indoor Studies. Boston: Houghton, Mifflin & Company.

Burroughs, John. 1904. *Literary Values and Other Papers, Works of John Burroughs*. Vol. 12. Boston: Houghton Mifflin.

Burroughs, John. 1904. *Indoor Studies, Works of John Burroughs*. Vol. 8. Boston: Houghton Mifflin.

Carson, Rachel. 1998. *Lost Woods: The Discovered Writing of Rachel Carson*, edited by Linda Lear. Boston: Beacon Press.

Emerson, Ralph Waldo. (1862) 1989. "Thoreau" in Baym, Nina, and Ronald Gottsman, eds. *The Norton Anthology of American Literature*. 3rd ed. Vol. I. New York: Norton, pp. 1033–1045.

Feldman, Paula R., ed. 1997. *British Women Poets of the Romantic Era: An Anthology*. Baltimore, MD: The Johns Hopkins University Press.

Fields, James T. (1872) 1883. *Yesterdays with Authors*. Boston: Houghton Mifflin Company.

Finch, Robert, and John Elder, eds. 1990. *The Norton Book of Nature Writing*. New York: Norton.

Haymaker, Richard E. 1954. *From Pampas to Hedgerows and Downs: A Study of W. H. Hudson*. New York: Bookman Associates.

Hudson, W. H. (1901) 1930. *Birds and Man*. London: Duckworth.

Hudson, W. H. (1904) 2007. *Green Mansions* with a foreword by John Galsworthy. New York: Overlook Duckworth.

Hudson, W. H. 1918. *Far Away and Long Ago: A History of My Early Life* New York: E. P. Dutton.

Hudson, W. H. (1918) 1939. *Far Away and Long Ago: A History of My Early Life*. With a foreword by John Galsworthy. 1941 reprint. London: J. M. Dent & Sons Ltd.

Hudson, W. H. 1923. *Hampshire Days*. London & Toronto: J. M. Dent & Sons Ltd.

Jefferies, Richard. (1883) 1922. *The Story of My Heart*. London: Longmans, Green and Co.

Jefferies, Richard. 1884. *The Life of the Field*. London: Chatto & Windus.

Jefferies, Richard. 1893. *The Open Air*. London: Chatto & Windus.

Keith, W. J. 1974. *The Rural Tradition: A Study of the Non-fiction Prose Writers of the English Countryside.* Toronto: University of Toronto Press.

Lynch, Deidre. 2000. "Homes and Haunts: Austen's and Mitford's English Idylls." *Publications of the Modern Language Association of America* 115 (5): 1103–08.

Marshall, Ian. 1998. *Story Line: Exploring the Literature of the Appalachian Trail.* Charlottesville, VA: University Press of Virginia.

Mitford, Mary R. (1824) 1893. *Our Village.* Vol. 1. London: MacMillan. With an introduction by Anne Thackeray-Ritchie.

Mitford, Mary R. (1824) 1870. *Our Village.* Vol. 1. London: Bell & Daldy.

Mitford, Mary R. (1826) 1870. *Our Village.* Vol. 2. London: Bell & Daldy.

Mitford, Mary R. 1828. *Our Village.* Vol. 3. New York: E. Bliss.

Mitford, Mary R., and Austin, Mary R. 1882. *The Friendships of Mary Russell Mitford: As Recorded in Letters from Her Literary Correspondents.* Ed. The Rev. A. G. L'strange. New York: Harper &Brothers, Franklin Square.

Muir, John. (1938) 1979. *John of the Mountains: The Unpublished Journals of John Muir.* Ed. Linnie Marsh Wolfe. Madison: University of Wisconsin Press.

Paul, Sherman. 1992. *For Love of the World: Essays on Nature Writers.* Iowa City, IA: University of Iowa Press.

Stringer, Jenny. Ed. 1996. *The Oxford Companion to Twentieth-Century Literature in English.* Oxford: Oxford University Press.

Thomas, Edward. (1917) 1937. *A Literary Pilgrim in England.* London: The New Library.

Thomas, Edward. (1906) 1909. *The Heart of England.* London: J. M. Dent & Co.

Thomas, Edward. 1909. *The South Country.* London: J. M. Dent & Co.

Thomas, Edward. (1909) 1972. *Richard Jefferies: His Life and Work.* New York: Kennikat Press.

Thomas, Edward. 1917. *Poems.* London: Selwyn & Blount.

Thomas, Edward. (1917) 2005. *A Literary Pilgrim in England.* Reading: Cox and Wyman.

Thomas, Edward. 1923. *New Hampshire Days.* London: J. M. Dent & Sons Ltd.

Thomas, R. George. 1985. *Edward Thomas: A Portrait.* Oxford: Oxford University Press.

Tomalin, Ruth. 1982. *W. H. Hudson: A Biography.* London: Faber and Faber.

White, Gilbert. (1789) 1926. *The Natural History and Antiquities of Selborne.* London: J. M. Dent.

White, Gilbert. (1789) 1996. *The Natural History of Selborne.* Ware, Hertfordshire: Wordsworth Editions.

Williams, Raymond. 1973. *The Country and the City.* New York: Oxford University Press.

# 6

# Female Writers in Nature Writing

## Introduction

Among the settlers in the American "New World," the natural environment a woman was in had traditionally been one that had been conquered and tamed by humans. However, the female nature writers introduced in this chapter all went against the conventions. They not only went boldly into nature itself, but also described their personal experiences of communicating with nature to the world. However, while pioneers of American nature writing such as Ralph Waldo Emerson found sources of thought and culture in nature, female nature writers focused more on the scenery as a reflection of their hearts. While nature was described as rough and unruly in some literary works, it had other characteristics in their eyes; to them nature was the mother, sister, friend or even lover. As contemporary American nature writer Terry Williams says, "Wilderness is a feeling. …if you know wilderness in the way that you know love, you would be unwilling to let it go" (Williams, [2001] 2002, p. 76).

Susan Cooper's *Rural Hours* was published in 1850, four years earlier than Thoreau's *Walden*. Thoreau went to the woodlands around Concord to "live deliberately, to front only the essential *facts of life*" (Thoreau, *Walden*, [1854] 2004, p. 77). Cooper's *Rural Hours*, on the other hand, was "the simple record of those little events which make up the course of the seasons in rural life" (Cooper, *Rural Hours*, [1850] 1998, p. 7). Reading this book, we follow Cooper through her hometown, the nineteenth-century Cooperstown, in spring, summer, autumn and winter, while at the same time also viewing the inner landscape in this nineteenth-century female nature writer's heart. And in these seemingly leisurely walks, in the admiration of flowers and birds, we also follow her and come to understand the relationship between nature and culture, and to understand that humans need a healthy culture based on nature. The value of *Rural Hours* also lies in the scholarly knowledge and rigorousness of the author's description of the natural world, perfectly combining environmental study and aesthetics.

Celia Thaxter, another female nature writer, described her "island life" and her "island garden." Thaxter was a poet, a writer, an artist and a gardener at the same time, and was once called by another female American nature writer "the most wonderful person" in American literature. Through her soft and feminine writing style, we see the unique beauty of a desolate island in the sea: white sails of the passing ships cast their light on the bare and gloomy rocks; dawn tints the wings of seagulls with a rosy hue; a ship sails slowly under the moonlight like a beautiful spectre floating by; stars like familiar faces of old friends from the awe-inspiring night sky. All this is recorded meticulously with loving care, imbued with the rich spiritual life of the author. Thaxter's *An Island Garden*, first published in 1894, not only describes the gardener's patience, loving care and hard work, but also reveals an inner contentment and joy. It is not surprising that readers often feel a kind of spiritual charm, for the garden is an externalization of the inner world of the gardener.

Mabel Wright, on the other hand, was the kind of true poet Emerson imagined, that is, she had the ability to see the miraculous in the ordinary. Her representative work *The Friendship of Nature*, also first published in 1894, is masterful. It starts with May Day in New England, then leads the reader through the "Song of Mid-summer" and the "misty autumn," and finally to "the mood of winter." *The Friendship of Nature* not only pulses with the rhythms of the season, but is also interwoven with the "Romance of Roses" and the "Story of the Gardens." In the writer's opinion, roses bring us not only beauty and fragrance, but also "tender memories" (Wright 1894, p. 39). Her garden is fused with humanity. It has a soul. Flowers, birds and human beings combine to make one feel the interdependence and friendship among the three. The power of Wright's writing and language commands admiration, urging one to protect the things that enrich and nourish the human spirit.

Contemporary female American nature writer Anne LaBastille, however, differs from those mentioned above. She regards the outdoor as a refuge for the heart. After a failed marriage, LaBastille built a cabin by Black Bear Lake in the Adirondack Mountains in northern New York state which could only be reached by boat or on foot. Her first book, *Woodswoman* (first published in 1976) tells of her first 10 years living in the cabin. In wild nature, she finds what her emotion and spirit can rely on: animals, lush trees, the mountains and the setting sun. Plants and animals in nature become her close and permanent companions. In addition, she describes a single woman's life in the woods in a uniquely feminine and romantic style. Her kitchen sometimes becomes a place where racoons gather for a party. The small forest in front of her cabin is her study full of "light and shadows" (LaBastille, *Woodswoman: Living Alone in the Adirondack Wilderness,* [1976] 1991. p. 45). The mountains in the setting sun are her friends to whom she can tell her innermost thoughts. She at last finds a sense of belonging, as everything in nature is part of the universe. She also hopes to be like the pine tree growing nearby when she grows old, spreading beauty among the grass and the plants around and with her head up resolutely, calmingly facing the sun, storms, snowflakes and spring swallows.

These female nature writers all show distinctive feminine characteristics, whether in their attitudes toward, relationships with or ways of describing nature. When facing nature, they show more care than control, more appreciation than mere taking, more harmonious coexistence than domination. To them, the natural world is an extension of daily life and can be accessed easily at any time. The gardens in front and at the back of the house provide the same direct and intimate contact with nature as the mountains in the distance. So, whether they are describing deserts or gardens, their works all naturally show a gentle affection in the emotional and spiritual communication between humans and nature.

## Lecture 19. Cooper: Beautiful Rural Hours

In American literary history, James Fenimore Cooper (1789–1851) was well known for his series of novels called *The Leatherstocking Tales. The Pioneers* (1823), *The Prairie* (1827) and other works in the series vividly describe American frontier life. His eldest daughter, Susan Fennimore Cooper (1813–1894), was less well known; but although she lived in her father's shadow, she wrote a beautiful book, *Rural Hours*, which was reprinted six times within five years of its first publication in 1850. It was written in the form of a diary, using spring, summer, autumn and winter as the organizing framework. The author describes the natural scenery, animals and plants, culture and history of the areas around her hometown, Cooperstown in the state of New York, as well as her understanding of the relationship between humans and nature. James F. Cooper loved his daughter's book so much that in letters to his friends he refers to his wife as "the old lady," his daughter as "young rural hours" and himself as "old rural hours" (quoted by Rochelle Johnson and Daniel Patterson, introduction to Cooper, [1850] 1998, pp. xiii–xiv).

However, like many other nature-themed literary works, Cooper's *Rural Hours* only enjoyed fame for a short period and then was forgotten. It was only toward the end of the twentieth century, with the worsening environmental crisis and the rising popularity of American nature writing, that works like *Rural Hours* were rediscovered and began to re-exert their influence. For example, the copy of *Rural Hours* that I own is a 1998 edition. The editor explains in the preface that this is the first reprint of the original edition since 1876. It is worth mentioning that this copy of *Rural Hours* was a gift to me from an American nature writing scholar and professor in the English Department at Brown University, and that he wrote on the front page of the book (I paraphrase) that he hopes I still have time to enjoy some "rural hours" in the fast-paced city of Beijing where I live.

## A Simple Country Life

When Cooper was only 13 she began to travel and study in France, Switzerland, Italy and England. In Europe she received a formal education in languages and literature, and also met some well-known people. In 1833, when Susan was about 20, the Cooper family returned to America and settled down in Cooperstown. Although life there was simple and completely different from the life she had had in Europe, Cooper saw infinite charm in the fresh natural environment, and in the simple life. Thoreau once said, "Why not begin his travels at home?" (Thoreau, Journal 6 August 1851) Cooper also understood that through the microcosm we can gain insight to the wider world. She walked around the fields and hills and along the paths and rivers of Cooperstown, and wrote *Rural Hours* with the eyes of a painter and the pen of a poet.

When talking about writing *Rural Hours*, Cooper affectionately recalled her "best friend in her childhood," her maternal grandfather De Lancey. She often drove with her grandfather in a two-wheeled carriage to his farms and to the forest, and it was he who taught her first lessons in botany. He explained the growth of all kinds of trees, starting from the bark and leaves, and taught her which was a birch and which an oak. She got to know all kinds of farming crops and other plants. Cooper called this early education in the fields "the beginning of my *Rural Hours* ideas" (quoted by Rochelle Johnson and Daniel Patterson, introduction to Cooper, p. x). She was also influenced by her other grandfather, William Cooper. The background of *Rural Hours* was the place where William Cooper had settled. He not only lived the life of a pioneer and created the Cooperstown settlement, but also wrote a booklet titled *A Guide in the Wilderness,* published in 1810. In this booklet, he describes his lonely experience of exploring in the silent wilderness away from home, and also provides guidance to future settlers on how to create a settlement in the American wilderness. This pioneering spirit and attachment to the land influenced the next two generations of Coopers.

Cooper's father James recorded in writing her grandfather's enterprising achievement. In his American frontier novel series he describes settling, tilling and farming in the wilderness as progressive and heroic human action. Cooper heard and saw her grandfather's and father's actions. In her preface to the reprinting of one of her father's works, she calls him "most thoroughly a pioneer in spirit" (quoted by Rochelle Johnson and Daniel Patterson, introduction to Cooper, [1850] 1998, p. xvii).

Cooper began writing *Rural Hours* in 1848, and completed it in the next two years, and the book was published in 1850. However, as a third generation Cooper describing Cooperstown, Susan Cooper had her own unique perspective. Despite her admiration for her ancestors who tamed and conquered the wilderness, she also saw the rapid disappearance of the wilderness during her time, and the consequences of this. Through her seemingly leisurely walks and her admiration of flowers and observation of birds, she realized that there is a close relationship between nature and culture, and that America needed a healthy, nature-based culture.

## *Rural Hours*

When *Rural Hours* was published in 1850, nobody was more excited than Cooper's proud father. In a letter to his wife he wrote, "Right and left, I hear of *Rural Hours.* I am stopped in the street a dozen times a day to congratulate me" (quoted by Rochelle Johnson and Daniel Patterson, introduction to Cooper, [1850] 1998, p. xiii). Excerpts from *Rural Hours* were included in John S. Hart's *The Female Prose Writers of America* (1851). In the collection, Cooper is described as

> a nature writer who avoids extravagant embellishments and ecstasies: Her work... contains an admirable portraiture of American out-door life, just as it is, with no coloring but that which every object necessarily receives in passing through a contemplative and cultivated mind. (quoted by Rochelle Johnson and Daniel Patterson, introduction to Cooper, [1850] 1998, p. xv)

Lawrence Buell, in his seminar work *The Environmental Imagination*, refers to Cooper and her *Rural Hours* many times, and especially mentions a pictorial titled *The Home Book of the Picturesque*, published in 1852. This beautiful coffee table book contains works of American writers and painters of the time, including Washington Irving, William C. Bryant, Asher B. Durand and Cooper, and in Buell's opinion it greatly influenced the development of nineteenth-century American nature writing.

However, regardless of how society commented on *Rural Hours*, Cooper's original reason for writing this book was very simple: it was out of love. As the author says in her foreword, it is

> the simple record of those little events which make up the course of the seasons in rural life.... In wandering about the fields, during a long, unbroken residence in the country, one naturally gleans many trifling observations on rustic matters, which are afterward remembered with pleasure by the fireside, and gladly shared, perhaps, with one's friends. (Foreword to Cooper, p. 3)

As a result, we have this natural and beautiful book that is like a lively conversation with friends by the fireside, a book of "rural hours" that follow the rhythm of nature, starting with spring and ending with winter.

On opening the book we are greeted by the spirit and the atmosphere of spring:

> Wednesday [March] 8th
> Very pleasant day; quite spring-like. The snow is melting fast. Spring in the air, in the light, and in the sky, although the earth is yet unconscious of its approach. (p. 5)
>
> Wednesday [May] 3rd
> Pleasant walk on the open hill-side. Sweet, quiet day; if the leaves were out, they would not stir, for the winds are all asleep. Walking over pasture-ground, we did not find many flowers, only a few violets here and there, and some young strawberry flowers, the first fruit-bearing blossom of the year.

> The valley looked pleasantly from the hill-side this afternoon; the wheat-fields are now very brilliant in their verdure, some of a golden green, others of a deeper shade. Nearly half the fields are ploughed this season, and the farms look like new-made gardens. As we stood on the quiet, open down, a sweet song, from a solitary bird, broke the stillness charmingly: it came from the edge of a bare wood above, but we could not see the little singer. (pp. 33–34)

Summer under Cooper's pen is a calm, green picture. In it there is not only modern people's yearning for the green, but also yearning for a quietness of the heart.

> Friday June 1st
> Beautiful day. Pleasant walk. The whole country is green at this moment, more so than at any other period of the year. The earth is completely decked in delicate verdure of varied shades: the fruit-trees have dropped their blossoms, and the orchards and gardens are green; the forest has just put on its fresh foliage, the meadows are yet uncolored by the flowers, and the young grain-fields look grassy still. This fresh green hue of the country is very charming, and with us it is very fugitive, soon passing away into the warmer coloring of midsummer. (p. 59)
>
> Saturday [July] 28th
> Passed the afternoon in the woods
> What a noble gift to man are the forests! What a debt of gratitude and admiration we owe for their utility and their beauty!
> How pleasantly the shadows of the wood fall upon our heads, when we turn from the glitter and turmoil of the world of man! The winds of heaven seem to linger amid these balmy branches, and the sunshine falls like a blessing upon the green leaves; the wild breath of the forest, fragrant with bark and berry, fans the brow with grateful freshness; and the beautiful wood-light, neither garish nor gloomy, full of calm and peaceful influences, sheds repose over the spirit. (p. 125)

Cooper's autumn is different from that of most conventional writers; where they tend to lament about the melancholy nature of autumn, she presents a colorful and glorious autumn scene.

> Tuesday [October] 17th
> In our walk this morning, observed a large stone farmhouse, with maples grouped about in most brilliant color; a party of men were husking maize in the foreground; a group of cows grazing, in one direction, and a cart with a pile of noble pumpkins lying in the other. It would have made a good picture of an American autumn scene. The coloring of the trees was just what one could wish for such a

purpose, and the contrast with the stone house and gray barns was all that could be desired. (p. 215)

Cooper's description of autumn scenes is closely connected with the cultural and spiritual independence and freedom of the new world. She thought that autumn under the pen of European poets was usually dark and gloomy: "cold autumn," "pale autumn," "autumn, a product of gloom." That was a symbol of a dead culture. Autumn in America was different, however. It showed the vitality of a new country, a new culture, the confidence of a new nation. She wrote:

> Mark the broad land glowing in a soft haze, every tree and grove wearing its gorgeous autumnal drapery; observe the vivid freshness of the evergreen verdure; note amid the gold and crimson woods, the blue lake, deeper in tint at this season than at any other; see a more quiet vein of shading in the paler lawns and pastures, and the dark-brown earth of the freshly-ploughed fields; raise your eyes to the cloudless sky above, filled with soft and pearly tints, and then say, what has gloom to do with such a picture? Tell us, rather, where else on earth shall the human eye behold coloring so magnificent and so varied, spread over a field so vast, within one noble view?" (p. 213)

For Cooper, winter turned from magnificence to grandeur, from busy to silent scenes. Winter turned people's enthusiasm to thoughtfulness and to reason, but without losing its vitality.

> Monday [December] 4th
> Charming day. Light sprinkling of snow in the night; but it has already disappeared. The grass on the lawn is quite green again. A light fall of snow, without a hard frost, always brightens the grass, perhaps more even than a spring shower. It often snows here without freezing. (p. 253)
>
> Saturday [December] 30th
> Still, half-cloudy day. Snow 18 inches deep; a fall of several inches during the night. The air is always delightfully pure after a fresh fall of snow, and to-day this sort of wintry perfume is very marked.... It is, indeed, a vast change from grass to snow; things wear a widely different aspect from what they do in summer. All color seems bleached out of the earth, and what was a few weeks since a glowing landscape, has now become a still bas-relief. The hills stand unveiled; the beautiful leaves are gone, and the eye seeks in vain for a trace of the brilliant drapery of autumn—even its discolored shreds lie buried beneath the snow.
> There is, however, something very fine and imposing in a broad expanse of snow: hill and dale, farm and forest, trees and dwellings, the neglected waste,

> and the crowded streets of the town, are all alike under its influence; over all it throws its beautiful vesture of purer white than man can bleach; for thousands and thousands of miles, wherever the summer sunshine has fallen, there lies the snow. (pp. 282–283)

Thanks to *Rural Hours*, we can follow Cooper and enjoy the nineteenth-century scenery of Cooperstown in spring, summer, autumn and winter, and at the same time also enjoy the scenery in the heart of this nineteenth-century female writer.

## *Watch over the Land*

As mentioned earlier, although Cooper was deeply influenced by her father and grandfather in her description of nature, her writing went beyond theirs and that of other writers of her time. In describing the same land that the pioneers had settled, William and James Cooper promote the heroic spirit of man in their struggle against nature and the progress humans made when they changed the wilderness into tilled fields. Susan Cooper, on the other hand, while acknowledging this human progress, focused more on the balance between the taming of the wilderness and the protection of nature. She also, before her time, associated this ecological view with the health of American culture and with literary aesthetics, regarding the maintenance of "a garden like America" as the moral duty Americans owed to the land. This is why today, more than 100 years after its first publication, *Rural Hours* is still of relevance and significance. To a certain extent, we can say that *Rural Hours* was written to make people adopt a respectful and admiring attitude to nature through the understanding and knowledge of the local land, thereby changing the greedy attitude of just taking without any regard for the consequences. On the one hand, Cooper praised the progressive action of changing the wilderness into productive fields. She exclaims in *Rural Hours*, "How pleasantly things look about a farm-house!" and describes in detail and with admiration the vivid scenes of farms, farmland and agriculture. For example,

> From the window of the room in which we were sitting, we looked over the whole of Mr. B—'s farm; the wheat-field, corn-field, orchard, potato patch, and buckwheat-field. The farmer himself, with his wagon and horses, a boy and a man, were busy in a hay-field, just below the house; several cows were feeding in the meadow, and about fifty sheep were nibbling on the hill side. (Cooper, p. 96)

On the other hand, Cooper was also critical of the irresponsible cutting down of pristine forests and hoped that nature would retain its original appearance. She made it very clear that "a large shady tree in a door-yard is much more desirable than the most expensive mahogany and velvet sofa in the parlor" (p. 133).

The value of *Rural Hours* also lies in the author's scholarly knowledge and rigorousness in her description of nature and in her perfect combination of ecology with aesthetics. For example, when she is describing a common local bald eagle, she quotes the observations and recordings of three famous nineteenth-century American ornithologists: Alexander Wilson, John J. Audubon and Thomas Nuttall. When Cooper describes the natural world, she not only uses scholarly insight, but also the heart of a writer. When she is describing the strength of the bald eagle, she emphasizes that "We Americans all have a national interest in this powerful bird as the emblem of our country" (p. 264). When she writes about American autumnal scenes, she lists the original poetic lines of British, French and Italian poets on autumn and vividly contrasts them with the American autumn. Under her pen, the white snow blown up by the wind falls back to the ground like "sand of gold." She praises the graceful postures of wild swans and their wonderful singing and then thinks, "Mr. Nuttall supposes that it was this note of theirs which led to the classic fancy of the song of the dying Swan" (p. 257).

Of course, some readers may feel that after reading *Rural Hours* they know too little about its author, or that nature under her pen, even though lively and vivid, lacks depth. However, this may be exactly where she, as a nineteenth-century female nature writer, was different from her contemporaries. Unlike Emerson, Cooper did not believe that "[n]ature always wears the colors of the spirit" (Emerson, "Nature", [1836] [1903] 1979, p. 11), but that "humans wear the colors of nature's spirit." Unlike Thoreau in *Walden*, the author of *Rural Hours* seldom uses the first person singular "I" in her walks in the country, but often uses the plural form "we." She deliberately places herself as a part of the surrounding natural environment, making the reader a participant in her exploration of the woodlands, prairies, valleys, streams and lakes. We cannot help but admire this nineteenth-century female writer's ecological view, which was quite ahead of her time.

Aldo Leopold, an American nature writer often called the "father of ecology," said in the foreword of his book *A Sand County Almanac* (1949):

> There are some who can live without wild things, and some who cannot. These essays are the delights and dilemmas of one who cannot. Like winds and sunsets,

> wild things were taken for granted until progress began to do away with them. Now we face the question whether a still higher "standard of living" is worth its cost in things natural, wild, and free. For us of the minority, the opportunity to see geese is more important than television, and the chance to find a pasque-flower is a right as inalienable as free speech. (Leopold, p. vii)

Nowadays, when we are gradually losing natural and wild scenery, picking up Cooper's *Rural Hours* again and re-reading Leopold's words give rise in our hearts not only to nostalgia for "rural hours," but also eagerness and passion for the wilderness. In the modernized industrial "forest," keeping a piece of green is not only our spiritual need, but also our inviolable human right.

## Lecture 20. Thaxter: Flowers on the Rocks

On the bookshelves in my study are two books given to me by American friends, both in hardcover by Celia Thaxter (1835–1894), an American female writer of the nineteenth century. One, with an ivory-colored cloth cover, is *Among the Isles of Shoals* (1873); the other, with a pea-green gilded cover, is *An Island Garden* (1894). Celia[1] was a contemporary of American female writers Louisa May Alcott (1832–1888) and Emily Dickinson (1830–1886), and they all lived and wrote in New England. Alcott is well known for her *Little Women* and Dickinson for her unique poems, while Celia seems to have been left in the "forgotten corner." So what kind of a writer is Celia, and what are her books like?

### *Queen of the Isles*

Celia was born in 1835 in the small coastal town of Portsmouth, New Hampshire, New England. When she was only 4 years old, her father moved his young family to White Island to be the lighthouse keeper there. At the time, this was quite sensational news, as her father had been a state congressman. He was still young and had a promising future, so becoming a lighthouse keeper seemed a kind of self-exile. White Island was one of the nine small islands of the Isles of Shoals, about

1 In this lecture I use "Celia" instead of her surname Thaxter to refer to this nineteenth-century female writer for two reasons. Firstly to avoid confusion, as I mention her husband by his surname Thaxter. Secondly, I feel the name "Celia" is beautiful visually and aurally. However, in the subtitle of this chapter I still use her surname, to be consistent with the other chapters.

10 miles off the New Hampshire coast which were obstacles to ships going into Portsmouth. Unlike the tree-covered islands along the coast of northern Maine, at the time the Isles of Shoals were bare, desolate, rocky places with only grass, low bushes, and wildflowers when spring came. However, Celia and her family stayed on the couple of acres of White Island for six years, the only people there.

> In *Among the Isles of Shoals*, Celia describes their arrival at White Island.
>
> It was at sunset in autumn that we were set ashore on that loneliest, lovely rock, where the lighthouse looked down on us like some tall, black-capped giant, and filled me with awe and wonder. At its base a few goats were grouped on the rock, standing out dark against the red sky as I looked up at them. The stars were beginning to twinkle; the wind blew cold, charged with the sea's sweetness; the sound of many waters half bewildered me. (1873, p. 120)

It was under such circumstances that "[w]e entered the quaint little old stone cottage that was for six years our home," and it was in this small stone cottage that Celia was for "the first time lulled by the murmur of the encircling sea." She exclaims, "It takes so little to make a healthy child happy!" (1873, p. 121).

Celia's childhood companions, apart from her two younger brothers, were rocks, flowers, grass and seashells. Many years later, she recalls in *An Island Garden*:

> Ever since I could remember anything, flowers have been like dear friends to me, comforters, inspirers, powers to uplift and to cheer. A lonely child, living on the lighthouse island ten miles away from the mainland, every blade of grass that sprang out of the ground, every humblest weed, was precious in my sight, and I began a little garden when not more than five years old. (1894, p. v)

In a collection of her poetry published later, there is a short poem describing her holding a lamp in the dark to guide her father to shore in a boat. This childhood seems to have determined her lifelong love of the islands and of flowers and plants.

Romantic love also came early. At 14, Celia fell in love with Levi Thaxter, a Harvard graduate. Thaxter was the son of a banker. He came to the island hoping to open a holiday inn together with Celia's father, and also to work as a teacher for Celia and her brothers. Thaxter was the first well-educated young man Celia had met. Some said that he had dramatic talent and was so good with words, he "could persuade a green lobster into a pot of boiling water, and he quickly sweet-talked the naïve young girl into marrying him" (quoted in Julia Older, ed., introduction to *Celia Thaxter: Selected Writings*, [1916] 1997, p. 12). Even though Celia's father was against the marriage, in September 1851 Celia, not yet 16, married Thaxter who was 11 years her senior, and became Celia Thaxter. Their wedding was held on Appledore, the largest island of the Isles of Shoals. A few years earlier, Celia's father had moved the whole family there and had jointly opened a holiday inn there with Thaxter. Celia's oldest son Carl Thaxter was born on the island of Appledore, the first baby born on the Isles of Shoals for nearly a hundred years.

Celia's married life, although romantic, was not uneventful. Her husband Thaxter once preached and taught on another small island of the Isles of Shoals. His chapel doubled as his classroom, and it had only one room. It is said that his church service was more like poetry recitation and dramatic performance, as Thaxter's interests were more in poetry and drama. He loved English poetry, especially the works of Robert Browning; and later, when Thaxter passed away, Browning wrote a poem for the epitaph on his tombstone. It has been said that Thaxter had substantial literary talent, but that he was overshadowed by the reputation of his wife. After nearly dying in a boat accident, Thaxter was so shaken that he swore he would never return to the Isles of Shoals. Celia followed her husband and lived in Massachusetts for a period of time; however, she could not stop yearning for life on the isles, and eventually went back with her eldest son to Appledore, where she spent most of her life, while her husband and her two younger sons remained on the mainland in Massachusetts.

During her stay in Massachusetts, Celia missed her parents and the islands very much. In 1861 she wrote a poem, *Landlocked*, which she only read to her son Carl and did not intend to publish. However, not long after, *The Atlantic Monthly* published this poem, making Celia famous overnight:

> O Earth! Thy summer song of joy may soar / Ringing to heaven in triumph. I but crave/ The sad, caressing murmur of the wave/ That breaks in tender music on the shore. (in Walker, Cheryl. Ed. *American Women Poets of the Nineteenth Century: An Anthology*, 1992, p. 294)

This sad and beautiful poem brought Celia into the circles of literature and the temple of poetry, and her poems were later published one after another in the *Atlantic Monthly*. James Fields, editor of the magazine and a famous publisher, introduced Celia to literary men in Boston such as Emerson, Hawthorne and Longfellow, and to the British writer Dickens, who was visiting Boston at the time. In 1872 Celia's first collection of poems was published, and in 1873 her first prose book, *Among the Isles of Shoals*, in which she presented the natural environment and life on the Isles of Shoals to the world. This book, tracing the author's footsteps and filled with her passions, demonstrates that her greatest talent lay in prose writing. Her last published book, *An Island Garden* (1894), was the fruit of her old age and the song of her life.

However, Celia is distinguished not only by her literary talent, but also by her difficult life experiences and charming personality and character. Celia's life was not easy; her husband suffered from chronic illnesses, had no steady job for many years, and also grew apart from her. Her eldest son was born with a slightly crippled leg and an intellectual disability, and his moods were unpredictable. The burden of life was on Celia's shoulders; for a period of time she had to write to support her family, like Louisa May Alcott. However, Celia's troubles did not disrupt her life's dream. Her warm personality shone like sunshine in all directions and attracted people. Her dreams, her talent and her will made her life fruitful and magnificent. In the nineteenth century, when female writers were few and far between, she embarked boldly upon a career of writing as a way of making a living. Her works were mostly focused on the intimate relationship between humans and nature, showing a consciousness of environmental protection and ecology ahead of her times. Furthermore, her vision extended way beyond writing. She studied painting under the American impressionist painter Childe Hassam, and gave collections of poems with illustrations she drew herself to friends and family as gifts. She also learned porcelain painting, and the porcelain objects on which she painted flowers and plants were admired and became a way of supplementing her income to support the family. Celia was also a born gardener, with more than 150 kinds of flowers in her garden, and a bird specialist. She joined the Audubon Society,[2] the aim of which was to protect birds, and wrote articles criticizing the custom of using birds' feathers as decorations.

Most unforgettable was The Isles Cultural Salon that Celia held every summer. Having grown up in the company of flowers, plants and birds on a desolate island, Celia had unique taste and appreciation and attracted Boston literary and art personalities to Appledore to spend their holidays there. Her guests included writer James Russell Lowell, poet John G. Whittier, publisher James Fields, artists Childe Hassam and Henry M. Hunt, and Norwegian violinist Ole Bull. Guests would usually spend time in Celia's lounge room, which faced the sea and was full of flowers from her island garden—blue delphiniums, red corn poppies, pink

2 An American nature protection organization named after the ornithologist and naturalist John J. Audubon (1785–1851).

oleanders—which she would pick herself and arrange in vases of various shapes. The most eye-catching feature was a big conch shell hanging from the ceiling in mid-air. In the conch were golden globeflowers, violets and other blue flowers. In the lounge there was always piano or violin music, and sometimes there were readings by writers and poets. In the sea of flowers Celia would sit quietly in a corner, either painting on porcelain or writing poetry, always dressed in white or grey like a simple background for her flowers. As Lyman V. Rutledge quotes in his *The Isles of Shoals in Lore and Legend* (1965, p. 124), "Those who were invited to Celia's parlor could feel that here as nowhere else their finest talent were felt and understood." Another friend wrote that

> Celia Thaxter was, I sometimes think, the most beautiful woman I ever saw, not the most splendid nor the most regular in feature, but the most graceful, the most complete,—with the suggestion of perfect physical adequacy and mental health in every look and motion. She abounded in life, it was like breathing a new life to look at her. (Rutledge, p. 90)

A female writer of the same period said, "In all our literature, she (Celia) is the most picturesque figure" (Rutledge, p. 124). Thus Celia, who had never received any formal education and was called the "flower on the rocks," also became the "Queen of the Islands" and a symbol of beauty.

At dawn on 26 April 1894, Celia passed away in her loungeroom on Appledore Island. To the east of the room was a grassy path leading to the top of a small hill, where her family burial ground lay. Celia used to go there before dawn with fresh flowers and place them in front of her parents' tomb, and she was also buried there, accompanied by her beloved flowers. The loungeroom which used to be filled with flowers and laughter became silent, though the windows were still open, facing the rising sun and admitting the slightly salty sea breeze.

Celia's family hotel and her cottage no longer exist, but fortunately, the two books she wrote can still lead us into her island life to experience the natural world there and the world of her heart.

## *Island Life*

English writer Virginia Woolf (1882–1941) once said that a female writer's life was hard. In *A Room of One's Own* (1989, p. 44) she wrote, "It was certainly an odd monster that one made up by reading the historians first and the poets afterwards—a worm winged like an eagle; the spirit of life and beauty in a kitchen chopping up suet." Celia's experience also showed the hard life of a female writer.

As early as 1862 James Fields, editor of the *Atlantic Monthly* magazine, had invited Celia to write about her life on the island for his literary monthly. However, three young children and housework forced her to write only intermittently and it was not until the autumn of 1869, seven years later, that her *Among the Isles of Shoals* came out in a four-part serial in the *Atlantic Monthly*, and another four years until it was published in book form (Older, p. 27). However, this book, written in the pauses between household tasks, presented to us another world, a realm beyond the worldly affairs and a spiritual home of one's dream.

Based on the personal experiences of the author, the book describes the Isles of Shoals. There are no chapters, nor section titles; it is more like a long piece written at one go. However, as soon as you open the first page you will forget yourself and follow the footsteps of the author from her childhood to wander around the Isles of Shoals and follow the rhythms of nature through spring, summer, autumn and winter.

The beginning of the book is original and fresh. Celia compares the Isles of Shoals with the Enchanted Isles described by Herman Melville (1819–1891). Melville first went to the Galapagos Islands in South America—also named *Las Islas Encantadas* (The Enchanted Islands) in Spanish—in the spring of 1841. The group of islands did not even have the very basic conditions necessary for life, with only uneven black volcanic rocks, split cliffs, and thousands of seabirds screeching like devils. Celia quotes Melville's comment on this group of islands, "It is to be doubted whether any spot on earth can, in desolateness, furnish a parallel to this group" (in Thaxter 1873, p. 7). She then comments that, although the Isles of Shoals appear to be as desolate and depressing, they are fascinating islands because they have a strange charm and their geographical position has a universally acknowledged—although difficult to express—effect on people. After living there for a while, people forget their worldly bustles, worries and anxiety. To imaginative people, everything on the islands is covered in dreamy poetic color. If Melville's description of the "Enchanted Isles" expresses a tragic view of nature, Celia's sees pastoral scenery in the seemingly tragic natural landscape.

The Isles of Shoals spread over a couple of square miles. At low tide you can see six islands, and at high tide eight or nine, as high tide separates islands that are linked together at low tide. This geographical feature gives the Isles of Shoals a confusing and dreamy aspect. The naming of these islands is also unusual. For example, Star Island because it is shaped like a star; Hog Island (also called Appledore) because the part that is above water looks like the back of a hog; White Island in memory of Captain White; Londoner's because a British ship once crashed onto the rocks of the island and sank there; Cedar Island because a few cedar trees, which are rare locally, grew there; Duck Island because there was a pond of fresh water where migrant wild ducks liked to stay; and Malaga Island because this island reminded Spanish sailors of the Malaga vineyards back home. The Isles of Shoals were so named because there were large groups of fish in the water around these small islands when in season. Thus, the desolate and awe-inspiring islands become vivid and lively in our eyes.

Under Celia's guidance, we approach the Isles of Shoals. "Landing for the first time, the stranger is struck only by the sadness of the place,—the vast loneliness; for there are not even trees to whisper with familiar voices,—nothing but sky and sea and rocks" (p. 14). Next Celia suddenly turns and presents the other aspect of the islands: "But the very wilderness and desolation reveal a strange beauty to him" (p. 14). Then, through the stranger's eyes, we see the loveliness of the islands. Waves of the Atlantic Ocean lap the shore lightly and accompany the stranger's sleep. The vivid life of a summer morning wakes him. The sea and sky are both colored rosy at dawn, passing white sails shine gentle light onto the bare rocks, even the wings of seagulls are tinged red. In order that we experience it more vividly, Celia also uses the second person narrative method to reproduce her sea scenery.

> While you sit watching the shifting splendour, uprises at once a higher cloud than usual; and across it springs a sudden rainbow, like a beautiful thought beyond the reach of human expression. High over your head the white gulls soar, gathering the sunshine in the snowy hollows of their wings. As you look up to them floating in the fathomless blue on the horizon the white sails shine. (p. 119)

Celia even finds the night sky in winter extremely romantic.

> When the clouds are swept away by the besom of the pitiless northwest, how the stars glitter in the frosty sky! What wondrous streamers of northern lights flare through the winter darkness! I have seen the sky at midnight crimson and emerald and orange and blue in palpitating sheets along the whole northern of the heavens, or rosy to the zenith or belted with a bar of solid yellow light from east to west, as if the world were a basket, and it the golden handle thereto. (p. 94)

On the island, no minute change would escape Celia's eyes: changes in the sky or the sea, comings and goings of coasters and of birds, sunrise and sunset, the waxing and waning of the moon, the blooming and falling of flowers—everything was lovingly recorded. People who live in busy, crowded cities could not have imagined these. Where there are no streets, no shops, no neighbors and no need for doorbells, the stars "seem like the faces of old friends looking down out of the awful blackness." She exclaims, "Never was life so exempt from interruptions" (p. 99).

On the desolate and even despairing Isles of Shoals, Celia finds joy and lives a spiritually rich life. She watches fishermen fishing and whales playing in the water and calls the scene "magnificent"; she observes the lives and experiences of the residents of the small islands and learns from them the attitude of finding joy in a hard life: in May she takes walks in fine rain and sees glistening green leaves opening, and exults that "the rainy days in May at the Isles of Shoals have seemed to me more lovely than the sunshine in Paradise could be"; she watches a ship sail slowly past under the moonlight like "a beautiful ghost stealing softly by"; she observes swallows dancing and flowers blossoming, listens to insects singing and takes in the scenery of all four seasons. (p. 161) She lives a life on the island that is interesting, rich and unique. She tells the reader very frankly that when she was watching flowers with curiosity and enthusiasm in her childhood, she did not know anything about the English poet John Keats. It was only later that she read Keats' poems and learned that the famous poet thought "the intensest pleasure in life had been to watch the growth of flowers." Celia then comments, "but I am sure he never felt their beauty more devoutly than the little, half-savage being who knelt, like a fire worshipper, to watch the unfolding of those golden disks" (p. 133). Besides, a flower in Celia's eyes was not only a flower, "it seemed more than a flower; it was like a human being" (p. 128), so when flowers fell, she would collect them and gently bury them together, a scene similar to that of "Lady Daiyu burying the flowers,"[3] as described in a Chinese novel on the other side of the world.

The islands, under Celia's pen, had fascinating natural scenery, but also a harsh and unpleasant side. Celia showed the pitilessness of nature: one stormy night, a boat manned by two Norwegians wrecked on White Island. The lighthouse keeper could do nothing to help, and had to watch one of them die in the storm, while the other man left to tell later generations of their tragic experience (p. 146). She also described the hard life on the islands: the small fishing village on Star Island, the small fishing boats tossed about in the gales and waves, the fishermen's wives anxiously waiting. She was sympathetic towards the women, who aged before their time due to the burdens of life, saying, "It seemed as if youth and joy could never have been theirs" (p. 66). There was also the

3 Lady Daiyu is a character in the Chinese classic novel *The Dream of the Red Chamber*.

difficult-to-endure winter on the islands, throughout which they were isolated from the rest of the world. Migratory birds had flown away, and grass and plants were all dead. Only snowy owls came and stayed, and their habitual silence was as depressing as ghosts (p. 107). The saddest incident was when two residents of Star Island came by boat to Appledore Island where Celia lived to make a coffin for a little girl who had died. Celia watched the making of the little coffin the whole day. When it was completed, she ran home and brought back a bunch of scarlet geraniums and put it in the coffin, then watched the two men take it back to Star Island through the waves in the dusk. She said with heartfelt gentleness,

> I never saw the little girl, but where they buried her I know: the lighthouse shines close by, and every night the quiet and constant ray steals to her grave and softly touches it, as if to say, with a soft caress, "sleep well." (p. 104)

Celia was unique in that, even when she was describing cold and brutal natural facts, there was still a tinge of romance and a light of humanity.

Celia's description of life on the islands is a perfect combination of the realistic and the fantastic: the realistic is to accept the natural situation of the islands and seek to co-exist with it, while the fantastic is the imaginative space which is free, independent and uninterrupted by the external world. Celia taught people, through her own personal experience, how to extract joy and pleasure under immutable natural conditions, how to survive together and how to share their fate with the natural environment.

## *An Island Garden*

While *Among the Isles of Shoals* shows us Celia's early life, *An Island Garden* is a song of her later life. The brilliance of this book lies in Celia's outpouring of her love of natural scenes, in her philosophy of life and attitude to life, and in her aesthetic tastes. She put all of these into her creation and appreciation of an island garden by the sea. The copy of *An Island Garden* I have, a (2001) reprint by Mifflin & Co., contains illustrations by the famous American impressionist painter Childe Hassam. Apart from the picture of Celia in her garden, there are 21 of Hassam's colored paintings which show the charm of Celia's island garden. It is said that Hassam, who admired Celia, lived for a time in her island salon, setting up his easel under the hop vines to paint these pictures. It has also been said that Hassam's pictures for *An Island Garden* put him in the same position in American impressionism as Claude Monet's garden paintings put Monet in

French impressionism and that Hassam's painting *Celia Thaxter in Her Garden* (1892) made him famous (in the Publisher's Note in Thaxter [1894] 2001).[4] The garden, rocks and glistening sea, and Celia in white with silvery hair, present a harmonious scene in this painting that often leaves a deep impression on the viewer.

In her old age, Celia managed to escape from her housework and the management of the island hotel and thereby had more time to look after her garden. When people asked her for gardening tips she said, "patience and love," and when they found her answer unsatisfactory she wrote the book.

"At the Isles of Shoals, among the ledges of the largest island, Appledore, lies the small garden which in the following pages I have endeavored to describe" (Thaxter [1894] 2001, p. v). *An Island Garden* begins with a simple and straightforward description, showing Celia's serenity in her old age. Then she turns to her childhood, when she had flowers and plants as company, and to the time when she was not even 5 years old and had started to plant a small garden. It was her childhood experience on the isolated island that led to a lifelong love of flowers and plants, her faithful companions throughout her life. To her, of all the wonderful things in the natural world, nothing was more joyful than sowing a flower seed in empty ground and then watching it grow.

> He who is born with a silver spoon in his mouth is generally considered a fortunate person, but his good fortune is small compared to that of the happy mortal who enters this world with a passion for flowers in his soul.... Like the musician, the painter, the poet, and the rest, the true lover of flowers is born, not made. (p. 5)
>
> Celia's garden was about 50 feet long and 15 feet wide, extending southward along the veranda, away from the north wind and facing the sun. It was on this piece of land that Celia planted from 50 to more than 100 kinds of flowers. She not only experienced the hard work of a gardener, but also harvested contentment and joy of the heart which ordinary people rarely experience.

The first half of *An Island Garden* described the details of gardening: how to bring in precious flower seeds by boat and to nurture seedlings in half an eggshell; how she had to get up at two o'clock at night to fight against the insects

4 It is worth mentioning that in the 1880s Hassam spent several summers on Celia's island and created several hundred paintings. More than 100 years later, Yale University held an exhibition of this batch of works by him, titled *An Island Garden Revisited*.

that would eat the flower seedlings, and use nets to protect the young seedlings against birds. Of course, the enemies of the nursery also include storms and winds, which could destroy all the work she had done. As a gardener and protector of her flowers, Celia often got up to work in her garden before the sun rose, and under the moonlight. This was hard work which called for strength, patience, alertness, perseverance and hope. Celia said frankly, "It is hard work beyond a doubt, but I do not grudge it, for great is my reward" (p. 8).

This reward was related to the heart. Celia believed what Francis Bacon had said, "God Almighty first planted a Garden. And indeed it is the Purest of Humane Pleasures; it is the Greatest Refreshment to the Spirits of Man" (quoted on p. 24). After nearly a year's hard work and wait, when finally the ice and snow melted and spring came, the slowly blossoming flowers gave the gardener a thousand rewards: violets, lilies, roses, mignonettes and sweet peas, with their layered petals and different shapes, their colors from red to dark brown, from light yellow to orange red, from white to light purple. Celia especially liked the honeysuckle, which, with its flowers like

> rays of light, has an effect of gladness and glory indescribable: the very sight of it lifts and refreshes the human heart. And for its odour, it is like the spirit of romance, sweet as youth's tender dreams. It is summer's very soul. (pp. 41–42)

In *An Island Garden* we see an interesting picture of harmony between humanity and nature: Celia is busy working in the garden with birds for company. The female birds nest while the males sing on the fence, and there are white rocks, blue seas and gentle waves. This small island in the sea has a paradise-like beauty and wonder and Celia conveys to us happiness. "Along the far-off coast the joyous hills seemed laughing in the sunshine, and the great sea rippled all over with smiles" (p. 55). She writes, "To stand by the beds at sunrise and see the flowers awake is a heavenly delight" (p. 83). Perhaps your eyes would follow the flight of a bird into the rosy clouds, but when you look back at the flower, the bud has already been released from the two light green shells. A moment later the petals would be unfolding in front of your eyes. They flutter out in the gentle breeze like silken banners and the color of the flower "warms the wind indeed!" (p. 84) The little garden cannot restrain the wandering grasses and flowers; they break its bounds as if "rushing down the sloping bank in a torrent of yellow," which, "lightly stirred by every breeze, sends a satin shimmer to the sun" (p. 75). Celia would also walk along the little island path lined with wildflowers and experience "fresh surprises" in meeting these unfamiliar flowers (p. 48). It was through this life close to nature that Celia was able to have an emotional communication with the animals and plants, as if they were humans. After a strong storm, the garden had "more green than red" left, and among the branches and leaves that had fallen, Celia discovered a frozen hummingbird. She put the little bird in her hand to warm its body until it had a slight pulse of movement, then she placed it in a basket cushioned with cotton and put a cup of honeyed water beside it. She hung the small cradle in the sunniest place she could find and let it move around in the blue spikes and white lilies. Gradually the bird opened its eyes and looked around quietly. Suddenly it rose into the air as if it were a sparkling emerald quivering among the blue flowers, "diving deep into each winged blossom for his breakfast of honey" (p. 111). From then on, the little hummingbird settled down in the garden. When Celia worked it would come and rest trustingly on her arms, hands or hair, and she believed that "he is the very most charming thing in the garden" (p. 112). In Celia's heart all flowers and plants have feelings. In the morning, "the fair face of every flower salutes me with a silent joy that fills me with infinite content; each gives me its color, its grace, its perfume, and enriches me with the consummation of its beauty." She feels the personality of each flower and greets them like friends, "Good-morning, beloved friends! Are all things well with you? …I lose myself in the tranquillity of their happiness. They seem like sentient beings, as if they knew me and loved me" (p. 113).

In *An Island Garden*, Celia uses beautiful language and interesting descriptions to represent Darwin's concept of "the survival of the fittest," demonstrating that she was aware of ecology before her time. The characteristics of plants also reflect various philosophies of life. Like human beings, they have astonishing survival instincts. "If you dig a hole in the ground and put into it a Rosebush, filling one side of the hole with rich earth and the other with poor soil, every root of that Rosebush will leave the poor half to inhabit the rich and nourishing portion" (p. 88).

She uses a wistaria as another example. Someone plants the wistaria on the south side of a house where it is dry and stony, only to find a few years later that the roots of the tree have crawled for about 30 feet underneath the house to get water from the well on its north side. (p. 89)

She cites a writer's interesting description:

> During a severe drought if you place a basin of water at night, say two feet to the left or the right of a strong vine, in the morning it will be found bathing in the basin! Is not this an indication of thought in the vine? Does it not indicate a knowledge in the vine analogous to human understanding? (pp. 90–91)

Celia concludes that "[c]limbing plants show often a surprising degree of intelligence, reaching out for support as if they had eyes to see" (p. 92). Every plant, therefore, is part of what we call life. Celia writes, "As I work among my flowers, I find myself talking to them, reasoning and remonstrating with them, and adoring them as if they were human beings.... We are on such good terms, my flowers and I!" (p. 92)

Celia's comments on the plant leaves and roots and the garden's enemy, weeds, are full of intelligent humour. She thinks about the seed, "how it is thrusting two hands forth from the imprisoning husk, one, the root, to grasp the earth, to hold itself firm and absorb its food, the other stretching above to find the light." Celia also leaves the reader with a puzzle: "It is curious that the leaf should so love the light and the root so hate it" (p. 25). Celia quotes widely to explain the definition of a weed.

> "What is a weed?" A lady says, "A weed is a plant out of place." A keen and close observer of nature says, "A better definition would be, 'A plant which has an innate disposition to get into the wrong place.'" (p. 35)

From Celia, we learn that in botanical books weeds are often called "troublesome," not because the weed is poisonous or ugly, but "it is being impertinent—thrusting itself where it has no business and hinders other people's business—that makes a weed of it" (p. 35). When reading this, apart from smiling, we might also feel that if everyone finds their own most suitable place and does their duty instead of yearning for other positions that do not belong to them, or becoming an obstacle to others, then our human society would be peaceful.

Celia was also a gardener before her time. When there was a pest in the garden she used the principle of the natural food chain, for example by introducing toads to the island from outside to eat up the slugs, instead of using pesticides. She talked to her plants while working in the garden to make them grow better and faster. Not until about a century later, did research discover that talking to plants or playing music is beneficial to their growth.

In a word, *An Island Garden* is a very interesting book with enduring charm. Madame Tasha Tudor,[5] who wrote an introduction for a 2001 edition of *An Island Garden,* recalled with affection that her mother gave her a copy of the book as a birthday present when she was 15 because she, like Celia, had started

5 This book was first published by Mifflin & Co. in 1894; Madame Tasha Tudor's introduction is in a copy published in 2001.

to garden when she was five. This book accompanied her in her enjoyment of gardening, through dreamy spring, summer, autumn and winter. When she was 86 she reread *An Island Garden*, and in her introduction she says that "with the increase of age, the words in the book, though the same, your experience would change its meaning" (Tasha Tudor, introduction to Thaxter [1894] 2001, p. vii). She praises *An Island Garden* as having a spiritual charm, life, beauty, wonder, and dreamy images—features that together create a sense of enchantment about growing flowers and plants, because a garden is a reflection of the gardener's inner world.

With the passing of Celia, her garden gradually became desolate and the island hotel and house burnt down in a fire; only legends about the Isles of Shoals and Celia are left for later generations.

## Lecturer 21. Wright: May Day in New England

American poet Robert Frost once said, "All literature begins with geography." When we talk about American literature, we tend to look at New England: Thoreau's *Walden* (1854), Hawthorne's *The House of Seven Gables* (1851), Harriet Beecher Stowe's novels set in the New England countryside, Celia Thaxter's *An Island Garden* (1894) and numerous others. New England is where American literature originated, and where its essence lies due to its unique geographical location. In this lecture we are going to discuss another American female writer from New England, Mabel Osgood Wright (1859–1934), and her representative work *The Friendship of Nature: A New England Chronicle of Birds and Flowers* (1894).

Wright was a nature writer, but also a nature protection activist. She published 25 books and tried to educate people, using novels, essays and popular natural science books, about how to be friends of nature; she took great care of her own garden, telling "the story of the garden" and asking people to start with gardens to cultivate friendship with nature and to experience the beauty of nature; she established the local Audubon Society (the aim of Audubon Societies is to protect birds) and served as its president for 26 years; she also helped to organize the first privately owned bird sanctuary in America, leaving behind a natural environment where birds sing and fragrant flowers bloom. Throughout her life Wright tried to do one thing: to change people's attitude from appreciation of nature to protection of nature. She showed via her own experiences how to become friends with nature and how to permanently retain the "New England May Day."

## *Birds and Flowers Complex*

Wright's interest in nature and her literary inclination came from her father Samuel Osgood, a pastor of the Church of the Messiah in New York who had studied with the well-known writer, ethicist and minister William Ellery Channing. Apart from fulfilling his duties as pastor, Osgood was also a writer, orator and editor of many theological and literary journals. As a member of the circle of William Cullen Bryant, an American poet, Osgood was in close contact with literary figures of nineteenth-century America. The only thing that could match his spiritual pursuit was his obsession with gardening.

In 1857, Osgood purchased eight acres of land in Fairfield, Connecticut and built a summer holiday home, Waldstein, which had 18 rooms. During the next 20 years he built several little cottages around the main house, as well as gardens, lawns, winding paths, waterlily ponds and sculptures of famous people. In the gardens of the holiday home he also planted many trees. It is said that he would plant a pine tree every time he presided at a baptism or wedding. In the gardens he also made clever use of rocks: simple stone walls, niches, caves, pavilions and memorial tablets were dotted among the flowers, plants and trees. On many of the rocks were also carved excerpts from the Bible, or from literary works. Culture and nature perfectly harmonized at the point where the country and the city joined. On January 26, 1859, Osgood's daughter Mabel was born, and around the same time his holiday home was also completed. Even though Mabel spent most of her childhood in Greenwich, New York, her early education at Waldstein left an indelible impression on her, as the gardens and trees there were wonderful, unforgettable memories. In *The Friendship of Nature* (1894) she recalls

> With my first consciousness, the days were filled with planting and with growth; the pines already hid the walls, and cattle tracks were widened into paths and wound among young maples, elms, and beeches. Then there grew in me a love that made the four garden walls seem like the boundaries of the world. (Wright [1894] 1999, pp. 94–95)

Four years after Wright married, her father passed away, and she and her husband inherited the right of use of Waldstein. Very little is known of the first few years of her married life, except that she was in poor health and her isolated life made nature her companion and nurse, helping restore her energy. She started to write essays about nature and showed them to her husband, who always urged her to wait till these green apples ripen. Six years into her marriage, Mabel Wright started to publish essays on nature in *The New York Times* and other journals. Later, Macmillan put these essays into a book which was published in April 1894—her first book and her representative work, *The Friendship of Nature.*

*The Friendship of Nature* was an instant success. Critics on both sides of the Atlantic praised its poetic language and vivid description of the local scenery. *The Dial*, the transcendentalist journal, described Wright as "a true poet in the Emersonian sense, namely, the power to see the miraculous in the common." Wright also received a warm letter from Oliver Wendell Holmes,[6] thanking her for her "charming little book which shall go into my shelves by the side of [Gilbert] White's [*The Natural History of*] *Selborne* (1789)." There were also newspaper and journal comments which said that *The Friendship of Nature* displayed the feelings of Henry D. Thoreau (quoted in Daniel J. Philippon's introduction to Wright [1894] 1999, pp. 6–7). Now Wright's husband was as happy as she was, because in his eyes his wife's green apples were finally ripe.

Gardens and birds were the themes of Wright's writing throughout her life. After *The Friendship of Nature*, she published *Birdcraft* (1895). This book mixed descriptions of the characteristics, songs and shapes of birds with the author's birdwatching experiences and her interesting anecdotes. John Burroughs, in a letter to Wright in 1895, called *Birdcraft* "[a] readable and interesting book," and also exclaimed, "I wish I could have had the help of such a book when I began my bird studies. I really wish I had a book like this to guide me when I first started to study birds" (Burroughs, quoted in Daniel J. Philippon's introduction to Wright [1894] 1999, p. 8). Another of Wright's books, *Flowers and Ferns in Their Haunts* (1901), complements *Birdcraft*. It shows the world that the wilderness where flowers and ferns can grow freely is their real home. These two books, like guides to outdoor life, combine science and literature and have abundant charm. Wright also wrote two semi-autobiographical novels, *The Garden of a Commuter's Wife* (1901) and *The Garden, You, and I* (1906), both of which show her love for flowers and birds.

Wright's literary works and environmental protection work all started with gardening, and this is closely related to her unique national and cultural background. America had been known as a "country of nature." In the nineteenth-century American frontier culture, America was a symbol of "virgin land" and "garden" and considered itself the "Garden of the World," a concept which later

6 In addition to being a doctor and a Harvard University medical professor, Holmes was also a well-known American poet, writer and orator of the nineteenth century. He named the journal *The Atlantic Monthly*, was also one of its main contributing writers, and was quite influential in the American literary circle of the time.

became the main theme of American cultural history. With the ending of the westward movement and the acceleration of industrialization, the city became another name for noise and pollution, and the middle classes started to move from the city to the suburbs so that their families could be away from the noise and bustle. What followed were the idealization of suburban life and the development of landscape gardening and garden design. Andrew Jackson Downing, a famous landscape designer of the time and an advocate of the idealized suburban and pastoral life, combined aesthetic principles with the moral precept of protecting the environment. He regarded the suburbs as the best place for people to improve their morals, since in the outdoors they could avoid the noise and vices of the city. It was against this cultural background that Wright placed the garden as the intersection between the wilderness and civilization, using it as a bridge between nature and culture. Gardens showed people nature, but not a nature that was terrifying in appearance like a long-fanged, green-faced monster, but a nature with which humans could maintain a close and friendly relationship. She started with tending her own garden in the suburbs, thus cultivating her feelings for nature through gardening, and based *The Friendship of Nature* on her own personal experiences, encouraging people to start with gardening and to enjoy a friendly coexistence with nature.

Wright then turned her eyes to the protection of birds. She was a self-taught naturalist whose knowledge of birds came from her observations in the forest, studying in natural museums, and advice given by ornithologists. At the beginning of 1898, Wright was elected president of Connecticut Audubon Society; she served in this position for 26 years, and became a leading advocate for the protection of birds. The Audubon Society's aim was to dissuade people from trading in bird feathers and using them as ornaments, and to stop people killing birds and destroying their nests and eggs. It also encouraged them to learn natural history, study birdlife in the forest and ways of protecting birds, and promote "Bird Protection Day" in the state schools of Connecticut. Wright also edited the Audubon Column in *Bird Lore* magazine and contributed essays to support the protection of birds and explain the characteristics and habits of various birds.

Wright's main achievement in protecting birds was setting up a bird sanctuary. She started with local cemeteries; deeply influenced by Downing, Wright believed that cemeteries should be beautiful quiet places with the atmosphere of the countryside and the landscape of a park, where the relatives of the dead could not only commemorate their loved ones, but also communicate with nature. Using cemeteries as bird sanctuaries could extend people's love for the dead to a love for the natural world, and make them appreciate the temporariness of human life and the everlastingness of nature. In 1907, Wright promoted this idea to the keeper of Fairfield Cemetery. In 1914 she persuaded a rich Fairfield lady to purchase 10 acres of grassland, transfer it to the Connecticut Audubon Society, and make it into a bird sanctuary. As the land was near Wright's holiday house, she could supervise the details of the construction of the sanctuary—from the fences to the stone signs and from the paths in the forest to the water baths for birds—and build an environment that was good for birds to live in. In honor of her contributions, the sanctuary was named after her book *Birdcraft*: Birdcraft Sanctuary.

Birdcraft Sanctuary became a model for protecting wildlife in America. In September 1923 Edith Roosevelt, the former first lady of America, was looking for a place to set up a memorial to her late husband. She came to Fairfield to have lunch with Wright and asked about the details of the establishment of Birdcraft Sanctuary, and she later set up the 12-acre Theodore Roosevelt Sanctuary in Long Island.

In July 1934, at the age of 75, Wright passed away. In 1993, Birdcraft Sanctuary became a national historic landmark in America. Wright's representative work *The Friendship of Nature* became a well-known work of nature writing, and it is the best monument to this brilliant woman who accomplished great achievements in nature study and bird protection.

## The Friendship of Nature

On opening *The Friendship of Nature*, chapters full of the atmosphere of nature and the scenery of all four seasons present themselves to our eyes: "A New England May Day," "A Song of Summer," "The Loom of Autumn," "A Winter Mood," "The Romance of the Rose," "The Story of a Garden" and so on. The 11 chapters of *The Friendship of Nature* vividly describe the four seasons and the scenery of Wright's hometown, Fairfield. As John Holman, Wright's successor as president of Connecticut Audubon Society says, "No one knew the older Fairfield better than she. Every road and woodland stream—every wild nook—every hill and meadow—all were to her places of enduring charm" (quoted in Daniel J. Philippon's introduction to Wright [1894] 1999, pp. 23–24).

Following the tradition, established by Thoreau and other American nature writers, of using the four seasons as the narrative frame, Wright's *The Friendship of Nature* has a strong natural rhythm, but also its own distinct characteristics. Thoreau's *Walden* ends with spring to indicate the circle of the four seasons, whereas *The Friendship of Nature* starts with spring, then guides the reader in listening to "A Song of Summer," followed by "The Loom of Autumn," and finally "A Mood of Winter."

"A New England May Day" is the first chapter of *The Friendship of Nature* and is also the song of spring. Wright starts brilliantly: "This year April has overslept, March has rudely jostled May, who in confusion takes up April's task, leaving its own for June." Then she explains, "Here in New England, we have no calendar of nature, no rigid law of season, or of growth." She urges the reader to "[c]ome into the garden. The wind blows sharply from the north, where the snow still lies, and the clouds hang low, yet it is May-day and a catbird is singing in the arbour" (p. 31). We then follow Wright to watch a pair of crows making a nest on top of a green pine tree and admire a patch of violets blossoming under the tree: white violets that her grandmothers liked, dark purple violets of English gardens, and also a Russian variety of pansy violets. Along the unique stone walls of New England are rows of hyacinths, and a group of bees are collecting nectar among the flowers. From the garden to the meadow, a concert of bird song greets us: the golden-crowned thrush, wood-pewee, phoebe-bird, the robins, the hermit thrush and other birds: a sea of songs. At the end of this chapter, the author advises us that it is time to go home because "we have gathered violets and daffodils, and we have heard the hermit thrush singing in the lane" (p. 35).

To Wright, spring is a restless season, while summer brings people the peace of fulfilment. One day in mid-July, Wright takes two dogs with her to explore the secrets of birds and to listen to the "song of mid-summer" when the little stream rushes past the roots of the trees. We follow Wright across hills covered with wildflowers, fields growing rye and fragrant herbal gardens, and we come to the land of oaks, locust trees, white birches and beeches. As Wright says,

> Entering the woods on a full summer day, it always seems as if another world of thought, speech, and sense lay open, that the mystic wood-sound, the creaking of interlacing branches, the snapping of twigs, and leaf lapping, might be construed as a language to tell of the tree's life and desires. (p. 67)

The singing of golden-crowned warblers fills the forest by the river, and the forest warblers try a few songs and then retreat into silence. Summer is singing its melodies. We lean against the tree and listen with Wright. "To-day let the world slip, and let us live in a summer reverie" (p. 65).

The earth married the sun and gave birth to four daughters: Spring, Summer, Autumn and Winter. Autumn, under Wright's pen, is a willful daughter and a busy weaver. During the daytime she spins silky threads from the earth, and at night she weaves beautiful brocades so that every dawn there will be different pictures. Autumn scenes are a banquet of beauty: the woodlands change into new clothes every day; small boats of leaves float down the river; the meadow puts on floral dresses of fallen leaves; vines of various colors patiently climb and cling onto the rough stone walls, forming uniquely beautiful scenes along the roads of New England. Flowers are about to go to sleep, leaving cheerful memories in the hearts of the lonely people. Birds are about to fly away, but as Wright hopes, "If only, like the birds, we may keep in our hearts the songs of another season" (p. 112).

Winter is usually a season when everything retreats and it symbolizes death. However, in the heart of Wright, "There is no dead season" (p. 129). She writes:

> Black and white winter, are you both flowerless and songless? It may seem so, but there is both music and color; for the tones of winter are as really distinctive as those of all other seasons. If you search, as you have done each day, in the spring, summer, or autumn, you will find constantly a new beauty, a fresh surprise. (p. 131)

We therefore see a varied and colorful scene: a snow-covered road and stony walls; however, through the white snow there glisten yellow flowers and several dry wild turnips. The barberries in front of the house are coral tinted and a few quails stand by the door. Looking along the small path, the vines of the wild roses are as red as if new blood has been infused into them; small pearly flowers grow on the pinkish vines of raspberries; the specked white beeches soar into the sky; the forest of grey-trunked oaks extend to the distance; on the iron black branches in the orchard are a few jays looking for frozen apples. Wright especially reminds us, "Stop! Here by the wall, in a sun streak, the witch-hazel shows its gold threads; pause as you pass, and listen to its story of the winter-opening flower, of youth in old age" (p. 133). From this we can see that there is no winter, only condensed scenes of grandeur, and the force of nature which reaches beyond the current time.

In Wright's garden, birdsong and flower languages are mixed; or perhaps different flowers have different language symbols. However, in Wright's view, all roses, whether red or white, pink-centered or yellow-centered, with multi-layered or single-layered petals, all sing of love. It is a "changeful melody of many keys" of the same song (p. 33). The wild rose along the edges of fields and rocks means pure love; the musk-rose that only exudes its fragrance at dusk means capricious love; the long-spiked Carolinian rose reminds people, "Dangerous is love"; the eglantine, also called the rose of sorrow, drops its fragrant leaves like tears as if whispering, "I wound to heal"; and "the lover's rose, the free gift of the sun, the air, and the dew, the marriage token of the earth to June." Wright invites the reader warmly, "

> Go into the garden now; do not delay! Roses are sweetest and keep longest when picked with the dew on them.... Here are the roses, here and everywhere; they will almost leap into your arms, fill your basket, your hat, your upturned gown."

Roses not only brings us beauty and fragrance, but also "tender memories" (pp. 47–49).

Wright's garden is different from other gardens. In her view,

> [i]n many of these gardens the flowers are only prisoners, forced to weave carpets on the changeless turf, and when the eye is sated and the impression palls, they become to their owners, who have no part in them, merely purchased episodes. (Wright, p. 91)

Wright's garden, on the other hand, is human and has a soul because nature created it and human wisdom, diligence and care nurtured it, changing it from a desolate wasteland to a place full of birds and flowers, evidencing the harmonious coexistence of humans and nature. This garden is the gardens in Waldstein, set among eight acres of wilderness. Although there is a chapter in *The Friendship of Nature* titled "The Story of the Garden," the whole book is actually the story of the garden: the endless tapestry of scenes of grass, flowers, moss, birds and trees in which humans and nature, city and country all blend into an inseparable eco-environment. The story also extends to Fairfield, and even the whole of New England, because it reflects a rich country life.

In Wright's book, you will see a traditional New England garden: the picket gate under the shade of a giant elm tree, the yellow and white house, and chairs on a pillared porch, a honeysuckle climbing along one side of the porch to meet a prairie rose; a little path through myrtle and green grass leading to the gate, lilac bushes beside the fence. (p. 95) In Wright's own garden, under the arch of spruces, there is a "hermit thrush lane," where hermit thrushes often gather in May; then there is the eye of the garden—a pond of azure water, filled by the stream. It not only helps to quench the thirst of birds in summer, but also coats the land in a light green. What is noticeable is the old queen apple tree in front of the house that "casts its petals every May against the window-panes, like snow blushing at its own boldness" (p. 95). Reading this, we cannot help but think of Thoreau's description of New England, "to shut every window with an apple tree" when taking solitary walks in New England (quoted in Foster, *Thoreau's Country: Journey through a Transformed Landscape*, [1999] 2001, p. 19). The elm also has a very strong symbolic significance. Buell, in his *New England Literary Culture* (1989, p. 307), calls the elm "the peculiar glory of New England."

New England is also well-known for its beautiful coasts. In Wright's heart, they are the garden of the sea. By the coast there are wetlands. A small path goes down gradually and disappears into the endless green marsh.

> What do you see? A cloth of soft warm green stretched toward the sea, a silver thread of water, fold by fold, embroiders it, framing with curious patterns the still, mirror pools. Long bands of dark oak woods lie on either side; before you are some tufts of sand grass, and a few dwarf trees are outlined black against the sky, while beyond the crest the boats slip through the golden haze. (Wright, p. 57)

The coast before dawn is extremely quiet and still: "no wind, not even a ripple, and the boats at anchor are motionless as sleeping swans. A fisherman poles his way up the creek, startling herons among the reeds." Wright exclaims: "Beauty still lingers and the sea is murmuring of its garden, and with its siren voices greets the returning gulls" (p. 60).

Perhaps the reader has already noticed the second person pronouns used in *The Friendship of Nature*. Wright used this narrative style to create an intimate atmosphere between the author and the reader, and between humans and nature.

> Look through the vista before you, over the field and up the hillside, where the tree tops meet the sky; there are the blooming apple orchards, foam-white, or rosy as aurora's fingertips. Flower, bird, man, what a triad these blossoming orchards typify; for man's dwelling are always near his orchards, and in the orchards are the birds. (Wright, p. 39)

From this intimate tone, we feel the interdependent relationship and friendship between man and nature. However, Wright does not lecture us but uses a poetic language and creates a dreamy artistic atmosphere to help us understand the lesson.

> Precious is the solitude and the song of the water thrush, for they soothe the spirit; precious are the orchards, the sunlight, and the home-going cattle, for they warm the heart. The red thrush perching high pours out his voluble song, while the lilacs sway over the wall. Still querying in an elm swings the oriole; is he bird, or flower, or cloud, or the transmigration of all. (p. 43)

Wright's language is brilliant, and charmingly poetic. Her pursuit was persistent and unusual, and her name and spiritual legacy shall live forever. Even now, many years after her passing, interest in her shows no sign of decrease. A new edition of *The Friendship of Nature* was published in 1999, and she was the only female writer included in Daniel J. Philippon's *Conserving Words: How American Nature Writers Shape the Environmental Movement* (2004). This first book on the close relationship between nature writers, their works and the environmental protection follows five influential writers and their works that spurred on the establishment of environmental organizations and the development of the environmental movement, to show how literature spreads the environmental concepts and thus leads to positive human action. Wright, through her words and actions, was successful in helping us realize the pressing need for our society to protect those sublime, healthy and beautiful things in nature.

## Lecture 22. LaBastille: Dreams and Reflections on Black Bear Lake

"The freeze-up on Black Bear Lake is a prelude to winter. The freeze-up is a prelude to hardship. The freeze-up is a prelude to loneliness." This is the beginning of Book I of *Woodswoman*, first published in 1976, by American nature writer and ecologist Anne LaBastille (1935–2011). *Woodswoman* describes her experience of survival in the Adirondacks in northeastern America.

I first learned of *Woodswoman* when I was researching for my PhD and writing about Thoreau. While searching for data on Thoreau's influence on modern American life, I learned that a modern American woman named Anne LaBastille left the modern world, built a log cabin beside a lake in the Adirondacks, and lived there for several decades. She wrote about her personal experiences, and this would become her book *Woodswoman*. By 2003, LaBastille had already published four books in her *Woodswoman* series. First published in 1976, then in 1987, 1997 and 2003, they chronicle the author's experiences, feelings, dreams and reflections while living in the wildness for more than four decades. This "modern Walden" series is very popular and has been reprinted many times.

LaBastille was born and grew up near metropolitan New York. She studied at Cornell University and received her PhD in wildlife ecology there. In 1964, after her divorce from her husband, she purchased 20 acres of land in the Adirondack Mountains and built a log cabin there, on the shore of Black Bear Lake, that could only be reached by a small boat or on foot. From then on, the cabin, without electricity or sewage system, became her new home. The copy of *Woodswoman* that I have—Book I of the series, subtitled "*Living Alone in the Adirondack Wilderness*"—describes the first 10 years of the author's life in the cabin.

However, LaBastille's life and her *Woodswoman* books are not simply a reproduction of Thoreau's life and his *Walden*. Thoreau only lived beside Walden Pond for two years, while LaBastille lived beside Black Bear Lake from the 1960s to the end of the twentieth century, and also worked hard to protect the wilderness. She did what many people only dream of doing and don't dare to put into practice. *Woodswoman*, as one commentator says, "*Woodswoman* is not simply a book about living in the wilderness, it is a book about living that contains a lesson for us all" ("Remembering" in *The Adirondack Almanack* 2001). Another writer who wrote passionately about wilderness says, "Following the publication [of *Woodswoman*], LaBastille became an instant role model for thousands of young women all across the country. Her story offered evidence that a lonely life in the forest can foster great confidence" ("Remembering"). There is no doubt that LaBastille's life and works convey the message that tranquility is priceless; or in other words how, in a restless and materialistic modern society, to find spiritual value in the wilderness.

## *Wilderness: Refuge of the Heart*

In the spring of 1964, LaBastille's marriage came to an end and her husband gave her his ultimatum to leave before the Fourth of July. This ultimatum fundamentally changed LaBastille's life. As early as in her childhood, living near New York city, LaBastille had been dreaming of living in a little log cabin like Thoreau's. In her high school years, she dreamed of hiking in the mountains and camping near clear lakes. At university she studied wildlife protection, and her professional aim was to observe and study wildlife. Her unsuccessful marriage gave her the impetus to realize her childhood dream: building a little log cabin to solve her problem of homelessness.

After looking around, LaBastille purchased 20 acres of wilderness on the edge of what she called "Black Bear Lake" in the Adirondack Mountains. It was a very isolated area with no electricity or telephone services or vehicular access, which suited her intention of going "back to nature." She called the land where she was building her cabin "West of the Wind" because one evening, after she had chosen the site of her cabin and was sitting in her small boat thinking, a wind blew from the west and made her feel fresh and clear-headed. Even though there was no habitable house yet, she felt she had found a place to settle down, and this led to a sense of relief and excitement. However, the experience of building the log cabin was not so poetic: she had to lay the foundations on the rocky land, and brought in logs via the waterway. She learned to use an axe to chop the wood, and saws to cut it to size. Working hard like the early American frontiersmen for several months, LaBastille's log cabin slowly came into shape. As she said in *Woodswoman*, "Suddenly I had a home. I had a door to open or close to the world. There were windows with peaceful views" (LaBastille [1976] 1991, p. 16). The cabin was about 12 feet by 12 feet, with a corridor going to the back veranda. One of the two windows faced south and the sun, and looked out on the lake, granite rocks and a small area of woodland. The other window faced the west and the fir forest.

LaBastille decided to move into her log cabin on the Fourth of July because to her America's Independence Day also symbolized her personal liberation and independence. She was giving up her former lifestyle and embarking on a completely new life. The first visitor, a single female writer, said to LaBastille, "Your log cabin is as softly outlined, as naturally brown, as proudly retiring, as completely indigenous in character as an Adirondack bear—a real, pre-people, independent bear" (p. 20). LaBastille called the cabin her own new home, the first home that really belonged to her in her adulthood.

However, when the author moved into the cabin, she had mixed feelings of concern and hope. She was concerned about whether living alone in the wilderness would be safe, and whether anybody could hear her calls for help in an emergency. If a thief or a wild animal got into her cabin, what should she do? On the other hand, she was looking forward to reading in the early morning at her desk without being interrupted, and to quiet evenings sitting thinking in her rocking chair in front of the fire, leading a simple life completely uninterrupted by the outside world. What made her happy was that she had moved from homelessness to a home of her own, and she felt joy in this new home.

The first dinner LaBastille made in her new home was rather romantic. She ate it at the boat dock in the fading light of the setting sun, and the surrounding scene was lovely and lively: a blackbird was singing merrily in the bush, two purple finches were singing in a tall spruce, barn swallows were playing across the water of the lake, and trout leapt breaking the surface of the lake from time to time. On the other side of the lake, a mother deer was leading her fawn through the shallow water, splashing around. "My wildlife neighbors seemed oblivious to my presence." That night, however, the author was woken up by the sound of beavers' tails in the lake, and in the morning she was woken by the "drums" of woodpeckers in front of her door. Sometimes a racoon would break in and make a mess of her kitchen, and red squirrels boldly challenged her as if questioning whose land it was. LaBastille felt a bit guilty, "How dare I erect this wooden square nest underneath their leafy round ones?" (p. 23)

LaBastille stayed in the cabin in the woods for decades. She believed that a woman who wished to survive in a cabin in the woods needed to have three qualities: work, health and love. During her years in the woods, she made a living in three ways, as: a guide in the mountains, an ecological consultant, and a freelance writer and photographer. These three professions made it possible for her to enjoy the peaceful tranquility of the mountains and communicate with the outside world at the same time. As she lived close to nature, nourished by the land, and supported by her life experience of wilderness, she needed good health. As for love, she had one short romance similar to the one portrayed in the film *The Bridges of Madison County*, in her seventh year in the cabin. Nick, a nature-loving male teacher, was traveling in the mountains and came upon her log cabin by chance in autumn when wild swans were flying south, a signal that "lonely" winter was about to arrive. Love was once more kindled in LaBastille's heart; however, although it was painful they separated, as LaBastille did not want to leave her home in the Adirondacks, and her lover did not want to leave his teaching position in Alaska. LaBastille devotes a whole chapter to this love: "A Man around the Cabin."

From her experiences in the mountains, it is evident that LaBastille had found a place for her heart in there. Her love had gone beyond the common love between man and woman and become a greater love for the natural environment in which she lived, including the animals and plants. This is because she understood that care of the natural environment—on which humans depended for survival—was not only her mission, but also where her emotions and spirit lay.

## *Forests: Natural Entertainment Park*

LaBastille's love was not confined to words, but rather could be found in a sincere engagement with her playmates and her closest friends. As she says, "My close and constant companions were trees. I became intimately acquainted with every tree inside a 400-foot radius" (p. 43). She also presented to us all kinds of trees with their different sounds, colors and fragrances, and the pleasure and vitality of life she found through her contact with them, all described with the observation of an ecologist, the subtlety of a fine writer and her own personal interest. She noticed that

> In wind, the spruces gave off a somber, deep, sad whoosh, while the pines made a higher, happier softer sough. Balsam firs made a short, precise, polite swishing; red and sugar maples gave an impatient rustling; yellow birches, a gentle, restful sighing. (p. 44)

The sound also varied with the size, shape and thickness of the leaves, and with the direction and force of the winds. All these constituted a beautiful symphony of the forest.

LaBastille found that in the long summer, firs, spruces and pine trees were like gigantic incense sticks, giving off fragrances which filled the cabin inside and outside. She also caught the play of light in the bushes which made the forest assume all kinds of vivid shapes. On a September evening, she watched the setting sun throw its light over her fir woods. The trunks of the firs cast their shadows on the glistening, golden bronze earth, the golden green moss and the bronze fallen logs. She exclaims, "My little forest had become a study in light and shadow, a stained glass window of gold and green panes with black bars, back-lit by the setting sun" (pp. 45–46).

More importantly, the author absorbed the essence of nature through her contact with the trees.

> I took more and more enjoyment from touching the great white pine. One morning, with my arms wrapped around the trunk, I began to feel a sense of peace and well-being. I held on for over fifteen minutes, chasing extraneous thoughts from my mind. The rough bark was pressed hard against my skin. When I stepped away from the white pine, I had the definite feeling that we had exchanged some form of life energy. (p. 56)

It is worth mentioning that in American nature writing, trees and communication with trees have been a constant hot topic. In the nineteenth century, Thoreau often went to the woods and describes a woodchopper who is "cousin to the pine" (Thoreau, Walden, [1854] 2004, p. 129). Annie Dillard says in her award-winning work *Pilgrim at Tinker Creek* ([1974] 1988), "My mind branches and shoots like a tree" (p. 96). Scott Russel Sanders, author of *Staying Put* (1993), describes how one stuffy summer night he could not fall asleep, and decided to get up and go to the yard. He embraced the big tree there and immediately felt consoled because the feeling was just like embracing his old grandmother. Poet Walt Whitman, who had suffered a stroke, said in his prose collection *Specimen Days* how he regularly went to a large oak to receive the medication nature gave him, and he praised the tree for giving humans the will to survive. In these nature writers' eyes, forests and trees are the embodiment of nature, providing a contrast to the restlessness of modern urban life. Trees have not only material value, but also priceless spiritual value. Seeing green and communicating with trees have become an inalienable human right.

Naturally, the change of seasons added infinite charm to the forest where LaBastille lived. One quiet early morning, from the frost-free window, warm light in the east and the smell of the air, she felt that spring was coming and was infinitely cheered. Spring was the best season for her to watch flowers and birds. In summer, the rolling mountains were all covered with green trees, looking like green waves under the sun, and the clear lake reflected inverted images of these beautiful scenes. Summer was the season when she went further into the mountains by boat to seek more rugged natural scenery, and it was also when she repaired the cabin. Autumn was the most glorious season in the Adirondack Mountains. Trees presented their varied colors, and their colorful leaves decorated the mountains. To LaBastille, this was not only a season to climb the mountains, but also to cut wood and to prepare for winter. In winter, as the lake froze, the log cabin was almost completely isolated from the world. Cold winter drove LaBastille inwards and tested her ability to entertain herself. At the small desk below the window facing the lake, she wrote down her thoughts on her life in the forest, including her first Christmas Eve in her cabin, taking walks in the forest accompanied by the moon, the stars, the frozen lake and her shadow on the snow-covered ground.

Animals and human visitors also brought pleasure to LaBastille's life in the woods. Guests to the cabin included nature-loving retired army officers from India, indigenous natural environment protection activists from Guatemala, and also a wildlife protection activist from Nairobi—those who shared her interests. However, the most frequent visitors to her cabin were uninvited and willful animals. Sometimes when the author came back to the cabin, she would discover there was a "farce" going on in her home. One day, a mother raccoon with her three babies had hollowed out the apple pie she had just baked, upturned her flour and sugar jars, left their floury claw prints all around the room, and were swinging merrily on the curtains. When the author swept these naughty visitors out with a broom they would throw her complaining looks, as if to say, "You broke up a good party." LaBastille not only accepted visits from animals, but also paid visits to their homes. She watched the foxes' holes, and watched beavers build dams. She also fell into a bear cave and had the most surprising experience. All this made her exclaim, "How much more exciting and varied my existence would be, and how much more balanced the ecosystem would be, with these true native animals in residence" (p. 200).

LaBastille named the book that recorded her life experience in the Adirondacks *Woodswoman* because her heart was linked to the forest. Fortunately, she found not only emotional support in the forest, but also infinite joy.

## *Loneliness: A Spiritual Stimulant*

Like most nature writers, when LaBastille went into wilderness and enjoyed the peace and quiet, she also experienced undeniable loneliness. One chapter of *Woodswoman* is titled "Endless Loneliness" and describes the first winter she spent in the woods. She writes, "Winter is a lonely season.... Loneliness is a pervasive quality in these mountains" (p. 130). In deep autumn when she saw wild swans flying south, she already felt the arrival of winter in the wind. In winter she tried to make herself busy, either writing or consulting on ecological protection. She says, "It's a conscious battle against the loneliness" (p. 130). However, when she had to go and live in Washington DC for a few months due to work obligations, she found the noisy city life annoying and missed her cabin in the wilderness, concluding that "[a] person can be just as lonely in a city as in the backwoods. Crowds of people are no more of a guarantee against loneliness than are forests of trees" (p. 268). This may explain, to some extent, why many American nature writers go into the wilderness without looking back. Pursuit of peace and quiet and struggle against loneliness are conflicting feelings common to LaBastille and other nature writers which they not only acknowledge, but even extract a certain pleasure from. Loneliness, which allows thinking and is also a spiritual stimulus, is the cost of freedom. In *Woodswoman*, the three hermits in the Adirondack Mountains described by LaBastille prove the value of loneliness.

In order to qualify as a mountain guide, LaBastille visited more than 30 different mountain lakes in the Adirondacks over two summers, under the guidance of an old guide, and also climbed to the top of two of the highest mountains alone. She discovered that three hermits had lived in the mountains before her, two of whom had already passed away, leaving behind the remains of their cabins. LaBastille describes the remaining old hermit thus: "I sensed that here was a man who had passively retreated from the world. Torpid as a woodchuck in winter" (p. 95). The second hermit had been described as "hard as Laurentian granite and tough as Adirondack spruce" when he was alive (quoted in LaBastille, p. 104). The third hermit had been called the "Hermit Mayor of Cold River City—population of one" (quoted in LaBastille, p. 109), because for many years he had lived alone by the Cold Water River. In his diary he recorded: "April 29 I walk about my Town Hall and City. Robins and Juncoes, Chipmonks and Sparrows. No gas stations, no dogs on every corner, no church bells in my ears. What freedom alone in natural primitiveness" (quoted in LaBastille, p. 108). In post-industrial America, high tech has brought convenience and comfort, but also spiritual depression and indifference, and people have become slaves of time. American nature writer Edward Abbey went into the deserts in the West alone. He wrote books expressing his wish that the quiet, empty desert become a counterforce against the blind development of humans, and calling on people to love space as they love time. Similarly, the love that LaBastille's three hermits showed is also a love of free space. LaBastille's own choice to live alone in the mountains is not only a love for free space; her experiences show that going out is also going inward to the self, and demonstrate why the heart needs the nourishment of wild nature.

One late autumn evening, after the painful ending of the deep romance with Nick, LaBastille sat quietly in her cabin as the sun set. Looking at the light of the September sunset on the log walls and on the surrounding mountains, a solemn feeling better than any you would have felt in an old European cathedral suddenly developed. She exclaimed:

> I felt as though I was sitting at the very core of something. From this core, this hushed cabin room, my mind was raying out like shafts of golden sun all over the Adirondacks. I could plainly "feel" them, their ancientness, solidarity, reliability....It was like living with a grandparent who could always be counted on. (p. 239)

At this time, LaBastille's loneliness was transformed into a sacred feeling, a feeling that was in cognizance with ancient and everlasting nature, a sense of belonging to this world. In modern society where "everyone is an island," this sense of belonging is precious and difficult to find.

In the last chapter of the book, LaBastille provides a clear explanation for living in the quiet cabin for so many years. The cabin was the source of her inspiration, the center of her existence. It gave her not only peace, health and the ability to be close to nature, but also resourcefulness and creativity. Most importantly, she concludes, "I share a feeling of continuity, contentment, and oneness with the natural world, with life itself, in my surroundings of tall pines, clear lakes, flying squirrels, trailless peaks, shy deer, clean air, bullfrogs, black flies, and trilliums" (p. 276).

The three remaining books in the *Woodswoman* series are all based on LaBastille's personal experiences during more than 4 decades living in the wilderness of the Adirondacks. In the second book, *Beyond Black Bear Lake*, published in 1987, she describes building a second, even smaller cabin in a place even wilder than Black Bear Lake. She uses recycled materials to reduce costs and for the sake of the environment, and names the second cabin "Thoreau II," hoping to experience the same kind of life as that of Thoreau at Walden Pond. However, you cannot go back in time, and LaBastille cannot find the quiet, uninterrupted life of the nineteenth century. In books III and IV of the *Woodswoman* series she tells of her conflict and stress at having to work as a freelance writer and an environmental protection consultant, and take on academic activities and teaching, despite being determined to retreat to a life in the wilderness. In the end she purchases a farm with a farmhouse, electricity and telephone, but still within the Adirondack Park, by the side of Lake Champlain.[7] Her original cabin near the lake becomes her seasonal retreat; however, she writes that she kept it because it was her spiritual refuge and she "wanted tranquility and to be close to the wellspring of life" ("Anne LaBastille"). In 2008 LaBastille left her cabin due to illness, and she passed away in July 2011. In an obituary in the *New York Times* she is described as "Woodswoman of Adirondacks" (Hevesi 2011). Although LaBastille has passed away, she and her cabin have, to a certain extent, entered eternity. Nature is everyone's final resting place. LaBastille, at the end of *Woodswoman* I, says of the hardship she experiences in the cabin when entering old age,

> these trials will give me depth and stature so that in old age I can be like my big white pines—dignified, lending beauty to the surroundings, and lifting their heads with strength and serenity to both sun and storms, snowflakes and swallows. (p. 277)

## References

Buell, Lawrence. 1989. *New England Literary Culture: From Revolution through Renaissance.* Cambridge: Cambridge University Press.

Buell, Lawrence. 1996. *The Environmental Imagination: Thoreau, Nature Writing, and the Formation of American Culture.* Second printing, Cambridge, MA: Harvard University Press.

7 Adirondack Park, built by the state of New York, is the largest park in America. It contains 3,000 lakes and 30,000 miles of river.

Cooper, Susan Fenimore. (1850) 1998. *Rural Hours*. Eds. by Rochelle Johnson and Daniel Patterson. Athens, GA: University of Georgia Press.

Dillard, Annie. (1974) 1988. *Pilgrim at Tinker Creek*. New York: Harper Perennial.

Emerson, Ralph Waldo. (1903) 1979. Centenary Edition, *The Complete Works of Ralph Waldo Emerson*. Vol. I: *Nature Addresses and Lectures*, 2nd edition of the Centenary Edition. New York: AMS Press.

Foster, David R. (1999) 2001. *Thoreau's Country: Journey through a Transformed Landscape*. Cambridge, MA: Harvard University Press.

Hevesi, Dennis. 9 July 2011. "Anne LaBastille, Advocate, Author and 'Woodswoman' of Adirondacks, Dies at 75." *The New York Times*. http://www.nytimes.com/2011/07/10/nyregion/anne-labastille-environmentalist-dies-at-75.html.

LaBastille, Anne. (1976) 1991. *Woodswoman: Living Alone in the Adirondack Wilderness*. New York: Penguin Books.

Leopold, Aldo. (1949) 1987. *A Sand County Almanac: And Sketches Here and There*. New York: Oxford University Press.

Older, Julia, ed. (1916) 1997. *Celia Thaxter: Selected Writings*. Hancock, NH: Appledore Books.

Philippon, Daniel J. (2004). *Conserving Words: How American Nature Writers Shaped the Environmental Movement*. Athens: University of Georgia Press.

"Remembering Anne LaBastille's Environmental Record." 4 July 2001. *The Adirondack Almanack*. http://www.adirondackalmanack.com/2011/07/rememberinganne-labastille.html.

Rutledge, Lyman V. 1965. *The Isles of Shoals in Lore and Legend*. Barre, MA: Barre Publishers.

Sanders, Scott Russell. 1993. *Staying Put: Making a Home in a Restless World*. Boston: Beacon Press.

Thaxter, Celia. 1873. *Among the Isles of Shoals*. Cape Cod: Wake-Brook House. Facsimile reprint of the first edition.

Thaxter, Celia. (1894) 2001. *An Island Garden*. Boston: Houghton Mifflin, with an introduction by Madame Tasha Tudor.

Thoreau, Henry David. (1851) 1993. *A Year in Thoreau's Journal: 1851*. Penguin contributor: H. Daniel Peck.

Thoreau, Henry David. (1854) 2004. *Walden*. Boston: Shambhala.

Walker, Cheryl. 1992. *American Women Poets of the Nineteenth Century: An Anthology*. New Brunswick, New Jersey: Rutgers University Press.

Wikipedia. 16 May 2013. "Anne LaBastille." http://en.wikipedia.org/wiki/Anne_LaBastille.

Williams, Terry Tempest. (2001) 2002. *Red: Passion and Patience in the Desert*. New York: Vintage Books.

Woolf, Virginia. 1989. *A Room of One's Own*. Orlando, FL: Houghton Mifflin Harcourt.

Wright, Mabel Osgood. 1894. *The Friendship of Nature: A New England Chronicle of Birds and Flowers*. New York: Macmillan and Company.

Wright, Mabel Osgood. (1894) 1999. *The Friendship of Nature: A New England Chronicle of Birds and Flowers*. Baltimore, MD: The Johns Hopkins University Press.

7

# Land Ethics in Nature Writing

## Introduction

When Mary Austin passed away in 1934, American society was facing the end of a long period of transition. A century earlier, America had been a pre-industrial country. Building of the railroad was just about to commence, and the vast wilderness was quietly asleep. However, by the end of the nineteenth century Thoreau's wish to "go westward" was no longer a dream, and the so-called pioneering or westward movement had drawn to a close.

By 1934, with rapid industrialization and the development of various methods of transportation, some American nature writers believed that the American wilderness had already become fragmented. If early Americans turned inward to self or to God to seek spiritual consolation or strength when faced with the harshness of nature and feeling helpless, modern Americans created another god—technology—a god that seemed to be omnipotent, even, as English professor Harold Fromm says in his 1996 essay "From Transcendence to Obsolescence," "putting nature in a cage" (in Glotfelty and Fromm, p. 35).

To illustrate the increasing distance between humanity and nature brought about by technology, Fromm (in Glotfelty and Fromm, p. 33) quotes a student who told him that she had no desire to venture out into the country to "enjoy

Nature" when she could see all the trees she wanted on color TV. Scott R. Sanders, in a 1996 essay titled "Speaking a Word for Nature" (in Glotfelty and Fromm, p. 192), describes a scene in which he climbs to the top of a mountain filled with wildlife and natural scenery with the sky shimmering with the colors of dusk, birds singing evening songs, and breezes bringing waves of fragrance of the pines; then he hears a noise that is in distinct disharmony with the natural scenes. It turns out to be a group of young people watching TV in their van. Sanders is amazed that when people take the trouble to go to nature, they are still incapable of being close to it.

Fromm says that with the disappearance of wild nature and the satisfaction of people's material needs, their minds seem to have reached a new autonomy and independence based on the control and wilful use of nature, and this has cut off the roots it relies on for absorbing nourishment from the earth. He explains the consequences of this situation thus: "Man's nurturing environment threatens to stop nurturing and to start killing" (Fromm, p. 34). The side effects of technology and humans' attitude that they are master of nature have not only caused spiritual harm to people, but also pose a direct threat to the earth that we rely on for survival. In the twentieth century, as technology advanced, the conflict between humanity and nature became all the more prominent. This forced people to think carefully about and try to deal with the relationship between humanity and nature, and as a result, American nature writing became more active.

In his *This Incomperable Lande* (1989), Lyon discusses the characteristics of the development of American nature writing in the twentieth century. He believes that from William Bartram's time to today, American nature writing has undergone no big changes in the subject matter. However, nature writers of the twentieth century have to respond to the everchanging environment; this requires of them "a depth of response that is fundamentally new," and they "have had to keep up with a tremendous increase in natural history knowledge" (Lyon, p. 74). If the thinking and focus of nature writing before the twentieth century was limited to nature and self, or nature, individual thinking and behavior, the nature writing of the twentieth century started to show the relationship between nature and humanity and the harmony between humanity and ecology.

We can say that twentieth-century nature writing has two distinctive features. Firstly, it includes not only pursuit of beauty in literature, but also an ethical view of land and nature—in other words, it describes the ethical relationship between humanity and nature from the perspective of aesthetics. Secondly, it particularly emphasizes a "sense of place." To the twentieth-century nature writers, without the geographical support of place there would be no spiritually supportive base; therefore, most nature writing focuses on lovingly engaging in life at a particular place and fusing into one with the land under one's feet.

As a result, nature writing in the twentieth century was practised by a large group of people from different geographical environments with a common topic. Its force and influence and the depth and range of the works were all unprecedented. Aldo Leopold had a deserted farm in the middle of Wisconsin as his background when writing *A Sand Country Almanac* (1949), which taught people how to treat the land well and take good care of it from a moral and ecological perspective. Rachel Carson's *Silent Spring* (1962) describes the threat that the abuse of chemical pesticides poses to both the ecological environment and humanity. Edward Abbey's *Desert Solitaire* (1968) conveys modern society's restless yearning for peace and quietness, as well as his thinking about wilderness and modern civilization. Annie Dillard's *Pilgrim at Tinker Creek* (1974) records the author's pilgrimage to nature around Tinker Creek in the Blue Ridge Mountains of Virginia. Barry Lopez's *Arctic Dreams* (1991) describes his experience of exploring the Arctic, and the human need for two kinds of scenery, external and inner. Terry Tempest Williams's *Refuge* (1991) describes three generations of women who, all suffering from breast cancer as a result of exposure to American nuclear testing, sought spiritual solace at Salt Lake. Scott R. Sanders's *Staying Put* (1993) raises a common theme in nature writing of how to find a piece of land of one's own in a crowded and restless world.

However, these twentieth-century nature writers also had close links with their predecessors. John Muir was called "the last disciple of Concord" and Edward Abbey a "modern Thoreau," while *Pilgrim at Tinker Creek* (1974) was regarded as a more courageous *Walden*. Naturally, they also went beyond their predecessors' views. Leopold introduced the concept of "land ethics" and called on people to cultivate an "ecological conscience"; Abbey presented a new model of harmony between man and nature: opposition, compromise and balance; Williams pointed out that "Wilderness is a feeling. If we love wilderness as we love our lovers, we will not let it leave us" (Williams, [2001] 2002, p. 76); and Sanders advocated connecting with the land under one's feet and encouraged people to rediscover their position in this world.

In a word, for contemporary nature writers the relationship between humanity and nature is no longer one between "me and it," but rather between "me and you." They no longer believe in a self, only a self that has merged into nature in a particular place. They no longer believe in the "survival of the fittest," but in "coexistence." This chapter will discuss Aldo Leopold, Edward Abbey and Annie

Dillard as representatives, and through them and their works show the trends, common features and characteristics of twentieth-century nature writing.

## Lecture 23. Leopold: Seeking Land Ethics

In the first half of the twentieth century in Wisconsin, the state that produced John Muir, another influential nature writer appeared: Aldo Leopold (1887–1948). Like Muir, Leopold had a lifelong fascination with the wilderness, and like Muir, he fought throughout his life for the protection of the natural environment. Unlike Muir, Leopold was an academic, a professor at Wisconsin University, and had an even richer background in scientific knowledge. He introduced the significant theory of "land ethics" and as a result has been called the "father of ecological ethics" and a "professional natural resources conservationist."

Leopold carried on Thoreau's tradition, wishing to "speak a word for Nature." In his representative work *A Sand County Almanac* (1949)[1] he introduced the concept of "land ethics," calling on people to assume mankind's obligation to protect the natural environment that we rely on for survival. His conduct criteria of land ethics were that anything that helps to protect the ecological environment is right and anything other than that is wrong. He wanted people "to think like mountains," that is to think from the perspective of ecology—of the relationship between man and nature, of maintaining the health of the land, and of cultivating an "ecological conscience." In the mid-nineteenth century, Thoreau had gone to the woods to seek his relationship with nature and to experience a simple life and spiritual purification. More than a century later, Professor Leopold of Wisconsin University purchased a farm that was almost completely desertified land due to inappropriate use and was nicknamed "Sandy County." His purpose was to take his family there to plant grass and trees on weekends and holidays, so that they could experience and demonstrate how to cure the land of its wounds and restore humanity's relationship with nature. He died of a sudden heart attack while helping his neighbor to fight a bushfire there in 1948.

A year after Leopold passed away *A Sand County Almanac*, with its main theme of land ethics, was published by Oxford University Press and attracted a

1 There are two Chinese translations of this book. The first, with the Chinese title of *Shaxiang Nianjian* (Sand County Year Book), was published in 1997 by Jilin People's Publishing House. The second, with the Chinese title of *Shajun Nianji* (Sand County Yearly Records), was published in 2000 by Joint Publishing.

great response. More than a million copies were printed, and in 1978 Leopold was posthumously awarded the John Burroughs Medal. *A Sand County Almanac* keeps Leopold alive in the heart of humanity. In his book *Speaking for Nature* (1980), Paul Brooks cites Roderick Nash describing Leopold's book as "the bible of the surging environmental movement of the 1960s and early 1970s," and cites another writer as saying that Leopold created in eloquent language "an entirely new way of defining conservation" (quoted in Brooks, p. 263).

## *The Father of Ecological Ethics*

Leopold was born in 1887 in Burlington, Iowa. From his youth he learned about nature in his hometown beside the Mississippi River. His parents loved outdoor activities and nurtured and inspired his interest in identifying the various birds that were native to his hometown. Perhaps it was this childhood interest in birds and his love of outdoor life that led him to the School of Forestry at Yale University, from where he graduated in 1908. Leopold's life was associated with two regions of America: the southwest, where he started work as a forester, and mid-Wisconsin, where *A Sand County Almanac* is based. In 1909 Leopold was employed by the newly established U.S. Forest Service, and subsequently spent more than a decade working for the Service in Arizona and New Mexico. During this time, he declined an attractive public relations position in the National Forestry Office in Washington DC, as he was determined to effect significant changes in the protection of wildlife in the southwestern mountain area. Leopold had both professional skills in forestry and the vision of an ecological scholar. He wrote wildlife protection manuals for his areas, worked with his colleagues to implement wildlife protection laws, and sought ways to increase the numbers of trees and wild animals. He also published articles on wildlife management, and soon attracted national attention. The Permanent Wildlife Protection Fund awarded him a medal, and President Theodore Roosevelt commended him for "setting an example to the whole country" (quoted in Nash 1968, p. 184).

Leopold's experiences in the southwestern mountains showed him that nature was far more complex than people had imagined: it was a gigantic organic body, and all living things were intricately interrelated. To him, a countryside without wild animals would be a spiritual vacuum. It was during this period of his life that the seeds of environmental protection and of the concept of land ethics that he later proposed were sown in his mind. In 1924 he was transferred to Wisconsin by the Forest Service. After working in the Forestry Office there for a few years, he became the first professor of game management at the University of Wisconsin. According to James Hart in *The Oxford Companion to American Literature* (1995, p. 368), Leopold's book *Game Management*, published in 1933, was regarded as "a landmark conservation textbook," while Brooks (p. 260) describes it as "a work whose literary and philosophic overtones have extended its influence beyond the confines of the profession, while it remains a standard text to this day."

Two years later, Leopold purchased an abandoned farm near Wisconsin River. On this land of about 100 acres, the only structure was a hunting lodge converted from a cattle shed that was usually referred to as "the shack." From then on, on weekends or holidays, he would take his family to this abused and ruined land to plant grass or trees there, staying in the shack while there. Susan Flader, in *The Sand County of Aldo Leopold* (1973), comments:

> The shack years were an experience in the slow sensitizing of people to land, the evolution of a sense of country. The shack originally acquired as a hunting camp soon became a "weekend refuge from too much modernity"...The more woodcock nests they discovered, the more trees and shrubs, grasses and flowers they planted, the more chickadees and nuthatches they got to know—in short the more familiar they became with the place—the more they found to anticipate, to ponder, and to marvel at.... *A Sand County Almanac* is eloquent testimony to the meaning and value of the experience. (Flader, p. 43)

The Wisconsin years were the critical period for Leopold in forming his wilderness values and his concept of land ethics. The Forestry Service's product lab, along with his teaching and research at Wisconsin University, provided opportunities for him to reflect on the value of wilderness and also convinced him that people must stop looking at nature only from the point of view of economic growth and at land only in monetary terms. "He felt that what was at stake in keeping some wild land was the quality of American life—the welfare of the nation beyond its material needs" (Nash, p. 187). Leopold's work on the farm made it possible for him to personally experience, explore and understand humanity's duties and obligations toward land, that is, land ethics. Leopold sacrificed his own life in the fight against a bushfire, which shows that he practised the land ethics and retained the ecological conscience that he advocated. As the "father of ecological ethics," Leopold taught us—through not only theory, but also practice—that humans are duty-bound to protect the health of the land and must cultivate an ecological or environmental conscience.

Leopold's representative work, *A Sand County Almanac*, has three parts. The setting for the first part is the Sand County farm in Wisconsin. The author follows the calendar months and describes the simple and leisurely country life there and how his family plants, sows and tries to rehabilitate the land. The second part is set against a broader traveling background. Leopold leads the reader out of the farm, first to other wilderness in Wisconsin, and to his hometown in Iowa, then to Arizona and New Mexico where he first worked, and southward as far as Mexico, then northward to Ohio and Utah, eventually crossing the border into Canada. In this part, the author gives an overview of the natural history of each area and the scenery there, also pointing out the damage humanity has done to the ecological system—the wilderness that the great majority of Americans had not had a chance to see before it disappeared. This loss causes sadness and anger; just as people have the right of free speech, they should also have the right to enjoy blooming plants. Following this is the third and final part of the book, which has a focus on land ethics. Here the author points out that the land is a complex organic body composed of an intricate food chain and circulating energy. The continued functioning of the land depends on cooperation and competition among its various components, and humanity is only one of the components. However, humanity often puts itself in the wrong position, regarding itself as the master of the land rather than a citizen of the land's community, and the result is abusive behavior toward the land with inevitable consequences. For this reason, we need new sets of values and concepts with regard to land health and the restoration of a harmonious relationship between humanity and the land, i.e. land ethics. These land ethics mean that humanity must have a sense of duty and obligation toward the land, and an ecological conscience.[2]

Sherman Paul, in his *For Love of the World* (1992), points out that the three parts of *A Sand County Almanac* can be seen as a Thoreau chapter, a Muir chapter and a Leopold chapter. The first part consists of records of nature following the seasons, reminding the reader of Thoreau's *Walden*; the adventures and conservation of nature in the second part bring Muir's work to mind; and in the third part Leopold synthesizes these previous two. Here he "brings both forward in the uncompromising upshot of the conclusion, where his divergence from the managerial conservation of Gifford Pinchot, in which he had been trained, also shows the extent of his education" (Paul, p. 46).

Leopold occupies an important position in nature writing because he proposed an unprecedented theory based on twentieth-century science: he regarded

2 For information on this book refer to: Fritzell, *Nature Writing and America*, pp. 193–200 and McClintock, *Nature's Kindred Spirits*, pp. 23–45.

humanity's need for wilderness as a human right; he examined humanity's relationship with nature from a biological point of view and asked people to think like mountains; he was the first to bring ethics into humanity's relationship with nature, And he himself practised what he preached on his Wisconsin farm, setting an example of ecological conscience for his fellow Americans.

In American nature writing, the view and understanding of wilderness have remained constant. Emerson believed that "[i]n the woods we return to reason and faith" (Emerson, "Nature" (1836) in Centenary Edition, *The Complete Works of Ralph Waldo Emerson*. Vol. I: *Nature Addresses and Lectures,* (1903) 1979, p. 10); Thoreau that "[i]n wilderness is the preservation of the world" (Thoreau "Walking" in *Portable Thoreau* [1947] 1977, p. 609); and Muir that "[i]n God's wilderness lies the hope of the world" (Muir, [1938] 1979, p. 317), while Leopold believed that wilderness offers a life of a higher quality than material enjoyment, and that access to it is an inalienable human right like that of free speech. In the foreword to *A Sand County Almanac*, Leopold speaks about his understanding of the wilderness in a light and intimate way: "There are some who can live without wild things, and some who cannot. These essays are the delights and dilemmas of one who cannot" (p. vii). He writes,

> Like winds and sunsets, wild things were taken for granted until progress began to do away with them. Now we face the question whether a still higher 'standard of living' is worth its cost in things natural, wild, and free. For us of the minority, the opportunity to see geese is more important than television, and the chance to find a pasque-flower is a right as inalienable as free speech. (Leopold, p. vii)

Leopold's regard of the enjoyment of wilderness as a human right has a twentieth-century flavor. During his lifetime, America experienced the economic depression and two world wars, and all sorts of sociopolitical terms appeared, such as "consume," "progress" and "living standards." However, Leopold saw that in the economic pursuit of consumption, progress and high living standards, people living in a highly materialistic society would face a yet more serious problem: a sick and groaning land. Paul (p. 39) compares Leopold's words on the issue (as recorded above) to President Lincoln's Gettysburg Address during the American Civil War, and believed that the question raised by Leopold as to "whether a still higher 'standard of living' is worth its cost in things natural, wild and free" effectively "reads in terms of the Civil War the present irreconcilable (irreconciled) conflict of man and land." Paul sharpened the focus of this issue. We can think of it like this: The difference between these two types of war is that the "civil war" with nature will not be resolved through bloodshed, but through humanity's cultivation of land ethics.

## *Think Like a Mountain*

Leopold developed his theory of land ethics—including learning to "think like a mountain," which means thinking from the perspective of ecology and of maintaining the health of the land—gradually. "Think like a Mountain" is the title of a chapter of *A Sand County Almanac* which starts with a description of a wolf's howling in the mountains.

> A deep chesty bawl echoes from rimrock to rimrock, rolls down the mountain, and fades into the far blackness of the night. It is an outburst of wild defiant sorrow, and of contempt for all the adversities of the world.
>
> Every living thing (and perhaps many a dead one as well) pays heed to that call. To the deer it is a reminder of the way of all flesh, to the pine a forecast of midnight scuffles and of blood upon the snow, to the coyote a promise of gleanings to come, to the cowman a threat of red ink at the bank, to the hunter a challenge of fang against bullet. Yet behind these obvious and immediate hopes and fears there lies a deeper meaning, known only to the mountain itself. Only the mountain has lived long enough to listen objectively to the howl of a wolf. (1949, p. 129)

Leopold then describes how, when he was young, he killed a mother wolf in the mountains due to ignorance. He could never forget "a fierce green fire dying in her eyes" and writes:

> I realized then, and have known ever since, that there was something new to me in those eyes—something known only to her and to the mountain. I was young then, and full of trigger-itch; I thought that because fewer wolves meant more deer, that no wolves would mean hunters' paradise. But after seeing the green fire die, I sensed that neither the wolf nor the mountain agreed with such a view. (p. 130)

It was from then on that Leopold began to look at the world from a different perspective. He regarded the wolf as a symbol of wildness, and felt a sense of mystery about the place the wolf held in the natural system. He realized that it was wrong to kill a certain animal just because it did not fit into the wilderness imagined by humanity. He saw that we humans, as visitors from an industrialized world to the natural world, needed to learn the ways of the land and of life, and to regard nature as a gigantic organism. We needed to let go of the mistaken belief that we could enter the natural world lightly and rebuild a system that we thought was ideal, resetting the rules of nature to serve man's own interests. We must think like a mountain, from the perspective of keeping the ecological balance, and relearn the rules of nature, cautiously entering the whole ecosystem and moving in synch with the pulse of nature.

Starting from "thinking like a mountain," Leopold gradually presented the concepts of ecological conscience and land ethics. He believed that building a road was much easier than thinking about what the country really needed; in a rapidly industrialized America, what people needed was a land ethic. Leopold used his ecological scientist's knowledge to explain the principle of the land pyramid and the food chain, and explained that humanity was only one of the components of the land community, which was made up of soil, rivers, plants and animals. All members of this community had their respective positions and were all interdependent. In the long journey of biological evolution, humans were fellow travelers with other biological beings, and regardless of man's designs, nature would always go its own way. In order to be in step with nature, man must merge with it, maintaining a strong sense of the whole community, and could not afford to deviate from the way of nature. In order to prove that man was part of the land community, we must first of all learn to respect and care for others in this community.

Thus, Leopold extended human ethics to the natural world and human survival to the way of nature. "A *land ethic,* then, *reflects* the existence of an *ecological conscience*, and this in turn *reflects* a conviction of individual responsibility for the health of the land" (p. 221). Through reading Leopold we realize that we abuse the land because we regard it as our commodity, and that when we regard it as belonging to a broader community of which we are a part, we will start to view and use it with love and respect. Humanity's relationship with the land is just like one human being's relationship with another: it involves ethics.

Leopold practised his land ethics on the farm in Wisconsin, on its wounded land. Two American scholars compared his practice with those of Thoreau and Muir. Lawrence Buell, in *The Environmental Imagination* ([1995] 1996), points out that for Thoreau, Walden Pond's natural environment provided him first and foremost with a laboratory in which to experiment with self-reliance and attempt to pursue purification of the self through living a simple life. In contrast, Leopold had no interest in self-reliance; for him, the most important thing was the environment. The difference between these two attitudes is also shown in their practical actions—Thoreau's at Walden Pond, and Leopold's on his farm. From the very beginning, Leopold regarded his practice at the farm as a long-term rehabilitation of the abused land. Buell comments with admiration, "Planting several thousand trees annually is not the usual kind of family outing" (p. 174).

Paul's collection of essays on nature writers includes one titled "The Husbandry of the Wild" in which he describes Leopold's theory and actions. Paul comments that when Thoreau went into the woods near Walden Pond, he was looking for a connection between himself and nature, seeking to establish a culture of the self and the independent spirit. More than a century later, as a trained scientist, Leopold went to his farm to restore both humanity's relationship with the land and the land's health. Although Leopold inherited the spiritual legacy of Thoreau and Muir, he went beyond these two predecessors. When observing nature, "he focused more on nature rather than self." In theory, he raised the issue of changing humanity's view of the land and the necessity of "changing from ego-thought to eco-thought." He was seeking a position of defence of nature, rather than of using nature. More importantly, he advised a correlative action, "not only the preservation of the wilderness but the husbandry of the wild, the wildering" (Paul, p. 52).

Leopold therefore put forward a theory that Thoreau and Muir did not raise, and took actions that they could not have taken. His writing and actions were great contributions to western civilization. His wilderness was no longer the terrifying and roaring wilderness of traditional western culture, but a joyful wilderness and a measure of whether human society was healthy or not; and his going to the wilderness went beyond the narrow range of curing personal physical and mental wounds of Thoreau's time, expanding into a much broader field to cure the wounds of the land and restore its health. His theories have helped us understand the importance of eco-balance and encourage us to turn from self to nature, from self to ecology.

Although Leopold developed land ethics and practised them, we cannot help but feel moved by his artistic eye and his poetic description when we open his *A Sand County Almanac.* Like many of his contemporary poets and literary writers, he held very clear beliefs and views about the land, with romantic imaginings and deep affection for it, as he believed that "the detection of harmony is the domain of poets" (Leopold, p. 153).

In the opening chapter of *A Sand County Almanac,* we immediately see his simple, clean writing style. In the chapter titled "Think like a Mountain," we see not only the solemnity of a philosopher, but also the imagination of a poet. However, the most impressive aspect of the book is still its beauty as a whole, the rhythm of which resonates with the pulse of nature, a feature of nature writing.

> A dawn wind stirs on the great marsh. With almost imperceptible slowness it rolls a bank of fog across the wide morass. Like the white ghost of a glacier the mists advance, riding over phalanxes of tamarack, sliding across bog-meadows heavy with dew. A single silence hangs form horizon to horizon. Out of some far recess of the sky a tinkling of little bells falls soft upon the listening land....
>
> ...At last a glint of sun reveals the approach of a great echelon of birds. On motionless wings they emerge from the lifting mists, sweep a final arc of sky, and settle in clangourous descending spirals to their feeding grounds. A new day has begun on the crane marsh. (p. 95)

The cranes have been arriving in Wisconsin every spring since the Ice Age. After Leopold's description of their arrival, he asks one question:

> To what end? Out on the bog a crane, gulping some luckless frog, springs his ungainly hulk into the air and flails the morning sun with mighty wings. The tamaracks re-echo with his bugled certitude. He seems to know.... Our ability to perceive quality in nature begins, as in art, with the pretty. It expands through successive stages of the beautiful to values as yet uncaptured by language. The quality of cranes lies, I think, in this higher gamut, as yet beyond the reach of words. (p. 96)

Leopold presents a picturesque and poetic scene to demonstrate wild animals' superior understanding of the ecosystem, and to show the separation of humanity from nature in the modern society.

Leopold's narrative strategy of following the steps of nature is also moving. In the first chapter of *A Sand County Almanac* he follows the calendar, recording outdoor observations by the month. These observations—like Thoreau's in *Walden* and Muir's in *My First Summer in the Sierra*—are based on his diary entries, full of scientific facts and poetic descriptions.

In February, early spring, finding an oak that has been split and felled by thunder and lightning, he traces its history from its rings and its many connections with its natural environment. In September, at misty dawn he listens to bird songs that are different from in spring. In the second chapter is an essay he wrote for a monument to the pigeon erected in Wisconsin State Park, and here the sense of season is especially strong.

> Perched like a duck hawk on this cliff, it will scan this wide valley, watching through the days and years. For many a March it will watch the geese go by, telling the river about clearer, colder, lonelier waters on the tundra. For many an

> April it will see the redbuds come and go, and for many a May the flush of oak-blooms on a thousand hills. Questing wood ducks will search these basswoods for hollow limbs; golden prothonotaries will shake the golden pollen from the river-willows. Egrets will pose on these sloughs in Augusts, plovers will whistle from September skies, hickory nuts will plop into October leaves, and hail will rattle in November woods. But no pigeons will pass, for there are no pigeons, save only this flightless one, graven in bronze on this rock. Tourists will read this inscription, but their thoughts, like the bronze pigeon, will have no wings. (Leopold, pp. 110–111)

Commenting on Leopold's writing style, Buell (p. 40) describes Leopold's strategy of letting the reader become completely immersed in an idyllic atmosphere, then making a controversial comment, then continue with the previous intimate affection for the land; a descriptive strategy that provides an experience of beauty, and as a result "[b]eauty becomes a form of action." As Buell says, when we read Leopold's works, we not only appreciate his rich philosophical points, but also feel his experience of beauty. Beautiful images leave a deeper impression than statistics, because they have a direct affective effect on the reader's heart. It is through descriptions of various moving experiences that nature writing seeks to gradually persuade people to adopt a new attitude toward nature.

Nash's (p. 182) statement that Leopold's ecological conscience and land ethics were "pointing the way to a new relationship between man and land as well as to a new significance for the wilderness" is a pertinent summary of Leopold's achievements. His works and ideas are a new, progressive interpretation of western civilization. Contrary to the tradition of man-centered culture, he puts forward a change from self-conscience to ecological conscience; he replaces the idea that man is the master of nature with the idea that it is an ordinary member of the land community, and that land ethics insist that humanity has moral obligations toward the land; he believes that the wilderness is the root from which human beings spring, and that it is a reminder to modern people of their relationship with the natural world; he links the wilderness closely with culture and advises that only through maintaining the health of the land can the health of human culture be maintained; he not only asks people to preserve the wilderness, but also called on them to rehabilitate abused land; and he regards the exploration of a harmonious relationship between man and nature as being in the realm of art, producing masterpieces of nature writing. All of this greatly influenced American nature writers in the late twentieth century, especially another influential nature writer who appeared immediately after him—Edward Abbey.

# Lecture 24. Abbey: Experience of the Beauty of Tranquillity

In the era in which Edward Abbey (1927–1989) lived, American industrialization was going ahead by leaps and bounds. During this period, the prevalence of consumerism led to the commodification of society. John Muir's forest cathedral in the mountains became the "empty chapel, only the wind's home" in T. S. Eliot's *The Wasteland* (Eliot [1922] 2005, p. 198). Confusion and anxiety haunted the world. The loss of spiritual belief and isolation from nature seemed to forebode the end of the world. Edward Abbey's representative work, *Desert Solitaire: A Season in the Wilderness,* was published in 1968. With it Abbey, understanding the complexity of modern society, sent a message from the desert of the American West that "out there is a different world, older and greater and deeper by far than ours, a world which sustains the little world of man as sea and sky surround and sustain a ship" ([1968] 1971, pp. 41–42).

With its new theory of harmonious coexistence between humanity and nature and its unique and grand artistic beauty, *Desert Solitaire* has had a great impact on modern American culture. According to Scott Slovic and Terry F. Dixon in their edited selection of nature writing titled *Being in the World: An Environmental Reader for Writers* (1993, p. 3), it is regarded as a classic of nature writing and a sign of a New American Renaissance, and has inspired a large group of nature writers.

Nature writer and critic Don Scheese,[3] in his essay "Desert Solitaire: Counter-friction to the Machine in the Garden" (1996), calls *Desert Solitaire* "one of America's great contributions to world literature" (p. 303). He tells of its influence on himself, saying: "After reading *Desert Solitaire* a new era began in my

3 It should be mentioned that Scheese, a professor of English and environmental studies, specifies the definition and source of nature writing in his book *Nature Writing* (1996), which systematically introduces the main representatives of the genre. In 2001 he published *Mountains of Memory: A Fire Lookout's Life in the River of No Return Wilderness,* based on his experiences while working as a summer fireman in the mountains of Idaho. He not only researched and taught nature writing, but also explored nature on foot as much as possible, to experience the wilderness himself. Readers commented that Scheese carried on the tradition of Thoreau, Abbey and Snyder and conveyed to his readers the value and significance of the wilderness, so it can be said not only that his own life entered a new era, but also that nature writing did so, as a result of his reading nature writing.

life: I made it my vocation both to study the nature writing tradition and devote a significant portion of time to living in the wild." He goes on to say that Abbey carried on the nature writing tradition that was started by Thoreau and expanded and developed by Muir and Leopold. However, although all four writers sought to establish a land ethic in the American people, Abbey differs from the other three in two aspects: his chosen place of residence was desert, "a natural environment closer to wilderness"; and he was "the most radical and iconoclastic figure" (pp. 303–304). James Bishop, in his biography of Abbey, *Epitaph for a Desert Anarchist: The Life and Legacy of Edward Abbey* (1994, pp. 228–229), cites Nash as identifying two aspects of Abbey's works that distinguish them from those of others: first, Abbey helped change societal attitudes toward the desert by calling on a new generation of Americans to see the real face of the desert, that is, the fact that the preciousness of the desert does not lie in what is present but what is absent. Secondly, he contributed to deep ecology by leading environmental theory into reality. The theme that runs through his works is that nature exists not for humanity, but for itself.

## Landscape Beyond Green

Edward Abbey was born in 1927 in the Allegheny Mountains of Pennsylvania. His father was a farmer and his mother a high school teacher. While still in high school Abbey hitchhiked from Pennsylvania to Seattle and Arizona, and began to take interest in the deserts and canyons of the West. It was a very different land from the green mountains and rivers, meadows and woods of the East, a landscape beyond green. In Arizona he experienced the charm of the desert for the first time,

> a land that filled me with strange excitement: crags and pinnacles of naked rock, the dark cores of ancient volcanoes, a vast and silent emptiness, smoldering with heat, color and indecipherable significance, above which floated a small number of pure and hard-edged clouds. For the first time I felt I was getting to the West of my deepest imaginings—the place where the tangible and the mythical became the same. (quoted in Scheese, p. 304)

During this trip Abbey formed an unbreakable tie with the desert in the West, and setting up a home on the desert by the mountains became a dream in his early years. In 1945, in his late teens and the last year of World War II, Abbey joined the army and was sent to front in Italy. After his return in 1947 he enrolled at the University of New Mexico to study philosophy. At university his literary talent became evident; he edited the university's literary journal and started work on his first novel. After graduating he went to live in Edinburgh, and traveled in Europe on a Fulbright scholarship. In 1954 he returned to New Mexico University, and received his master's in philosophy in 1956, then took up a position with the National Park Service, leaving his wife and young child behind to go to work in the desert and canyons of Utah as a seasonal ranger.

Over the following 15 years, Abbey experienced various kinds of solitude amid grand natural scenery while working as a seasonal ranger in various nature preservation areas in the West. His understanding of and affections for the wilderness in the West increased and deepened, while at the same time his antipathy toward science and technology, urban life and the industrialized East of twentieth-century America grew. He disliked blind development, regarding it as false, dishonorable and destructive behavior. Abbey's dislike of the negative effects of modern civilization and his love for the wilderness in the West can be seen throughout his diaries and works. Every time he returned to the desert and canyons of the West he felt a surge of new life, a restoration of health and a heart filled with joy and contentment.

Edward Abbey wrote more than 20 books, including novels based on the wilderness in the West and also essays on nature. Garth McCann, in his critical essay "Edward Abbey" (1977, p. 20), describes Abbey as "more at home in the contemporary West than in the West of myth and history." As a modern man, Abbey embodied contemporary American culture, vividly exemplifying the American dream. When Americans faced the threat of increasing consumerism and urbanization and angrily asked, "who gave you the right to rob me of the blueness of my sky," Abbey became a spokesperson for the wilderness and the freedom of humanity, and his works emphasized American citizens' right to inherit this priceless treasure.

Literary critic Harry Middleton comments that "when it is asked how we came to know the desert, surely the most lasting and valuable images left us will be those of Abbey—words set on paper, durable as lines cut in stone" (quoted in Bishop, p. 188). It can be said that Abbey defended the wilderness throughout his life. In *Down the River*, a collection of autobiographical and philosophical essays written over the four years preceding its publication in 1982, he excitedly declares, "We are obliged, therefore, to spread the news, painful and bitter though it may be for some to hear, that all living things on earth are kindred" ([1982] 1991, p. 13). A few days before his death, Abbey completed his last book *A Voice Crying in the Wilderness* (1990). When he left this world on 14 March 1989, his last words were that he felt he had done what he could. Abbey's death, instead of making people forget him, turned him into a legend and he continues to live on in people's hearts; like Mary Austin before him, he achieved permanent fame through his prose work *Desert Solitaire*; since his death the book has been reprinted many times and has also been translated into German, French, Japanese and many other languages.[4]

*Desert Solitaire* describes Abbey's experiences when working as a ranger one summer in a canyon in the West. The 18-chapter book describes in detail his spiritual and physical adventures in the canyons, his reflections on the relationship between humanity and nature, his concerns about the blind development American society was pursuing, and his new definitions of human civilization and modern culture, even though in the foreword to the illustrated 1988 edition of *Desert Solitaire* he says,

> *Desert Solitaire*, I'm happy to add, contains no hidden meanings, no secret messages. It is no more than it appears to be, the plain and simple account of a long, sweet season lived in one of the world's most splendid places. (Abbey [1968] 1988, p. xii)

And his "plain and simple" straightforward narration follows the tradition of the simple and concise nature writing of Thoreau, Muir and others.

*Desert Solitaire* is based on diary entries Abbey made when working as a national park ranger. He presents these original records as discrete and yet continuous stories, demonstrating a unique style. He stands alone in the mountains, facing the rocks and clouds, the sky and the vast land under a rising sun on the first amazing morning; plays solitary games in the empty desert, beside a campfire under starry skies; fights with a snake which he dubs the Satan in the desert paradise; wanders in what he calls a "rock garden"; recalls and reflects on the legends of western cowboys and American Indians; discovers that he can find water through smell in order to survive; ponders upon nature and life when faced with the silent valleys under the hot sun at noon; deliberately floats down the Colorado River without a lifejacket on a risky adventure; chases a horse that has escaped into the mountains and tries to have a conversation with it; gets lost in the canyons one unforgettable night, falls into a pond under a cliff, then struggles out and sleeps in a mountain dog's cave. In a word, the features of the desert and the eternity of nature show humans feelings and urge them to think. Abbey tells

4 For Abbey's life, please refer to: Lyon, ed. *This Incomperable Lande*, pp.84–85; McCann, *Edward Abbey*, pp. 6–10; Bishop, *Epitaph for a Desert Anarchist*, pp. 6–71.

his restless contemporaries that there is wilderness in the world and it can soften the conflicts of modern society, purify the soul, call on the affections between one human and another, and maintain the primitive relationship between humanity and nature.

*Desert Solitaire* has become a classic of American nature writing because the author, while following the nature writing tradition of Thoreau and others, raises the new theory of conflict-compromise-balance coexistence between nature and humans expanding the self-centered concept of nature to the ecology-centered concept of nature and regarding blind development at the expense of nature as a betrayal of modern civilization itself. He has inherited Leopold's idea of land health from *A Sand County Almanac* and regards entirety, stability and beauty as the measures of ecological health of the desert. *Desert Solitaire* is itself a demonstration of desert beauty.

Abbey has been called "the Thoreau of the American West," (McMurtry, Larry 1975) and says himself that "Thoreau's mind has been haunting mine for most of my life" (Abbey [1982] 1991, p. 13). In the preface to the 1988 edition of *Desert Solitaire* he writes, "The few such writers whom I wholly admire are those, like Thoreau, who went far beyond simple nature writing to become critics of society, of the state, of our modern industrial culture" (p. xi). Like Thoreau's *Walden*, Abbey's *Desert Solitaire* has the diary form as its framework and the season as its index. Thoreau went to Walden Pond before the industrial era in order to face the reality of nature, study nature's course, and explore self-reliance, whereas Abbey went to the desert after industrialization and declared, "if it is possible, directly face the naked facts of survival, which support us" (Abbey [1968] 1971, p. 6). Just as Thoreau, with the foresight of a prophet, had said, "In wildness is the preservation of the world" (Thoreau "Walking" in *Portable Thoreau* [1947] 1977, p. 609), Abbey—living in a modernized society and realizing the great importance of wilderness to it—put forward a completely new view of wilderness as "not a luxury but a necessity of the human spirit and as vital to our lives as water and good bread" (Abbey [1968] 1971, p. 6).

Abbey and Thoreau are not only kindred spirits in their thinking; they are also similar in terms of the style of their representative works, written as prose essays in language that is close to the land. Thoreau would rather his essays retained some naivety than attained ornamented formality, while Abbey hoped for naturalness, preferring surface "freshness" to profound "deadness." While *Desert Solitaire* is Abbey's spiritual autobiography and also the mythology of his self-education and self-fulfillment, *Down the River* shares the sad, mourning tone of Thoreau's autobiographical *A Week on the Concord and Merrimack Rivers*

(1849), although Thoreau was mourning the passing of his brother John while Abbey was mourning the death of the river, the valley and the wilderness, damaged by a dam. Not long before his death, Abbey went down the Green River in Utah with a copy of the 33rd edition of *Walden*, communicating spiritually with both nature and Thoreau, and it is evident that Thoreau's thoughts have been close to him all his life. However, Abbey lived in a more modern society, and his views were of his time, going beyond the tradition he inherited.

Compared with the nature writers of the nineteenth century, Abbey's views were closer to being eco-centered rather than self-centered. His emphasis of the harmony and balance of the ecosystem in the desert is particularly evident in Chapter 3 of *Desert Solitaire*, "Satan in Paradise." One morning, when drinking coffee on the steps of his mobile house, he sees a rattlesnake on the ground behind him. Faced with this direct threat, Abbey intuitively thinks of the revolver he has inside the house; however, he feels that as a ranger his duty is to protect all living things in the park without exception. He cannot bring himself to kill the snake, as he holds a love for all living things: "I prefer not to kill animals. I'm a humanist," so he adopts an ecofriendly method to deal with it: first he uses a spade to throw the snake out of the house; after it comes in again he catches an earth snake, which is not poisonous, and keeps it as a pet, so the earth snake can drive away the rattlesnake. His way of dealing with the rattlesnake is the first ecological lesson Abbey had in the desert, and his personal experience of engaging with the eco-environment. He thus develops a living-things-centered view and a deep ecological knowledge, writing with deep feelings, "We are obliged, therefore, to spread the news, painful and bitter though it may be for some to hear, that all living things on earth are kindred" (Abbey [1968] 1971, pp. 20, 24).

In the desert there are no people, grass, or flowers, only rattlesnakes, cacti and exposed red rocks. With his own caring behavior toward all living things, Abbey shows people around the world how, as humans, we should admire and care for a land that is not short of loveliness even though it lacks a conventional pastoral landscape; its loveliness perhaps lies precisely in that absence and in its vastness.

Where Leopold, in *A Sandy County Almanac*, linked the health of the land with that of humanity, Abbey, in *Desert Solitaire*, links the eco-health of the desert with human culture and civilization. On the isolated and vast desert, Abbey escapes from a modern society obsessed with quick profits and instant success and is able to ponder on the problems of modern civilization. He believed that civilization at the expense of what remains of damaged nature is, in fact, a betrayal of the fundamental principles of civilization itself, and that blind development and growth will lead to the death of society and civilization, and in *Desert Solitaire* he uses one noontime in the desert as a backdrop against which to describe his view of nature and civilization:

> Noon is the crucial hour: the desert reveals itself nakedly and cruelly, with no meaning but its own existence.
>
> My lone juniper stands half-alive, half-dead, the silvery wind-rubbed claw of wood projected stiffly at the sun. A single cloud floats in the sky to the northeast, motionless, a magical coalescence of vapor where a few minutes before there was nothing visible but the hot, deep, black-grained blueness of infinity.
>
> Life has come to a standstill, at least for the hour. In this forgotten place the tree and I wait on the shore of time, temporarily free from the force of motion and process and the surge toward—what? Something called the *future*? I am free, I am compelled, to contemplate the world which underlies life, struggle, thought, ideas, the human labyrinth of hope and despair. (Abbey [1968] 1971, p. 155)

The contrast between the stillness in the desert and the motion of modern society makes Abbey realize that many of the things we are busy with are meaningless. People become slaves to time, and lose the freedom they should have. Facing the vast, quiet desert, he exclaims, "We are preoccupied with time. If we could learn to love space as deeply as we are now obsessed with time, we might discover a new meaning in the phrase 'to live like men'" (1971, p. 66).

Precisely because of this, Abbey sees the importance of desert wilderness to humans and their culture. He believes that blind growth and development threaten human society just as cancer does. To pursue an end regardless of means forces people to a state near madness, separating them from the roots they need to survive. "A civilization which destroys what little remains of the wild, the spare, the original, is cutting itself off from its origins and betraying the principle of civilization." He goes on to use other people's various experiences on the sea, in the mountains and in the desert to explain the special meaning of desert to human civilization. When sailing the ocean we reach the other shore and when climbing a mountain, if we persevere, we reach the summit, but the desert is different: it is hard for us to find similar feelings there, all we can feel is the indifference of the endless sands (1971, pp. 192, 271). It is precisely this "endless indifference" that is appealing: this is a unique landscape, a special land where people can gain insight into the universe, a buffer zone where they can resist the frantic behavior of culture. The desert, for its own reason, leads modern people out of the valley of self-contradiction; in its vastness restless people find spiritual balance, and its generosity of scale gives modern civilization new inspiration.

Abbey acquired a new understanding of the value of wilderness.

> We need wilderness whether or not we ever set foot in it. We need a refuge even though we may never need to go there. I may never in my life get to Alaska, for example, but I am grateful that it's there. We need the possibility of escape as surely as we need hope; without it the life of the cities would drive all men into crime or drugs or psychoanalysis. (1971, pp. 148–149)

Abbey links wilderness closely with the American spirit. As Bishop points out, he warns people that "once the wilderness is gone…freedom will be endangered, too.… We can have wilderness without freedom; we can have wilderness without human life at all, but we cannot have freedom without wilderness." He believes that "[t]rue human freedom, economic freedom, political freedom, social freedom, remain basically linked to physical freedom, sufficient space, enough land." Bishop explains that to Abbey, unless people go to the desert and the rivers, unless they walk into the ancient, primeval, vast and democratic vistas, they "will never understand the secret essence of the word 'freedom'" (Bishop 1994, pp. 13, 188). Abbey therefore considers it his duty to protect the "Garden of Eden" in the last wilderness of America, and he raises the attitude toward wilderness to a religious level, "Original sin, the true original sin, is the blind destruction for the sake of greed of this natural paradise which lies all around us—if only we were worthy of it" (Abbey 1971, p. 190).

Abbey also sees inspiration for humanity and civilization in animal behavior. One night beside the pond, he hears frogs singing happily and starts to look for the source of their happiness.

> Why do they sing? What do they have to sing about?…It may nevertheless be the case that these small beings are singing not only to claim their stake in the pond, not only to attract a mate, but also out of spontaneous love and joy, contrapuntal choral celebration of the coolness and wetness after weeks of desert fire, for love of their own existence, however brief it may be, and for joy in the common life. (p. 143)

From the singing of the frogs, Abbey draws a philosophical lesson.

> Has joy any survival value in the operations of evolution? I suspect that it does; I suspect that the morose and fearful are doomed to quick extinction. Where

> there is no joy there can be no courage; and without courage all other virtues are worthless. (p. 144)

McCann points out that "[p]hilosophically, *Desert Solitaire* centers on the concepts of opposition, compromise and balance" (p. 21). Modern people's need for wilderness and their pursuit of modern, material life form a confrontation. Throughout the book, we can feel the author's confusion and inner conflict that comes from humanity's dual needs for civilization and for nature. The desert, to humans, is both dangerous and hostile, yet also relaxing and comforting; silence is both depressing loneliness and also a psychological necessity; we need economic growth, and also need social stability; we yearn for a comfortable material existence, but also dream of a simple life. Abbey makes these unavoidable contradictions and competing needs particularly clear in the last chapter of the book, "Bedrock and Paradox."

> After twenty-six weeks of sunlight and stars, wind and sky and golden sand, I want to hear once more the crackle of clamshells on the floor of the bar of Clam Broth House in Hoboken. I long for a view of the jolly, rosy faces on 42 Street.... I grow weary of nobody's company but my own—let me hear the wit and wisdom of the subway crowds again, the cab drivers' shrewd aphorisms, the genial chuckle of a Jersey city cop.... Unlike Thoreau, who insisted on one world at a time, I am attempting to make the best of two. (pp. 298–299)

He believes that the way to solve this conflict lies in compromise and balance. "Balance, that's the secret. Moderate extremism. The best of both worlds" (p. 299).

After living for half a year in the desert, Abbey realizes that nature will last longer than humanity. He writes:

> Men come and go, cities rise and fall, whole civilizations appear and disappear—the earth remains....
>
> ...I sometimes choose to think, no doubt perversely, that man is a dream, thought an illusion, and only rock is real. Rock and sun. (p. 219)

As a writer, Abbey took up the challenging task of helping humanity survive, as he loved not only wilderness and freedom, but also humans. About to leave the desert, he sighs,

> The finest quality of this stone, these plants and animals, this desert landscape is the indifference manifest to our presence, our absence, our coming, our staying

> or our going. Whether we live or die is a matter of absolutely no concern whatsoever to the desert. Let men in their madness blast every city on earth into black rubble and envelope the entire planet in a cloud of lethal gas—the canyons and hills, the springs and rocks will still be here, the sunlight will filter through, water will form and warmth shall be upon the land and after sufficient time, no matter how long, somewhere, living things will emerge and join and stand once again, this time perhaps to take a different and better course. (pp. 300–301)

Here Abbey shows that in a rapidly urbanized modern society, the worst victim is not the natural environment, but humanity itself. If humans want to break through the contradictions that surround and trouble them, then they must seek compromise and balance.

In fact, Abbey believed that industrial technology itself was not evil; the key was to control it and prevent it from running wild and blindly developing into a monster. Abbey truly saw a way out of the conflict between nature and culture, believing that fair and reasonable compromise by humanity would bring about a balanced equilibrium. How can this balance be achieved? We need to keep the existing wilderness and the industrial areas as they are, as far as is possible: since we need the freedom and beauty of wilderness, we must keep its original conditions; and since we need the products and services of industry and the city, we must seek them within the industrial areas. Abbey's balance through compromise will avoid damaging either the wilderness or the city, while maintaining the best conditions of both worlds.

## *Real and Magic Desert*

Examining Abbey's contribution to American nature writing, Scheese (1996, pp. 305–306) points out that Abbey not only put forward a new theory and described particular ways of behavior to protect wilderness, he also established a different and appealing desert aesthetic. Indeed, under Abbey's pen, *Desert Solitaire* is imbued with a rough and rugged beauty. It has become a nature writing classic because he described the desert in an unprecedented way, capturing the elusiveness of the desert and the paradox of its being both beautiful and terrifying at the same time.

The very first sentence of *Desert Solitaire* is, "This is the most beautiful place on earth" (p. 1). The mountains, the wondrous architecture created by nature, the colorful and variously shaped rocks and the vast, sacred wilderness stir up a strong desire and passion in him. He writes,

> Standing there, gaping at this monstrous and inhuman spectacle of rock and cloud and sky and space, I feel a ridiculous greed and possessiveness come over me. I want to know it all, possess it all, embrace the entire scene intimately, deeply, totally, as a man desires a beautiful woman. (p. 6)

However, the desert is not a soft and beautiful woman; Abbey soon discovers that it is "a land of surprises, some of them terrible surprises. Terrible as derived from terror" (p. 132). Abbey experiences these 'terrible surprises'; it is very difficult to find water in the desert, and even if he finds water which looks extremely beautiful on the surface, it turns out to be bitter and undrinkable, so that in the end he has to learn to find water by smell, just as the animals do. His desert is "[a] sculptured landscape mostly bare of vegetation—earth in the nude" (p. 187).

Elsewhere in the book, when Abbey writes about his experience of falling into a pond under a cliff, we can experience with him the mixed feelings of beauty and terror. Abbey gropes around in the dark pond and, realizing in desperation that nobody will come to help him, tries again and again to climb the slippery cliff, only to fail again and again. Looking up in terror and despair, he sees the patch of blue sky between the dark canyons, and a small white cloud floating by over his head. The scene is "so lovely and precious and delicate and forever inaccessible that it broke the heart and made me weep like a woman, like a child. In all my life I had never seen anything so beautiful." At sunset, he at last manages to get out of the pond and climbs to the top of the cliff. A storm arrives in the canyon, and he has to take shelter and sleep in a coyote's cave. Later he admits that it was a very long, painful night. He has pains all over his body, has to endure wetness, cold and hunger, and is haunted by nightmares. However, in the end he concludes, "It was, all the same, one of the happiest nights of my life" (pp. 228–231). *Desert Solitaire* is a moving story, perhaps precisely because of such unusual experiences and writing. The desert, under Abbey's pen, has a cold and haughty beauty, a rough, desolate and arid beauty, a mysterious beauty that is difficult to grasp. All these kinds of beauty are accompanied not by pastoral scenes, but by wild terror and adventure.

Slovic, analyzing Abbey's desert aesthetics from a psychological perspective, believes that "to write about nature is to write about how the mind sees nature, and sometimes about how the mind sees itself." He points out that "[s]uch writers as Annie Dillard, Edward Abbey, Wendell Berry, and Barry Lopez are not merely, or even primarily, analysts of nature or appreciators of nature—rather, they are students of the human mind, literary psychologists" (Slovic 1992, pp. 3–4). In *Desert Solitaire*, Abbey often stops to comment and explain how the heart views nature. Commenting on rock arches in the desert, he says that their significance lies "in the power of the odd and unexpected to startle the senses and surprise the mind out of their ruts of habit, to compel us into a reawakened awareness of the wonderful—that which is full of wonder" (Abbey 1971, p. 41). This natural and wonderful power

> has the curious ability to remind us—like rock and sunlight and wind and wilderness—that out there is a different world, older and greater and deeper than ours, a world which surrounds and sustains the little world of men as sea and sky surround and sustain a ship. The shock of the real. For a little while, we are again able to see, as the child sees, a world of marvels. (Abbey 1971, pp. 41–42)

In the materialism and soullessness of modern society, people need this sensible reminder and truthful shock. Thoreau once awakened people's awareness of nature like a rooster awakening the neighborhood; now Abbey wanted—through contact with real nature in the desert and through the power of nature—to awaken their awareness of their roots and of the real, and also of the need to protect the land that they rely on for survival, not only physically but also spiritually.

In *Desert Solitaire*, we also experience a tranquil beauty that is a part of the isolated landscape. This peaceful stillness is necessary to the spirit of a restless modern society, although it may not initially be acceptable to some people; as Abbey says of the desolate and uninhabited desert "a man could live out his life in this place, once he had adjusted his nervous system to the awful quietude, the fearful tranquility" (p. 183). Abbey himself had adjusted his nervous system, and could therefore play 'solitaire' in the desert, transforming the terrifying quietude to a wonderful realm.

After his first dinner in the desert, with a campfire for company, Abbey watches the dusk sky.

> Dark clouds sailing overhead across the fields of the stars. Stars which are unusually bold and close, with an icy glitter in their light—glints of blue, emerald, gold. Out there, spread before me, to the south, east and north, the arches and cliffs and pinnacles and balanced rocks of sandstone (now entrusted to my care) have lost the rosy glow of sunset and become soft, intangible in unnamed unnamable shades of violet, colors that seem to radiate from—not overlay—their surfaces. (p. 13)

After the fire dies out, Abbey starts to walk around in the deepening darkness. He has a flashlight, but will not use it unless absolutely necessary. He explains,

> like many other mechanical gadgets it tends to separate a man from the world around him. If I switch it on my eyes adapt to it and I can see only the small pool of light which it makes in front of me; I am isolated. Leaving the flashlight in my pocket where it belongs, I remain a part of the environment I walk through and my vision though limited has no sharp or definite boundary. (p. 14)

Thus Abbey teaches us that walking in the darkness without a flashlight not only allows us to become more a part of the natural environment, but also put us in a different realm, where human imagination, like the endless darkness of night, is hidden, boundless and free, and belongs completely to the self.

When Abbey finishes his walk and returns to the place where he is staying, it is already late night.

> I wait. Now the night flows back. The mighty stillness embraces and includes me; I can see the stars again and the world of starlight. I am twenty miles or more from the nearest fellow human, but instead of loneliness I feel loveliness. Loveliness and a quiet exaltation. (p. 15)

Barton Levy St. Armand [1984] 1989) quotes English writer Walter Pater, "The perfection of culture is not rebellion but peace, only when it has realized a deep moral stillness has it really reached its end" (p. 259). We can say that Abbey has, at this moment, obtained artistic peace and reached cultural perfection.

In summary, it is no accident that Abbey is regarded as both an outstanding representative of American nature writing and the embodiment of contemporary American culture, or that *Desert Solitaire* is regarded as a classic of nature writing. Abbey strikingly highlights the main characteristic of nature writers: they are people troubled by the conflict between wilderness and modern civilization. Nature beckons Abbey, and civilization prevents him from going forward. He goes into the desert with the aim of solving his conflict with society, in other words the conflict between modern civilization and wild nature. On the bare land without much greenery he ponders the issue and comes up with a new view of harmonious coexistence between humans and nature: opposition, compromise and balance. He links ecological health with modern civilization and gives modern civilization a new definition: blind development at the expense of nature is a betrayal of modern civilization itself. He puts forward the idea that protection of wilderness is a moral matter, based on the belief that all living things are kindred. He points out that the original sin is the greed and blind damage of the natural paradise. He links wilderness closely with human freedom, warning humans that their freedom will be in danger once wilderness is lost. He emphasizes the inalienable right of humans to inherit the priceless wilderness. He analyzes the relationship between nature and culture and puts forward the belief of nature writers: nature gave birth to culture. He promotes a simple life close to wilderness, calling on modern people to love space as they love time, and to pursue the beauty of peace in both nature and the heart.

Edward Abbey passed away in March 1989, and a memorial gathering was held in Archie National Park,[5] his favorite place, in May of the same year. The contemporary American nature writer Terry Tempest Williams gave a moving eulogy, saying,

> Ed is here and will always be. His words reverberate on canyon walls, his voice being carried by desert winds on the open skies of the American West. He is Coyote, a dancer upon the desert. (published in Williams's *An Unspoken Hunger: Stories from the Fields*, 1994, p. 73)

## Lecture 25. Dillard: Pilgrim at Tinker Creek

In the mid-1970s, *Pilgrim at Tinker Creek* (1974) by Annie Dillard (1945–) caused a big stir on the American literary scene. It was regarded as "the most influential and widely imitated works of contemporary nature writing" (Finch and Elder in their introductory note of Annie Dillard 1990, p. 816) and won the Pulitzer Prize (General Nonfiction) in 1975. The success of the book—based on the author's observations and reflections around a creek near her home in Virginia's Blue Ridge Mountains—lay in its depiction of how Dillard both followed and rebelled against the transcendentalist masters like Emerson and Thoreau. And her interpretations of nature contained religious mysticism, but also comedic popularity.

Dillard, in the very essence of her thinking, inherited the tradition of American transcendentalism. Her year at Tinker Creek has a very strong flavor of Thoreau. *Pilgrim at Tinker Creek* was not only a pilgrimage into nature, but also to the saints of Concord, and was considered a "kind of gutsy *Walden*"' It is worth mentioning that the focus of Dillard's master's thesis had been Walden and Thoreau. Unlike postmodernists but like Emerson, Dillard found beauty, ideas and spirit in nature; Unlike T.S. Eliot in *The Wasteland*, she wandered along Tinker Creek on a Thoreau-style spiritual pilgrimage. Emerson emphasized

5 Archie National Park is located in eastern Utah and contains more than two thousand natural rock arches.

the order, harmony and uniformity of nature, while in Dillard's eyes nature is consistent yet also diverse, orderly yet also chaotic, sublime yet also destructive. Emerson was an idealist, while Dillard looks at nature from not only an idealistic perspective but also a scientific one. She describes nature more objectively and more reasonably, observing that it has insurmountable power. In *Walden* Thoreau urged people to "Simplify, simplify" (Thoreau, *Walden*, [1854] 2004, p. 78), while in *Pilgrim at Tinker Creek* Dillard reminds them of the complexity beneath the apparently simple surface of nature. More importantly, Dillard wants, through her experiences and works and as a witness of the relationship between humans and ecology, to send people the message that human culture is only part of nature and that humans have intricate connections with animals and plants. While Leopold wanted to "think like a mountain" and Abbey to "feel like a river", Dillard hopes that her life runs along freely as the creek, "new every minute" (Dillard, p. 2).

Annie Dillard was born in Pittsburgh, Pennsylvania in 1945 and spent her childhood there. As a child she loved playing the piano, dancing, and collecting rocks and insects. In *Pilgrim at Tinker Creek* she regards collecting rocks, watching the clouds, and collecting drops of water from the sea—all activities that show traces of her childhood—as the greatest pleasures of life. In high school Dillard was a little rebellious, and the excitement of youth turned her interest to poetry writing. At this time the person she admired most was Emerson. She later studied English, theology and writing at Hollins College, and obtained her master's in English with a thesis titled "Walden and Thoreau" in 1968. She went on to teach poetry and writing at Western Washington University at Bellingham, then taught at Wesleyan University in Middletown, Connecticut from 1979 until she retired. Since 1973, she has also been a staff writer at Harpers, one of the five biggest publishing houses in America.

In 1971 Dillard, then suffering from pneumonia, decided to go to nature to experience life, and chose Tinker Creek in the Blue Ridge Mountains of Virginia. She spent four seasons there, among the mountains, rivers, and many kinds of animals and plants. Like Thoreau, she recorded in her diaries what she observed and experienced, then sorted the more than 20 diaries into several hundred cards, and based the astounding *Pilgrim at Tinker Creek* on these. Dillard also wrote many other books, including: a poetry collection, *Tickets for a Prayer Wheel* (1974); essay collections *Holy the Firm* (1977), *Teaching a Stone to Talk* (1982) and *The Writing Life* (1989); a literary criticism collection, *Living by Fiction* (1978); a novel, *The Living* (1992); a memoir, *An American Childhood* (1988); and also a collection of random notes titled *Encounters with Chinese Writers* (1984). In

1982 Dillard was invited to visit China, where she met a group of contemporary Chinese writers who, in turn, later visited America and were warmly welcomed by Dillard. She accompanied them to a conference and to visit various places, took them to Walden Pond, and treated them to a Chinese meal at home. *Encounters with Chinese Writers* describes these mutual visits (Slovic & Dixon 1993, p. 27; Thompson & Bryson 2003, pp. 107–120; Lander 1997).

## *Beautiful and Terrible Nature*

In its 15 chapters, *Pilgrim at Tinker Creek* vividly records the four seasons the author spent at Tinker Creek, from January to December. Although this way of following the seasons is a tradition of Thoreau, Muir, Leopold and Abbey, Dillard also shows her uniqueness and difference.

At the beginning of the book, the author describes the surroundings where she is living.

> I live by a creek, Tinker Creek, in a valley in Virginia's Blue Ridge....
> ...It's a good place to live; there's a lot to think about. The creeks—Tinker and Carvin's—are an active mystery, fresh every minute. Theirs is the mystery of the continuous creation and all that providence implies: the uncertainty of vision, the horror of the fixed, the dissolution of the present, the intricacy of beauty. (Dillard, pp. 2–3)

The nature that is ordinary in the eyes of other people becomes mysterious and new under Dillard's pen, attracting people to explore and develop together with the author. Thus Dillard leads readers on a pilgrimage into nature.

On her way she finds not only the brilliant sky of January, but also the romance of a wanderer. She writes,

> Today is one of those excellent January days partly clouded in which light chooses an unexpected part of the landscape to trick out in gilt, and then shadow sweeps it away. You know you're alive. You take huge steps, trying to feel the planet's roundness arc between your feet. Kazantzakis[6] says that when he was young he had a canary and a globe. When he freed the canary, it would perch on the globe and sing. All his life, wandering the earth, he felt as though he had a canary on top of his mind, singing. (p. 3)

6 Nikos Kazantzakis (1883–1957) is a Greek writer.

Dillard herself also went on her journey in such mood.

From January to June, Dillard sees the flourishing beauty of nature and tries to grasp the essentials of how to observe it. In this part of her description beauty is accompanied by horror, yet on the whole she looks at the world in a positive light; the natural world under her pen is generally good. However, Chapter 9—"Flood"—is a turning point. From Chapter 10 "Fecundity" onwards, Dillard starts to look at the world with suspicious and negative eyes. She describes the horror of reproduction, the cheapness of life and the inevitability of death. This landscape, with the arrival of autumn and the falling of leaves and the death of plants, becomes a vacant world, a world that makes the heart empty. The book ends on Christmas Eve, returning to the first chapter of the book in January to form a cycle of life and seasons. Buell (p. 239) calls this kind of structure, following the seasons and joining the ending to the beginning, "The serpent with its tail in its mouth," an image which vividly describes Dillard's unique way of using the seasons to symbolize the cycle of nature.

From this, we can see that Dillard's *Pilgrimage at Tinker Creek* has a transcendentalist color, but at the same time rebels against the tradition. She went to nature with a pious attitude, only to come to a puzzling conclusion: that in nature, mystery and beauty, purification and destruction, flourishing and horror coexist. While Emerson and Thoreau assumed a worshipping attitude toward Nature, Dillard, while worshipping, also had doubts and confusion. Comments on *Pilgrim at Tinker Creek* vary. Most critics see Dillard as being closely connected with American transcendentalism. Margaret Reimer, in her article "The Dialectical Vision of Annie Dillard's *Pilgrim at Tinker Creek*" (1983, p. 183), points out that although Dillard, in her attitude toward nature, inherits the ancient optimism of transcendentalism, she also has her own uniqueness. For Marc Chenetier, while Dillard's many references to Thoreau and Emerson show her connection with the tradition, he describes her book as "an improved formula" ("Tinkering, Extravagance: Thoreau, Melville, and Annie Dillard", p. 158), richer and more profound. Buell agrees, calling *Pilgrim at Tinker Creek* "[a] major reinvention of an ancient genre" (p. 242).

Like Emerson, Dillard looks at nature with an optimistic attitude, finding providence, beauty, surprise and spirit. The title of the book, *Pilgrim at Tinker Creek*, already shows the author's adherence to tradition—she goes to nature piously. In her pilgrimage she says,

> Then one day I was walking along Tinker Creek thinking of nothing at all and I saw the tree with the lights in it. I saw the backyard cedar where the morning

> doves roost charged and transfigured, each cell buzzing with flame. I stood on the grass with the lights in it, grass that was wholly fire, utterly focused and utterly dreamed. It was less like seeing than like being for the first time seen, knocked breathless by a powerful glance. The flood of fire abated, but I'm still spending the power. Gradually the lights went out in the cedar, the colors died, the cells unflamed and disappeared. I was still ringing. I had been my whole life a bell, and never knew it until at that moment I was lifted and struck. I have since only very rarely seen the tree with the lights in it. (Dillard, pp. 32–33)

This experience of being struck by the glance of God and of communicating with the spiritual being is so moving that Dillard comes to regard such experiences as the aim of her life. During this incident she experienced the wonder, the power of nature, and the electrifying shock of burning and singing together with nature, the primitive connection of the individual body with the ecological whole. In this moment of Dillard's we can see Jonathan Edwards under the light of God in nature in the eighteenth century; Emerson becoming "a transparent eyeball" in the nineteenth century; and Austin who "revived" together with heaven and earth, trees and grasses in the light at the turn of the century.

Emerson believed in looking at nature with an innocent eye. In *Nature*, he explains with vivid images, "The sun illuminates only the eye of man, but shines into the eye and heart of the child" (Emerson, "Nature" [1836] in Centenary Edition, *The Complete Works of Ralph Waldo Emerson*. Vol. I: *Nature Addresses and Lectures,* [1903] 1979, p. 8). In *Pilgrim at Tinker Creek*, Dillard advocates a similar childlike innocence, by which she means a wholehearted and concentrated engagement with natural scenes. The precondition of experiencing nature is to get rid of all other thoughts and have an empty mind. "You catch grace as a man fills his cup under a waterfall." She believes that only children can hear the singing of the mice, as only children's eyes are always open. What they have is "sense" and therefore they have developed "input systems." "When we lose our innocence—when we start feeling the weight of the atmosphere and learn that there's death in the pot—we take leave of our *senses*" (Dillard, pp. 81, 90). Therefore, keeping our innocence is the precondition of experiencing nature and learning through the experience.

It is precisely due to this undying, childlike innocence that Dillard sees the beauty and the freshness. She believes that there is beauty in nature and finds beauty there. She spends one night at a "paradise enow" and in the morning, seeing the scenery on both sides of Tinker Creek, she says,

> This is what I had come for, just this, and nothing more. A fling of leafy motion on the cliffs, the assault of real things, living and still, with shapes and powers under the sky—this is my city, my culture, and all the world I need.

In this "garden in the wilderness", she feels like she has "just been born" (Dillard, p. 213).

Reading *Pilgrim at Tinker Creek* reminds us of Thoreau's *Walden*. Both the author's year spent experiencing life in nature and the structure, which follows the seasons of nature, exhibit the influence of Thoreau. In the book, "Winter," "Spring," "Walk," "Night Watch" and other chapter titles echo *Walden*'s structure; and more importantly, Dillard inherits the spiritual legacy of Thoreau.

Like Thoreau, Dillard considers walking to be a life-long undertaking; she believes that "our whole life is a stroll" which can be regarded as a "spiritual wandering." Like Thoreau, this walker along Tinker Creek uses not only her eyes but all her five senses to closely observe, experience and collect nature. Her eyes capture almost all the changes in the natural scenery, animals, and plants, from the beautiful fig tree by the side of the creek to the grasshoppers jumping happily in the grass, the pitiable little frog being sucked to death by a water bug, and the egg-laying white moth which led to her horror of reproduction. Hearing a group of restless bulbuls in the sky, watching them flying past in an arch formation and return to the forest, she feels as if they have flown past in her heart, and this makes her tremble all over. She walks into the clear creek, thinking about what changes would be happening in the world at this moment as the water flows under her feet. The nature around her is closely connected with her life. She writes, "Living is moving; time is a live creek bearing changing lights." "I go down to the creek again, and lead a creek life. I watch and stalk" (Dillard, pp. 83, 85, 82, 182). As Thoreau has become the incarnation of Concord, Dillard has become the incarnation of Tinker Creek. Buell explains the similarity between Dillard and Thoreau from another angle: she bought "a pet goldfish for her companion at Tinker Creek and named it Ellery Channing,[7] thereby 'becoming' Thoreau" (Buell, p. 155)?

From the above, we seem to find Dillard's *Pilgrim at Tinker Creek* to be a return to the tradition of Emerson and Thoreau after Muir, Austin, Leopold and Abbey; yet it is not a simple return, but more an improvement of the tradition and a modern return. Her attitude toward the tradition is a dual attitude of

7 Ellery Channing was an American poet, a good friend of Thoreau and also his first biographer.

inheriting and rebelling. Under her pen, nature has two sides. Unlike Emerson and Thoreau, who look at nature in a simple, idealistic and optimistic light, Dillard looks at it in a more complex and more objective light, seeing not only the bright side but also the dark, not only the beautiful side but also the horrible.

At the beginning of the chapter titled "Intricacy" Dillard describes "a rosy complex light" which gradually becomes red, moving through the forest of pines, oaks and hawthorns and onto the valley on the other side of the Blue Ridge in the west, forming a mysterious and scattered fog. From here, gradually, the theme of the complexity of the natural world is introduced. Dillard uses the ecological knowledge she has learnt and practical examples to explain that "[e]volution, of course, is the vehicle of intricacy." She then further explains that landscape itself is a complex structure which "consists in the multiple, overlapping intricacies and forms that exist in a given space at a moment in time." Unlike Thoreau's advocacy to "simplify, simplify," Dillard lets us see that "wherever there is life there is complexity and chaos" (Dillard, pp. 132, 138).

In the following chapter titled "Fecundity," Dillard describes a terrible scene in which the natural world goes astray due to its complexity. At the beginning of this chapter Dillard recounts how she had a nightmare after watching a white moth laying eggs. In her dream, the eggs all hatched in front of her eyes and became weird-shaped fish wriggling on her bed. This frightened her and she woke up with a scream. She says, "I don't know what it is about fecundity that so appals. I suppose it is the teeming evidence that birth and growth, which we value, are ubiquitous and blind, that life itself is so astonishingly cheap, that nature is as careless as it is bountiful, and that with extravagance goes a crushing waste that will one day include our own cheap lives" (p. 160). Dillard then explains that the force behind the breeding is the pressure of growth.

With these two images of intricacy and fecundity, Dillard tries to show that although beauty exists in nature it was not a simple beauty, but a complex and sometimes even terrifying and conflicting beauty.

> The texture of the world, its filigree and scrollwork, means that there is the possibility for beauty here, a beauty inexhaustible in its complexity, which opens to my knock, which answers in me a call I do not remember calling, and which trains me to the wild and extravagant nature of the spirit I seek. (p. 139)

To her, beauty was the fruit that nature had produced from the wonderfully horrible flowers growing freely in chaos. She points out insightfully, "I saw how freedom grew the beauties and horrors from the same live branch" (p. 180).

The root of Dillard's dual attitude toward nature lies in the fact that her nature is no longer the nature that Emerson and Thoreau faced. The environment in her time includes not only the mountains, rivers, forests and desert, but also modern cities and other high-tech products; nature had become an environment in which things conflict. Dillard's heart is full of the confusion and contradictions of modern life. She loves the pure mountains and rivers, but she also cannot live without a modern study filled with books. She wants to return to Emerson's time, but this exists only in her spirit and her desire; the natural environment can never return to the nineteenth century, and she cannot rediscover the simple nature of Emerson's time.

## Master of Natural Description

With a romantic mood of the nineteenth century and the confusion and contradictions of the twentieth century, Dillard cast her eyes to Tinker Creek again. She writes, "I could use some calming down, and the creek is just the place for it. I must go down to the creek again. It is where I belong" (Dillard, p. 178). Given this premise, *Pilgrim at Tinker Creek* cannot be just a dry record of wilderness, but rather a spiritual journey of experiencing and re-recognizing wilderness. Through this process, time, water and trees all became the focus of Dillard's thoughts. As a witness of the coexistence of humans and nature, she is able to present to readers the notion that humans have intricate connections with the natural world.

A chapter titled "The Present" in *Pilgrim at Tinker Creek* shows how people view the world and its complex images at a particular moment. The "present" is the moment when the dormant soul is awakened to the beauty of nature, and when human consciousness communicates with nature. Dillard regards the time when "the glistening pine tree" early in the morning, and the mountain ranges reddened by the sunset evoked various feelings and thoughts in her as those precious "present" moments in which she is burning together with nature and feels being reborn. She reminds readers that they need to capture such "present' moments, because they are like lightening, and will be gone in a flash.

Tinker Creek is the center of Dillard's natural world. To her, the water in the creek is a great saviour and is the personification of Christ. It is the flowing water and light that come from a mysterious place carrying the latest news, and continually updating it to renew the world again and again. When she looks down at the creek, she captures its glistening moment. When she looks over the creek, she sees the future reflected in its transparent waves. In the end, she wants to live like it and become part of the eco-environment. As she says, "The creek is the mediator, benevolent, impartial, subsuming my shabbiest evils and dissolving them, transforming them into live moles, and shiners, and sycamore leaves" (p. 101).

In the picture of nature that Dillard has woven if we see Tinker Creek as forming the horizontal lines, then the trees by its side forms the vertical lines. Dillard regards trees as symbols of eternity and sincerely admires their elegant stability on the land, with their branches reaching out toward the sky and their dignity of being rooted in the earth. In her view, trees have height and depth, the space for unlimited free growth, and the ability to integrate and synthesize the specifics, just as the human brain does. Trees with flourishing branches and leaves prod her into spiritual pondering. She thinks that "[t]he trees especially seem to bespeak a generosity of spirit," and that her "mind branches and shoots like a tree." Under a beautiful fig tree, she scoops up a handful of soil and feels the world is moving in her palm, because she, with the biological knowledge she possesses, knows that there are on average 1,356 biological traces in the soil of every square foot of forest land. She feels, therefore, that in the forest beside Tinker Creek she had blended into the eco-environment, becoming herself a part of nature. She writes, "I was the lip of a fountain the creek filled forever; I was ether, the leaf in the zephyr" (pp. 112, 95, 32).

Dillard's contribution to nature writing lies not only in her awareness of the dual characteristics of nature and in her description of the relationship between humans and the eco-environment, but also in her ability to "see" nature in two different ways. In a chapter titled "Seeing" she explains vividly:

> Nature is like one of those line drawings of a tree that are puzzles for children:—Can you find hidden in the leaves a duck, a house, a boy, a bucket, a zebra, and a boot? Specialists can find the most incredibly well-hidden things. (Dillard, p. 17)

Dillard then describes the two kinds of seeing she had experienced. One is a kind of "concentrated seeing," an "active," analytical and inquisitive seeing, like people taking a camera on a walk and taking photos of what they see. The other kind is a "letting go," "casually looking around," "passive, fusing-oneself-with-nature way of 'seeing'" (Dillard, pp. 31–32).

In Glotfelty and Fromm's *The Ecocriticism Reader* (1996), professor of American studies Vera Norwood calls Dillard "a master of natural description" (p. 339). The nature she describes has wondrous charm. In an intimate tone and with a gentle rhythm, she describes the contrast between the movement and the stillness of nature, its romance and philosophy. This greatly attracts readers. She catches the souls of the seasons and concludes that "summer is hidden" and "winter is exposed," making readers imagine the dense leaves and branches in summer and the desolate winter scene. One winter evening she records the stillness of nature,

> It's the most beautiful day of the year. At four o'clock the eastern sky is a dead stratus black flecked with low white clouds. The sun in the west illuminates the ground, the mountains, and especially the bare branches of trees, so that everywhere silver trees cut into the black sky like a photographer's negative of a landscape. (Dillard, p. 10)

In early spring, she captures the movement of Tinker Creek.

> Long before I could actually see the creek, I heard it shooting the sandstone riffles with a chilled rush and splash. It has always been a happy thought to me that the creek runs on all night, new every minute, whether I wish it or know it or care, as a closed book on a shelf continues to whisper to itself its own inexhaustible tale. (p. 68)

Dillard's description of snow is also unique.

> Yesterday I watched a curious nightfall. It was like dying, this watching the world recede into deeper and deeper blues while the snow piled; silence swelled and extended, distance dissolved, and soon only concentration at the largest shadows let me make out the movement of falling snow, and that too failed. It was like dying, that growing dimmer and deeper and then going out. (pp. 43–44)

To Dillard, nature is a product of beautiful fabric that has hidden beauty in its rough threads. However, beauty is also always accompanied by death. Dillard's description of the cloudless sky is also extraordinary, "It was a clear, picturesque day, a February day without clouds, without emotion or spirit, like a beautiful woman with an empty face" (p. 56). Her description of how the features of the mountains differ from those of the river, on the other hand, is philosophical.

> I look to the mountains, and the mountains still slumber, blue and mute and rapt. I say: it gathers; the world abides. But I look to the creek, and I say: it scatters, it comes and goes.… (p. 201)
>
> …The mountains—Tinker and Brushy, McAfee's Knob and Dead Man—are a passive mystery, the oldest of all…You can have your spirit into a mountain and the mountain will keep it, folded, and not throw it back as some creeks will. The creeks are the world with all its stimulus and beauty.… But the mountains are home. (Dillard, p. 3)

Sandra H. Johnson, in the preface of her book *The Space Between: Literary Epiphany in the Work of Annie Dillard* (1992), describes how she felt when she read *Pilgrim at Tinker Creek* for the first time.

> I was astounded by its condensed, gemlike language that not only considered the phenomenon of illumination as a central theme, but offered up these fiery moments page after page with poetic power and biting energy. Dillard's writing seemed at once ancient and contemporary. (p. ix)

Perhaps this is precisely why *Pilgrim at Tinker Creek* is such a thought-provoking book, exemplary nature writing that helps readers to explore tradition and modernity.

Dillard is a representative of American nature writing in the twentieth century. In her works, we can clearly see the clashing and merging of the old tradition and the new waves of nature writing. Dillard inherited Romanticism and transcendentalism. She sang of the beauty and loftiness of nature, and regarded experiencing nature as a way of communicating with God. On the other hand, she felt uncertain and sometimes even doubtful about nature; she loved it, but her degree of piety was less than that of the Romantics and transcendentalists, showing some influence of post-modernism. We can say that Dillard's works have two sides because she witnessed not only the beauty but also the horror of nature. However, she courageously accepted this duality of nature and integrated them into one through compromising, developing her own unique characteristics. In a certain sense, Dillard's works are a mix of transcendentalism and post-modernism, with the dual qualities of nature fusing into one in the mix.

However, Dillard was still faced with a dilemma: she went to nature because she wanted to become part of it, just like a drop of water, a leaf or a fish. Yet her self-awareness was so real and definite that it was impossible for her to be part of nature in the long term. The dilemma she faced was representative. Even though twentieth-century American nature writers advocated a view that changed from self-centered to eco-centered, they could not escape themselves; they could not solve the problem of being different from nature and at the same time also part of it. Therefore, perhaps we can say that one of the characteristics of American nature writers is their obsession with this contradiction, their description and analysis of this contradiction, and even their play with and enjoyment of this contradiction.

As a major representative of contemporary American nature writing, Dillard adopted a dual method in her description of nature: one is objective, the other subjective. When being objective, through her detailed and vivid description of nature she made what seems like drawing after drawing of her adventures in the natural world. When being subjective, she magnified her personal feelings about the natural scenes and personified nature, which revealed its spiritual force. These two methods are the products of her conflicting attitude toward nature.

Buell describes Dillard's prose in *Pilgrim at Tinker Creek* as "a rushy kaleidoscope of perceptual and intertextual fragments, precariously contained by a basketry of image motifs" (p. 237). We can say that Dillard would inevitably look at nature from multiple perspectives because she lived in a diverse modern society. As a result, *Pilgrim at Tinker Creek* becomes a 'mysterious journey.' In the process, Dillard is immensely drawn and moved, and her spirit is sublimated. In her conceptual world, nature is beautiful; however, to modern readers mysterious profoundness is itself immensely attractive. What *Pilgrim at Tinker Creek* shows them is more wonders of nature, thereby attracting them to continue to explore nature's mystery.

## References

Abbey, Edward. (1968) 1971. *Desert Solitaire: A Season in the Wilderness.* New York: Ballantine Books.

Abbey, Edward. (1968) 1988. *Desert Solitaire: A Season in the Wilderness.* An illustrated edition, University of Arizona Press.

Abbey, Edward. (1982) 1991. *Down the River.* New York: First Plume Printing.

Bishop, James, Jr. 1994. *Epitaph for a Desert Anarchist: The Life and Legacy of Edward Abbey.* New York: Macmillan.

Brooks, Paul. 1980. *Speaking for Nature: How Literary Naturalists from Henry Thoreau to Rachel Carson Have Shaped America.* Boston: Houghton Mifflin.

Buell, Lawrence. (1995) 1996. *The Environmental Imagination: Thoreau, Nature Writing, and the Formation of American Culture.* Second print, Cambridge, MA: Harvard University Press.

Chenetier, Marc. 1990. "Tinkering, Extravagance: Thoreau, Melville, and Annie Dillard." *Critique: Studies in Contemporary Fiction* 31 (3): pp. 157–172.

Dillard, Annie. (1974) 1988. *Pilgrim at Tinker Creek.* New York: Harper Perennial.

Eliot, T. S. (1992) 2005. *The Wasteland and Other Poems.* New York: Barnes & Noble Classics.

Emerson, Ralph Waldo. (1903) 1979.Centenary Edition, *The Complete Works of Ralph Waldo Emerson.* Vol. I: Nature Addresses and Lectures, 2nd edition of the Centenary Edition. New York: AMS Press.

Finch, Robert, and John Elder, eds. 1990. *The Norton Book of Nature Writing.* New York: Norton.

Flader, Susan. 1973. *The Sand County of Aldo Leopold.* San Francisco: Sierra Club.

Fromm, Harold. 1996. "From Transcendence to Obsolescence." In Glotfelty and Fromm 1996, pp. 30–39.

Glotfelty, Cheryll, and Harold Fromm, eds. 1996. *The Ecocriticism Reader.* Athens, GA: University of Georgia Press.

Hart, James D. 1995. *The Oxford Companion to American Literature.* New York: Oxford University Press.

Johnson, Sandra Humble. 1992. *The Space Between: Literary Epiphany in the Work of Annie Dillard*. Kent, OH: Kent State University Press.

Lander, Cheryl. 1997. "Earth Saint," *Earthlight Magazine*, # 24, Winter 1997.

Leopold, Aldo. 1949. *A Sand County Almanac: And Sketches Here and There*. New York: Oxford University Press.

Lyon, Thomas J., ed. 1989. *This Incomperable Lande: A Book of American Nature Writing*. Boston: Houghton Mifflin Company.

McCann, Garth. 1977. "Edward Abbey." *Western Writers Series* 29.

Muir, John. 1979. *John of the Mountains: The Unpublished Journals of John Muir*. Ed. Linnie Marsh Wolfe. Madison: University of Wisconsin Press.

Nash, Roderick. 1968. *Wilderness and the American Mind*. Second print, New Haven, CT: Yale University Press.

Norwood, Vera. 1996. "Heroines of Nature: Four Women Respond to the American Landscape." In Glotfelty and Fromm 1996, pp. 323–350.

Paul, Sherman. 1992. *For Love of the World: Essays on Nature Writers*. Iowa City, IA: University of Iowa Press.

Reimer, Margaret Loewen. 1983. "The Dialectical Vision of Annie Dillard's *Pilgrim at Tinker Creek*." *Critique: Studies in Contemporary Fiction* 24 (3): pp. 182–191.

Sanders, Scott Russell. 1993. *Staying Put: Making a Home in a Restless World*. Boston: Beacon Press.

Sanders, Scott Russell. 1996. "Speaking a Word for Nature." In Glotfelty and Fromm 1996, pp. 182–195.

Scheese, Don. 1996. "Desert Solitaire: Counter-friction to the Machine in the Garden." In Glotfelty and Fromm 1996, pp. 303–322.

Slovic, Scott. 1992. *Seeking Awareness in American Nature Writing*. Salt Lake City, UT: University of Utah Press.

Slovic, Scott, and Terrell F. Dixon, eds. 1993. *Being in the World: An Environmental Reader for Writers*. New York: Macmillan.

St. Armand, Barton Levi. (1984) 1989. *Emily Dickinson and Her Culture: The Soul's Society*. Cambridge: Cambridge University Press.

Thompson, Roger and Bryson, J. Scott, eds. 2003. *Twentieth Century American Nature Writers: Prose. Dictionary of Literary Biography Series 275*. Farmington Hills: Gale Group.

Thoreau, Henry David. (1864) 2004. *Walden*. Boston: Shambhala.

Thoreau, Henry David. (1947) 1977. *The Portable Thoreau*. Ed. and with introduction by Carl Bode, The Viking Press, Penguin Books.

Williams, Terry Tempest. 1994. *An Unspoken Hunger: Stories from the Fields*. New York: Pantheon Books.

Williams, Terry Tempest. (2001) 2002. *Red: Passion and Patience in the Desert*. New York: Vintage Books.

8

# Refuge of the Heart in Nature Writing

## Introduction

Having entered the twenty-first century, even though humanity is continuing to push ahead in many areas, the prospect is not so optimistic, especially in America. First, there was the terrorist attack now known as 9/11, which destroyed the dream of homeland security, then the long war in the Middle East, and then came the global financial crisis. The new book *Finding Beauty in a Broken World* (2008) by contemporary nature writer Terry Tempest Williams reflects the mentality of writers in this field when faced with multiple challenges and crises. We may as well say that in the gloomy reality of this changing world, contemporary American nature writers are seeking beauty and consolation of the soul, and a return to the sacred scenery.

In the twentieth century the First World War not only destroyed human lives, but also alienated and marginalized nature. T. S. Eliot's poem *The Wasteland* is often used to indicate the pessimism that resulted from the spiritual injuries suffered by humanity and the death of nature. At the beginning of the twenty-first century, with humanity once again facing endless wars and an unstable world, people feel confused and perplexed anew: how to survive in a changeable and unrecognizable world? Williams quotes from Eliot's *Four Quartets*: "Turning shadow into transient beauty" (in Williams 2008, p. 343), and declares that "if humans want to survive, they must belong to a natural Mosaic" (p. 384). While the jigsaw puzzle of modernism and the emptiness and "void" of post-modernism reflect the desperate pessimism brought about by the death of nature, contemporary American nature writers must find support from—and re-establish belief and confidence in—the natural world, where a multiplicity of different and changeable varieties of living beings coexist. They suggest that humanity should pick up old, useless, fragmentary things and put them together to reconstruct entirety and beauty.

The three writers and their works that this chapter focuses on show how contemporary American writers find the strength and determination to respond to today's challenges and find both love for the land and love between people in a complex and changeable world. Some of them have experienced the cruelty of war; others have borne the sorrow of losing family; and still others feel perplexed by the quick-profit-oriented modern society, and therefore join the ranks of the "beat generation" of the 1950s. It is Mother Nature that is healing the wounds in their hearts and giving them spiritual support.

Henry Beston's *The Outermost House: A Year of Life on the Great Beach of Cape Cod*, first published in 1928, reproduces the beauty of this unique beach on the Atlantic Coast in epic language. Beston had gone to fight twice during the First World War, and witnessed terrible scenes on the battlefields of Europe. After the war he tried to heal his spiritual wounds and his inner despair through writing children's stories, but this had little effect. In the end he found solace on Cape Cod, "the end of the land" (Thoreau, [1865] 2008, p. 22), and thus began his pilgrimage to the land. In a modern world made insipid by the lack of nature, Beston reconnects us with primitive nature and awakens in us a vitality as vibrant as the wild natural world itself.

The other two writers commented on in this chapter, Terry Tempest Williams and Gary Snyder, straddle the twentieth and twenty-first centuries and are both still alive. Williams's *Refuge: An Unnatural History of Family and Place* (1991) presents her family history against the background of the Great Salt Lake in the American West, a landscape that is realistic yet also surreal: the undrinkable green water of the Great Salt Lake in the desert, with birds refuges dotted around the lake like emeralds, and the ecological harmony there, all contribute to healing the human heart. Williams herself relied on the scenery of the Great Salt Lake to accompany her mother on the last stage of her life and to accept the fact that she herself also had cancer. The Great Salt Lake became her spiritual support and the solace of her heart. Snyder belonged for a period to the category of the "beat generation." Even though he himself did not believe so, his work shows his confusion and perplexity when faced with the materialistic world. However, in the end Snyder escaped from the trap of modern civilization. He regarded the mountains on the West Coast as the place of "re-inhabitation" in the high-tech era, drawing ideas from the wilderness as he tried to communicate with the ancient Chinese poet Han Shan (which literally means Cold Mountain) across time and space. His writing is not only very rational, linking an individual's place with the whole world, but also full of the practice of wilderness as he shows through his own experiences what "re-inhabitation" means.

Beston, Williams and Snyder are writers from the "workshop of nature," each of them unique in their respective nature writing style and language. From his experience of watching the sea, Beston developed a style that is synchronized with the rhythm of the waves. Williams said that the birds in the migrant bird conservation area near Bear River shared her natural history. This unusual experience helped her find a unique direction and create a new style and text structure. Snyder has been called Mountain Saint and "an American Han Shan." His personal experience has also influenced his language. He believes that good writing is written in a "wild language," and the wild language he has been exploring is a combination of western and eastern cultures. His writing contains both the western wild imagination and the calm simplicity of the eastern Zen. He is also a practitioner of bioregional literature; reading his works is like living his experiences and feeling the glistening of the wilderness. Generally, the themes expressed in their works are positive and healthy and show a calm and accommodating mentality and a sense of ecological obligation.

When looking at twenty-first century nature writing, we may gain some insights from Scott Slovic's book *Going Away to Think: Engagement, Retreat, and Ecocritical Responsibility* (2008). In Slovic's opinion, America lost a sense of security after the 9/11 incident, and no nature writer can escape from terrorism and social upheaval. In this place that lacks a sense of safety, people have to not only become world citizens, but also join the ranks of the species of this planet so as to obtain a sense of belonging. In a situation where fear and hope coexist, Slovic calls on people to start from small things near them and protect this world while appreciating life. There needs to be a balance between "appreciation" and "protection," and between "engagement" and "retreat" (Slovic, pp. 3, 12). He calls on people to take up the responsibility of protecting ecology, just as Snyder says in an interview, "Live a life burdened with ecological responsibility, a life committed to protecting the areas of their residence and a life as if they are going to live there for ten thousand years" (Slovic 2008, p. 67). Mary Oliver's poem Williams quotes at the beginning of *Refuge* is also an inspiration for contemporary people: cast away confusion and despair, believe in an eternal land and brilliant sunshine, listen to the call of the wild geese, return your heart and body to the home on the land of the planet. The works of these three writers covered in this chapter show the changing characteristics of nature writing in today's society. Their literary call is for obligations toward and responsibilities for the land, as well as society and the family, and is the great love that links human love with love of the land.

## Lecture 26. Beston: The Outermost House

Jutting out from the East Coast of America, extending for about 30 miles from the inner coast of Massachusetts, there is a uniquely shaped piece of land that looks like an arm encircling the west shore of the Atlantic Ocean. This is Cape Cod, much talked about in American nature writing. On 11 October 1964, a special day for Henry Beston (1888–1968), about 1,000 people gathered on Cape Cod to attend the ceremony of naming The Outermost House as a "national literary landmark." About 40 years earlier, Beston had lived alone on the desolate beach on Cape Cod for a year, and his book *The Outermost House* is based on this experience. Now Beston, aged 76, came back to the hut he had lived in many years earlier and read paragraphs from his book to his admiring readers. "Touch the earth, love the earth, honour the earth, *her* plains, *her valleys, her* hills, and *her* seas; rest your spirit in *her* solitary places." Even though most had read the book before, many in the audience had tears in their eyes; Beston's words once again moved them.

Cape Cod is separated from the sea by glacially formed steep cliffs. There the sea met the last barrier between the two worlds. The Cape, over 20 miles long, is known not only for its natural beauty, but also for its cultural history. The well-known representative of seventeenth-century immigrants from Europe, William Bradford (1590–1657), in his *Of Plymouth Plantation* (1651), described Cape Cod as a "hideous and desolate wilderness" of "a wild and savage hue" (Bradford [1651] 2002, p. 62). This became one of the perennial themes in writings about Cape Cod and awakened Americans to the scenery of the North American coast. In the nineteenth century, Thoreau visited Beston's Great Beach several times, and exclaims movingly in his collection of essays on the area *Cape Cod* (1865), "A thousand men could not have seriously interrupted it, but would have been lost in the vastness of the scenery as their footsteps in the sand" (Thoreau [1865] 2008, p. 54). Nature writer John Hay (1915–2011), lived on Cape Cod and published many collections of essays about the Cape, including *In the Company of Light* (1998), and these have been ranked in the same category of the works of Thoreau, Beston and Carson. Contemporary American poet Mary Oliver (1935–) has lived on Cape Cod for more than 40 years and the Cape has provided much inspiration for her prize-winning works. Another nature writer, Robert Finch (1943–), attended that 1964 ceremony of naming The Outermost House a "national literary landmark." Finch, who not only published many works about Cape Cod but also edited *The Norton Book of Nature Writing* (1990), wrote a long introduction for the 1988 edition of *The Outermost House* (of which the 1992 copy that I have is a reprint).[1] In it Finch writes,

> Few, if any, natural areas of comparable size on the continent have been written about so extensively and so well, perhaps in part because this meeting place of land and sea—of the gentle, human-scaled contours of the Cape's soft sands and the protean, formless wilderness of the open ocean—allows us to see what we need to see, providing us with the raw materials of personality and voice. (in Beston [1928] 1992, p. xii)

American writer Nan T. Waldron says in her *Journey to Outermost House* (1991) that, of all the books written about Cape Cod, *The Outermost House* is the most outstanding and is widely regarded as the most definitive. It shows the reader a journey with endless water and mountains and confirms a restless life force and a possible harmony between the heavens and humanity. More importantly, as Don Wilding says in *On Its Solitary Dune* (a special publication that commemorates Henry Beston and *The Outermost House*), when the reader closes the book, they still feel it (Wilding 2001, p. 26).[2]

## A Writer Blended into the Landscape

Henry Beston was born in 1888 in Quincy, Massachusetts (also the birthplace of two former US presidents, John Adams and his son John Quincy Adams). Beston's father was a well-known doctor of Irish descent and his mother a devout Catholic of French descent. Beston describes his early experience as "a New England boyhood of sea and shore, enriched with a good deal of the *French spirit,*

1 A Chinese translation of the book *The Outermost House,* translated by Hong Cheng, was published in 2007 by Joint Publishing.

2 The title comes from a sentence in Chapter One of *The Outermost House,* "On its solitary dune, my house faced the four walls of the world."

from a French mother" (quoted in Wilding 2001, p. 3). He grew up in a bilingual environment (English and French), and thought in French as comfortably as in English. Beston obtained his bachelor's degree in 1909 and master's in 1911, both from Harvard College. After graduation he taught at the University of Lyon in France. Beston served in the First World War, and wrote two books based on his wartime experiences, *A Volunteer Poilu* (1916) in French and *Full Speed Ahead* (1919) in English. However, he was traumatized by his war experiences and after returning from the war he turned to a different writing genre—fairy tales and children's stories—to heal the wounds of war in his heart. After publishing two fairy tales he worked as an editor for a magazine in New York, continuing to seek themes and styles that would give full play to his writing potential, and finally turned to the beach on Cape Cod.

In 1925, in middle age, Beston purchased a piece of land near the beach on Cape Cod, designed a small cottage, and had it built on the beach facing the sea. He named the cottage "The Fo'castle." At first, he only wanted to live in the cottage for a couple of weeks of autumn in the following year, and did not intend it to be his permanent home. However, after two weeks Beston did not leave because the land, its beauty and the mystery of the sea fascinated him. He had finally found a way of expressing himself as a writer, and wrote in his diary in French about the themes and goals of his work.

> La Nature—voilà, mon pays.
> L'œuvre—cé lé brer, révé ler la mystère, la beauté, et la mystique de la Nature, du Monde visible
> Attacher ce sentiment mon nom.
>
> Nature—there is my country.
> Work—celebrate, reveal the mystery, the beauty and the mysterious of Nature, of the visible world.
> Bind this sentiment to my name. (Beston [1928] 1992, p. xv)

Beston lived on Cape Cod for a whole year and recorded vivid images of nature: the high and low tides, the waves rolling toward the beach, all kinds of birds coming, passers-by on the sea, winter storms, the grandeur of autumn, the spirituality of spring and the burgeoning of summer. He found nature's celebrations incomparable, and wanted to become a nature writer; however, when he left the beach in autumn he did not have a publishable manuscript, although he had several notebooks filled with materials. The fact that Beston did get down to writing the book is credited to his then fiancée Elizabeth Coatsworth, also a writer, who firmly believed that Henry's talent indicated a great future. One story says that when he asked Elizabeth to set a date for their wedding she answered, "No book, no wedding." *The Outermost House* was published in autumn 1928, and the couple married in June the following year and spent two weeks in The Fo'castle during their honeymoon.

*The Outermost House* has 10 chapters, structured according to the rhythm of nature, starting from autumn and ending in autumn, going a full circle. The author uses poetic language to describe the house he lives in, the beach and sand dunes he explores and all kinds of birds and plants he observes on them, the scenery of the four seasons, and occasional passers-by on the beach. He not only praises the beauty of nature but also reveals its cruelty. The most moving part is the spiritual enlightenment and insight he experiences when he is alone on the beach, encountering nature and communicating with it in solitude. Beston wrote many books on nature, but *The Outermost House* was the peak of his writing career. As his widow Elizabeth says in one of her introductions in *Especially Maine: The Natural World of Henry Beston from Cape Cod to the St. Lawrence*, a selected collection from Beston's works, "Sand dunes can appear and disappear, humans live and die. But he felt that his writing reached the top and had no regret in his life" (in Beston 1970, p. 16).

At first sales of *The Outermost House* were slow, but gradually its readership grew; by 1949 it had gone through 11 reprints, and it has been reprinted a further 26 times since then, with four audio editions of the book in 2007 alone. In 1991, the well-known American composer Ronald Perera composed music for *The Outermost House* which was sung like hymns. Perera, who had composed music for Charles Dickens's and James Joyce's works, had a special place for *The Outermost House* in his heart and he called his musical edition of the book "the perfect combination of words and music" ("Ronald"). Finch, a long-term resident of Cape Cod, thought extremely highly of *The Outermost House*, saying in his introduction,

> With the emergence of a post-World War II environmental literature, it began to achieve something of a cult status. Rachel Carson said that it was the only book that influenced her writing. In 1960 *The Outermost House* was cited by federal officials as one of the motivating forces in the creation of the Cape Cod National Seashore. It is now generally recognized as one of the classics of American nature writing, and each succeeding generation seems to have discovered anew the book's unique attraction. (in Beston [1928] 1992, p. xvi)

Barton Levi St. Armand says that *The Outermost House* must be read from the perspective of musical tonalism, poetical imagism and artistic luminism so as to experience the aesthetic pleasure and conceptual sublimity (Quoted in Hong Cheng's translation of *The Outermost House*, 2007, p. 3). In modern society, with cultural fast food everywhere and many writers publishing books at a speed similar to that of an assembly line, Beston's writing is intense and meticulous in a way that is rarely seen nowadays. According to his widow Elizabeth, "*only when he had become part of a landscape* could *he write* about it" (quoted in Wilding 2001, p. 28). She also describes how Beston wrote *The Outermost House*

> with a pencil or pen—on typewriter paper, except when taking notes—he never typed, for the sound of a machine would have interfered with the rhythm of his sentences, which meant so much to him. As he worked, the floor became littered with sheets of discarded paper. He sometimes spent an entire morning on a single sentence. (Beston 1970, p. 4)

Precisely because of this, we can feel the poetic, picturesque and musical beauty that St. Armand refers to.

The small lone house that the author built, although simple, did not lack romance: its wall boards and the window frames were painted light yellowish brown, the typical color of a sailor's cabin. The author named it "The Fo'castle" because it was built on a sand dune jutting into the sea, and was like a small boat on the sea, exuding a remoteness and mystery. The house had many windows; as the author says, "Seven windows in one room perched on a hill of sand under an ocean sun—the words suggest cross-lights and a glare" (Beston [1928] 1992, p. 7). It was a semi-outdoor house; sunshine came inside, and the sea came to his door. He could lie down with his head on his pillow and see the sea; watch the stars over it, the lights of berthed fishing boats and the white waves; and listen to the waves echoing among the sand dunes.

Nature under his pen had an epic, grand beauty, and Cape Cod was shown in a powerful picture, "East and ahead of the coast of North America, some thirty miles and more from the inner shores of Massachusetts, there stands in the open Atlantic the last fragment of an ancient and vanished land" (p. 1). On this piece of land, a battle raged continuously between the sea and the land.

> Age by age, the sea here gives battle to the land; age by age, the earth struggles for her own, calling to her defense her energies and her creations, bidding her plants steal down upon the beach, and holding the frontier sands in a net of grass and roots which the storms wash free. (p. 2)

This natural phenomenon of waves has a certain moving and tragic beauty: "Sound of surf in these autumnal dunes—the continuousness of it, sound of endless charging, endless incoming and gathering, endless fulfilment and dissolution, endless fecundity, and endless death" (p. 44).

Then we follow the author to repeatedly watch the waves coming from the outer sea of the Atlantic Ocean. They go over obstacles, and after breaking up and reforming, become a gigantic wave using its last energy and beauty to reflect the blue sky and break into pieces on the beach. In the repeated pattern of filling up, gathering together, achieving and breaking up, rebirth and re-death, we feel deeply the progress of human beings, one succeeding another generation after generation, and the sorrowful grandeur and poetry that accompany this process.

The sea and the sand on the dunes also had unspeakable charm in the author's heart, as their color tones changed with time and seasons. Hence, we see a picture painted by Beston,

> The winter sea was a mirror in a cold, half-lighted room, the summer sea is a mirror in a room burning with light. So abundant is the light and so huge the mirror that the whole of a summer day floats reflected on the glass. (p. 192)

We follow the author's eyes and also see, "a beautiful white gull sailing along the volute of a breaker accompanied by his reflection in the wave" (p. 54). Under Beston's pen, the sands also change with the seasons. In summer the warm, bright color is beautiful beyond words. At dusk, a mysterious light purple would cover the place where the beach met the sea. In winter, the warm golden-yellow color is replaced by a cold silver grey which cannot return the sunshine.

Beston summarizes the three great elemental sounds in nature: the sounds of rain, wind in a primeval wood, and of ocean waves on a beach. He believes that the sound of ocean waves on a beach is the most beautiful and awe-inspiring. He says in an interview,

> I like the rhythm of English prose. Sentences should be like waves in a sea, coming one after another, and each maintaining its own individuality. A sentence should have appropriate rhythm, wave coming, wave breaking up and wave retreating and then should be a short pause before the next wave. (quoted in Wilding 2001, p. 28)

In *The Outermost House* Beston uses the tonal variability in western music to describe the harmony of music on the sea.

> Listen to the surf, really lend it your ears, and you will hear in it a world of sounds: hollow boomings and heavy roarings, great watery tumblings and tramplings, long hissing seethes, sharp, rifle-shot reports, splashes, whispers, the grinding undertone of stones, and sometimes vocal sounds that might be the half-heard talk of people in the sea. (p.43)

No wonder the author exclaims that he would listen, "never wearying of the sonorous and universal sound" (p. 46); and no wonder Finch calls *The Outermost House* "a book that begs to be read aloud" (quoted in Wilding, "Finch Introduces 'A Sense of Place' Exhibit," 2008).

However, under the author's pen, not everything is rosy colored; we also see the cruelty of nature in the book. Beston reminds us that

> [t]o understand this great outer beach, to appreciate its atmosphere, its "feel," one must have a sense of it as the scene of wreck and elemental drama....
>
> ...Go about in the cottages, and you may sit in a chair taken from one great wreck and at a table taken from another; the cat purring at your feet may be himself a rescued mariner. (p. 121)

After describing a shipwreck which caused the death of nine people Beston directs our attention to the sea, filled with the fragments of the shipwreck, and to the seagulls circling overhead. "Gulls were milling over the surf and spume...and gulls were flying back and forth between the breakers and the marsh. From their point of view, perhaps, nothing had happened" (p. 89). This last short sentence speaks volumes about the indifference of nature. In another tragic shipwreck, Beston describes a moving scene. In the cabin of the five sailors who have died, among the bits and pieces they had left behind, he finds a booklet titled "If You Were Born in February." From its mouldy pages he reads, "Those who are born in this month have a particular affection for home.... They will go through fire and water for their loved ones." Beston makes the thought-provoking comment, "Who brought this thing aboard? One wonders. Whose curious hands first opened it in the lamplight of this tragic and disordered space?" (p. 120) Natural tragedy at this moment highlights the human impact.

Even though there is a cruel indifference in nature, what we see under Beston's pen is scene after scene of beautiful landscape paintings, the scenes of an imagist poet: "The ocean thunders, pale wisps and windy tatters of wintry cloud sail over the dunes, and the sandpipers stand on one leg and dream, their heads tousled deep into their feathers" (p. 18). He describes the sunset: "The afternoon sun sinks red as fire; the tide climbs the beach, its foam a strange crimson; miles out, a freighter goes north, emerging from the shoals," the lonely late autumn: "Mid-October and the land birds have gone. A few sparrows linger in the marshes. The plum bushes have lost their leaves. Walking the beach, I read winter in the new shapes of the clouds," and the sea at night: "It is dark to-night, and over the plains of ocean the autumnal sky rolls up the winter stars" (pp. 22, 25, 34).

## A Philosopher Who Respects the Land

In the twenty-first century, when harmonious coexistence between humanity and nature has become a major theme, *The Outermost House* is not only of literary interest but also a source of thoughts and realizations about the relationship between humanity and nature.

American scholar Sherman Paul, commenting on Beston and his representative work *The Outermost House*, compares Beston with Han Shan, a Chinese poet of the Tang dynasty, pointing out that both places of residence were unique in that they were close to nature and to the cosmic world. (Paul 1992, p. 119) More than 1,000 years ago, Han Shan lived in the mountains in order to be "away from the troubles of the world" and lived a life close to nature. He wrote in one of his poems, "Han Shan has a house, Without beams or walls. Six doors are all open left and right, the hall sees the blue sky" (in Xiang 2000, p. 440; *Han Shan you yi zhai, zhai zhong wu lan ge. Liu men zuo you tong, tang zhong jian tian bi*). Similarly, in the 1920s Beston went alone to the isolated beach on Cape Cod, built a house with seven windows, and kept company with the sea, the sky, the beach and the seabirds in order to live a life close to nature. In their poems and writings, both narrated their experiences of nature and their spiritual sublimation, showing us a grand world view of humanity merging with nature. Beston felt the problems of modern society deeply, and the contrasting vitality of nature. He writes,

> The world today is sick to its thin blood for lack of elemental things, for fire before the hands, for water welling from the earth, for air, for the dear earth itself underfoot. In my world of beach and dune these elemental presences lived and had their being, and under their arch there moved an incomparable pageant of nature and the year. (p. 10)

The main character of this pageant is the sun; the book advocates a "sun worship and worship for the burning year" from beginning to end. This worship for the burning year is a basic form of nature writing. Beston says, "A year indoors is a journey along a paper calendar; a year in outer nature is the accomplishment of a tremendous ritual" (p. 59). He reminds us that we need to worship the sun if we wanted to join in the ritual, because the exciting experience of the sun is the drama of Mother Nature, on whom we rely for survival. If we do not appreciate the sun, respect it, join it, we close a heavy door to the everlasting and poetic spirit of nature. This worship of the sun and nature presents a sharp contrast to modern society's worship of material wealth. In fact, Beston urges us to walk out of the cement forest full of modern technology, control our desire for growth and development at any cost, and come close to the ancient nature we have long forgotten, to experience the deep and everlasting rhythm of the land and absorb a secret remedy and sustainable energy. We can say that in a modern society which is eager for instant success and quick profit, reconnecting with primitive nature and letting it support us has become necessary for human survival.

Beston, looking at the starry sky over the sea and comparing day and night, reflects on the relationship between humanity and nature and expresses his respect and love for the night, writing, "By day, space is one with the earth and with man—it is his sun that is shining, his clouds that are floating past; at night space is his no more." Night dominates and shortens the distance between man and nature, the human spirit is sublimated to it, and with the night sky "a new door opens for the human spirit." He regards watching the Milky Way as a spiritual journey, and on this short journey "a poetic mood" and an exulted sublimation arise from this experience (pp. 173–174). Therefore, our reliance on nature is not only necessary for survival, but also for our spiritual needs; as Beston says, "the ancient values of dignity, beauty, and poetry which sustain it are of Nature's inspiration; they are born of the mystery and beauty of the world" (p. 218).

Beston's ethics, which foreground nature, show that his sense of ecology was ahead of his time. In his book he expressly points out that we humans need to have a new, wiser or perhaps more mystical view of animals and comments that, remote from nature and surviving on resourcefulness, in modern civilization we view animals through distorted lenses. We patronize them for their wrong reincarnation, their lower position and their tragic fate, and this is where we are wrong, as animals should not be judged by humans. In a world more ancient and more complex than our living environment, animals have grown and evolved more perfectly and with more complexity than us. They have various sensitive organs which we have lost or never had; they communicate in sounds that we have never heard; they are not our fellow humans, nor are they our underlings; in the long river of life and time, they are other species drifting around with us, imprisoned by the splendour of the world and burdened with its trials (pp. 24–25). It was through this understanding of the animal world that Beston could watch with respect as white swans flew across the blue sky in October or wild geese flew along a streak of golden light at the edge of the horizon in the March dusk. He could listen in the quiet April moonlight to the sound of "a river of life" flowing "across the sky," that is, large groups of wild geese flying north. Observing a marsh hawk calmly deal with a provocative tern, he thinks of the sculpted eagle on the granite wall in ancient Egypt. He praises the Egyptians for having captured the spirit of the eagle, as the animals they carved had no human characteristics but were calm, proud members of the rough and wild world.

We know that birds migrate, but in Beston's book, he describes in detail the little-known migration of fish and sees that animals have certain abilities we humans don't understand. He notices that every April, a type of grey herring leaves the sea to swim up a stream in Weymouth, Massachusetts, lay eggs in a freshwater pond, then swim over dams and dykes back to the sea. The baby herrings born in the pond follow them in 10 months or a year's time, and return to the pond the following spring. Beston puzzles over how, in the vastness of the sea, every fish born in the pond in Weymouth remembers its birthplace and crosses the sea without any navigational marks to return there. Beston cannot help asking,

> What stirs in each cold brain? What call quivers as the new sun strikes down into the river of ocean? How do the creatures find their way? Birds have landscape and rivers and headlands of the coast, the fish have—what? (p. 160)

Yet the fish very quickly find their way back to their birth pond. Beston's description of the herrings' migration not only fills us with respect and curiosity for these creatures, but also makes us think about the eagerness and passion of nature to spread life everywhere, filling the land, the sky, the sea and every corner of the world with life. As we know, Beston's era was that in which Eliot's *The Wasteland* was published, and among humans a pessimistic view that nature was already dead pervaded. However, in *The Outermost House* what we read is optimism. In the summer of his one year on Cape Cod Beston writes,

> During the months that have passed since that September morning some have asked me what understanding of Nature one shapes from so strange a year? I would answer that one's first appreciation is a sense that the creation is still going on, that the creative forces are as great and as active today as they have ever been, and that tomorrow's morning will be as heroic as any of the world. *Creation* is here and now. (pp. 216–217)

From his book, we learn that Mother Nature is trying her hardest to fill every empty corner and forgotten place with life, giving the dead new life and the living more vitality. Mother Nature's passion for life is endless, unstoppable and also relentless. Beston exclaims,

> And all these her creatures, even as these thwarted lives, what travail, what hunger and cold, what bruising and slow-killing struggle will they not endure to accomplish the earth's purpose? And what conscious resolution of men can equal their impersonal, their congregate will to yield self life to the will of life universal? (p. 160)

This is thought provoking: humans are only part of the whole ecosystem. Shouldn't we reset our position by learning from the collective consciousness of animals, changing our behavior so as to obey the will of life of the whole universe?

*The Outermost House*'s year-long life experience starts and ends in autumn. The author calls the autumn he spends at the beach on Cape Cod the highlight of the year, and the book proves to be the highlight of his writing career. After Beston married, he moved with his wife to a small farm in Maine and spent the rest of his life there. On the farm, apart from work and writing, Beston was often invited to give speeches on nature writing and regularly went to Dartmouth College to lecture on nature writers. He later published nine more works on nature, including *Herbs and the Earth* (1935), *The St Lawrence* (1942) and *Northern Farm* (1948). However, none was never able to reach the enduring heights of *The Outermost House* again.

Beston and *The Outermost House* were awarded honor after honor. Beston was awarded two honorary doctorate degrees, and also a medal from The American Academy of Arts and Sciences for his outstanding contribution to literature—an honor that only Robert Frost and T. S. Eliot had received before him. In the autumn of 1964, when his little house became a national heritage building, Beston returned to the beach on which he built it for the last time to give a speech, and he died in 1968. With erosion from sea water, The Fo'castle—which he had donated to the Audubon Ornithology Society of Massachusetts—had to be moved back from the sea, then moved again a few years later, becoming the only mobile heritage building. However, what has remained unchanged is the spiritual symbol The Fo'castle had become. Its site still attracts numerous nature lovers, among them the contemporary American writer Nan T. Waldron (1922–2000), who stayed in The Fo'castle from time to time in the last 17 years of her life, during which time she wrote *Journey to Outermost House*.

In the autumn of 2004, during my time as a visiting scholar in America, I went to the original site of *The Outermost House*, the beach facing the Atlantic Ocean on Cape Cod, which I had read about in the book numerous times. It was deep autumn. A red brick house with white windows, the old coastguard station, stood alone at the top of a sand dune covered in plants. Near the station, a board had been erected with information about Beston and *The Outermost House*. By the time I visited, the house itself no longer existed; in February 1978, a winter storm had swept it into the sea and buried it about one mile from where I was standing. I looked around for the things I had read about in *The Outermost House*: the dunes with waist-tall grasses and other plants at the back of the beach; further back, the blue ponds formed by accumulated seawater; and lower down, the desolate beach with rolling waves crashing in. I went down the sand dune and walked along the beach—on which there were very few tourists—trying to experience how Beston felt when he carried food and other necessities from the Norseta coastguard station along the beach, stepping through the waves on his way back to the Outermost House more than 80 years earlier. I imagined the poetic scene one autumn day when Beston walk the beach and "read winter in the new shapes of the clouds" (p. 26). I sat down on a pale piece of driftwood, watching the sea rise and fall and the wind whipping the waves in a final act of "destruction on this solitary beach" (p. 34). I turned my eyes to the sea, knowing that about one mile away from me was the "burial place" of the Outermost House. From the time Beston left the house to the time when it was buried by the sea, only a number of decades, the sea had already encroached onto the narrow land by a mile; perhaps before long, the beach I was sitting on would also be swallowed up by the sea. However, at that moment, the material things were no longer important; what was important was that Beston had already left the spirit and poetry of the Outermost House to humanity. We can say that the invisible, at that moment, was more than visible and the absence was more than the presence. *The Outermost House* remains in our memory not as a material form, but as a sign of respect for the ancient natural environment and as a rich imagining of a simpler, more poetic life.

## Lecture 27. Williams: Refuge of the Heart

Contemporary American writer and naturalist Terry Tempest Williams (1955–), in her book-length essay *Refuge: An Unnatural History of Family and Place* (1991), writes,

> A heron stands on the edge of the lake, solitary and serene. The wind shinnies up her back, raising a few feathers, but her focus remains steady. This is a bird who knows how to protect herself. She has weathered the changes well. Throughout the high water and now its retreat, the true blue heron has stayed home. Perhaps this is a generational stance, the legacy of her lineage. (p. 266)

Using a heron as the reference object, Williams reveals in a few sentences the determination that humans should have to engage body and soul in nature in a changeable modern society. She reminds us that this inner force is, in fact, the cultural heritage handed down from generation to generation, although it has been weakened and covered over by modern colors.

Williams is committed to describing the natural world through its shapes and manners from an aesthetic perspective, and her writing has attracted both scholars and general readers alike. She has received many honors, including the Wallace Stegner Award and the National Wildlife Federation's Conservation Award for Special Achievement, and has been inducted to the Rachel Carson Honor Roll, and is an influential figure in American nature writing, eco-criticism and environment protection activities.

Although Williams was born in Corona, California, she grew up in Salt Lake City, Utah and has lived there ever since. As a child of Mormon parents she inherited a spiritual legacy and a love for the land, the indigenous culture and her family which have supported her life and her writing. She reminds people that in Mormon culture there is one thing you must be familiar with, that is, historical geography. In order to have religious freedom her ancestors walked 1,200 miles to the Salt Lake Valley in the 1850s, and the family has lived there ever since. Her forebears' experience helped Williams to realize that "[a]s a people and as a family, we have a sense of history. And our history is tied to land." In Mormon religion the natural world has spiritual values, and the author says very frankly, "We learned at an early age that God can be found wherever you are, especially outside. Family worship was not just relegated to Sunday in a chapel" (p. 14).

Williams obtained her Bachelor of Arts in English in 1978 and her Master of Science in Environmental Education in 1984. She has worked as a resident naturalist in the Utah Natural History Museum and part-time professor at the University of Utah, and since 1996 she has been focusing on writing and environmental conservation activities.

Williams is a prolific writer and has published many books, including *Pieces of White Shell: A Journey to Navajoland* (1984), *Coyote's Canyon* (1989), *Refuge: An Unnatural History of Family and Place* (1991), *An Unspoken Hunger: Stories from the Field* (1994), *Desert Quartet: An Erotic Landscape* (1995), *Leap* (2000), *Red: Passion and Patience in the Desert* (2001), and *Finding Beauty in a Broken World* (2008). These works make the Great Basin in the American West the backdrop to stories in which land and humanity are integrated. Scholars comment that Williams adopts the genre of "story essays," uses a "lyrical language" and is "a writer of prose poetry" (in Chandler and Goldthwaite eds. *Surveying the Literary Landscapes of Terry Tempest Williams: New Critical Essays*, pp. xi, 95, 96). *Refuge: An Unnatural History of Family and Place*, Williams's representative work, has been called an American nature writing "classic." It is about how humans, when faced with a variety of unpredictable factors—even disasters and life tragedies—seek refuge and spiritual sublimation in nature. In this "unnatural history of family and place," the title, structure, chapter names, content and language are all singularly extraordinary. Every chapter is named after a specific bird and then followed by the record of the water level of the Great Salt Lake, the rise and fall of which are closely linked with both the condition of her mother's cancer and with the survival of the migrant bird conservation area. The lake, birds and humans, forming an inseparable whole, are the protagonists of this book, and nature is no longer a backdrop or background. The book demonstrates a great sense of eco-vision and land ethics. After its first publication in 1991 it was published as audiobooks by two different publishers (in 1994 and 2002), and has also been extracted in collections and textbooks used in universities.

## *The Story of Land and Family*

Williams grew up by the Great Salt Lake in Utah, an expanse of undrinkable water in the desert which presents a surreal scene. However, this seemingly useless lake is closely connected with America's largest waterbird refuge, Bear River Migratory Bird Refuge. The wetlands that surround the Great Salt Lake like green diamonds provides an important natural breeding place for waterbirds and shore birds alike. There are 208 kinds of birds there, and it is the resting place for millions more birds in their spring and autumn migrations. It is these birds and the author who share a natural history. The Mormon Williams family has lived there for six generations. However, as their homes were located downwind from the American nuclear testing base used from the mid-1940s to the 1990s, most of the female members of the family have suffered from breast cancer. Williams's grandmothers, mother and six aunts all had mastectomies and seven of them died of breast cancer, prompting her to refer to her family as "the clan of One-Breasted Women." At the same time, Williams found that the water level of the Great

Salt Lake was increasing and that as a result, some of the birds in the Bear River Migratory Bird Refuge could disappear; human tragedy and tragedy in the natural world were happening in parallel. This phenomenon made the author boldly adopt a unique writing style to express her ecological views, integrating human inner feelings with specific natural scenery. Her beautiful diary-like essays record her experience of accompanying her mother in the last stage of her life beside the Great Salt Lake. In other words, the book illustrates how the mother and daughter watched as the animals and plants in the natural world by the side of the lake dealt with cruel fate while facing their own personal tragedy and trying to heal the wounds in their hearts.

This is a very moving book. Around the time that the author started to sort out her diary entries about the Great Salt Lake and prepare the book, she herself was also diagnosed with breast cancer, at the age of only 34. When she opened the diary notebook in which she had recorded the last days of her mother, a bird feather, sand from the shore of the Great Salt Lake and sage leaves all fell out, triggering great grief and sorrow in her heart. As she explained,

> Perhaps, I am telling this story in an attempt to heal myself, to confront what I do not know, to create a path for myself with the idea that "memory is the only way home." I have been in retreat. This story is my return. (Williams 1991, p. 4)

At the beginning of the story, the water level in the Great Salt Lake was rising and reached 4204.7 feet above sea level, almost reaching the Bear River Migratory Bird Refuge's 4206 feet above sea level, meaning that the refuge would be flooded and the birds there would become "refugees fleeing in panic." At the same time, the author's mother's health was deteriorating, and the cancer had spread to her stomach. However, her mother did not seek medical treatment and instead she went to the Grand Canyon because, to her, days spent in the Grand Canyon and on the Colorado River were a meditation and a spiritual revival. Passing through the ancient rocks gave her hope and enabled her to endure whatever she must face, and she found strength in solitude to confront the illness. After undergoing an operation, and with very difficult conflicting thoughts, she decided to undergo chemotherapy. From a conversation between mother and daughter, we can see their unique way of dealing with illness. The mother says,

> "Perhaps you can help me visualize a river—I can imagine the chemotherapy to be a river running through me, flushing the cancer cells out. Which river, Terry?" "How about the Colorado?" I said. It was the first time in weeks I had seen my mother smile. (Williams 1991, p. 39)

Mother Nature gives the mother and daughter strength and resilience to conquer their illness. In her struggle against the cancer, the mother feels her life has concentrated and she lives every day consciously. The way she absorbs energy from nature inspires admiration in the author. She sees her mother obtain optimism and openness from nature, her life rhythm accelerating, and describes her mother as "the bird touching both heaven and earth, flying with newfound knowledge of what it means to live" (p. 136).

However, as the water level rises to the highest point of 4211.85 feet above sea level, the wetland is flooded, the Bear River Migratory Bird Refuge office is closed, and her mother's condition further worsens. Even though seriously ill, her mother prepares and experiences her last Christmas dinner with her family, then passes away at home. However, as the author reports, her mother's passing is a calm, spiritual process. Keeping watch over her dying mother, she senses the light in the room dimming, and suddenly thinks that her mother will wait till the sun has set. The sunset that day is magnificent, with apricot yellow glistening over the purple mountains in the distance. She tells her mother that the sunset is beautiful—she remembers how her mother once applauded a wonderful sunset, and that she and her mother walked hand in hand among sunflowers as tall as their shoulders.

Her mother passes away smiling. The following night, she goes outside and looks at the night sky, and condenses all her feelings into two sentences, "A full moon hung in a starlit sky. It was Mother's face illumined" (p. 232). This reminds me of a Han Shan poem about a night scene at a cliff, "Stars line up deep in the night, / A lone lamp at the cliff is the moon still there, / Full, round and crystal bright, / hung in the dark sky is my heart" (Xiang 2000, no. 200; *Zhong xing luo lie ye ming shen, ya dian gu deng yue wei chen. Yuan man guang hua bu mo ying, gua zai qing tian shi wo xin*). The similarity here between the eastern and western cultures is remarkable. Only when humans and human feelings are integrated into nature can they reach sublimity and eternity. No wonder Williams says, "My physical mother is gone. My spiritual mother remains" (p. 241).

However, the family tragedy does not stop there. A year after her mother's death, Williams's maternal grandmother, who has been suffering from breast cancer and Parkinson's disease, also dies. The author and her maternal grandfather kept vigil for her, and it is a very moving scene. "He held her hand and I held his. Mother felt near. Death has become a familiar landscape. I can smell it" (p. 261). Then her paternal grandmother and soulmate, who taught her to watch birds, dies of cervical cancer. Another tragic scene: "Jack sits quietly beside Mimi's body. A single candle burns on her dressing table. Reflected in the mirror, it appears as two." This scene echoes one that Li Shangyin, a Chinese Poet of the Tang Dynasty, describes in one of his poems, "A spring silkworm will not stop spinning silk until death, / a candle's tears dry only when it is burned down to ashes. (Li Shangyin, "Untitled" in (1705) 2008 *Quan Tangshi* Vol. 539, No. 130; *Chun can dao si si fang jin, la ju cheng hui lei shi gan*).

In the last chapter of the book, the water level of the Great Salt Lake is at last controlled and falls back to the level in the first chapter: 4204.7 feet. When the spring bird migration starts, Williams and her husband row a canoe from Crescent Bay to the Great Salt Lake. The author stands at the bow in the wind, with all the things that have happened over the last seven years before her eyes—the love continues. She remembers the unforgettable experience of going to Mexico for *Dia de Muertos* (the Day of the Dead). In a cemetery she had met an old lady who had five or six relatives buried there. The old lady, who was holding a bunch of marigolds, talked to her in Spanish, waving her hand at the sky, and the author understood her to mean, "Very beautiful—this sky above us. With clouds like roses. The Dead are among us" (p. 279). When saying goodbye, the old lady gave her a marigold. Marigold is the flower her mother planted every spring. While the author is recalling these memories, the little boat comes to the center of the Great Salt Lake. The author carefully takes out a packet from her pocket, and from the packet takes the petals of the marigold, which she had kept. Carefully, with her husband, she sprinkles the petals on the lake. Her family members who have passed away would live together with the mountains and waters in the natural world. At the end of her story, the author calls the Great Salt Lake "My basin of tears. My refuge" (p. 280).

## *Listening to the Call of the Wild Geese*

Opening *Refuge*, the first thing that appears before our eyes is the poem *Wild Geese* by Mary Oliver (1935–). This poem sketches a vivid picture: the earth is moving, the sun is passing the beautiful land, and the brilliant sunshine is falling on it, geese flying high in the sky are on their way home. A whole world is before you, calling you like the geese. It repeatedly reminds people that they are a member of the family. This poem indicates the theme of the book: nature is an interconnected ecosystem, a big family in which all members coexist. Modern civilization should reawaken humans' love for family, for their connection with the land and with the whole ecosystem, and encourage them to find a balanced life, thus guiding them from a personal, emotional world to a realm of great love that accommodates all things.

However, in modern society the rhythms of life are accelerating, and blind development makes people restlessly pursue instant success and quick profit. Among the tall cement forests, even familial love is fading, let alone the close relationship with nature, because we have forgotten our love for the land. *Refuge* uses family stories and bird stories to guide people in how to reawaken their memories of the land and of family, to understand the meaning of being "from the same line and intimately connected," and to keep an inner, steady force in a restless modern society. Nature writing advocates not retreat from society, escape from responsibility or indulgence in self-admiration, but a greater love that links human kinship with land kinship.

From this perspective, *Refuge* is a very moving exploration of the importance of kinship, family and kindness. At night, the author is awakened by a nightmare of her mother dying and is left with a lingering anxiety that she cannot get rid of: "Without a mother, one no longer has the luxury of being a child" (p. 202). Her heartache makes her exclaim, "An individual doesn't get cancer, a family does." There is deep, painful anxiety. "Her anguish over us—the living watching the dying—the dying watching the living" (p. 214). While members of her family try everything to make her mother happy, her mother still tries to create a peaceful environment at home, and cooks the last Christmas brunch for them, even though she is very sick.

Previous night at the Christmas Eve dinner, with four generations gathered together, the author's husband's toast is significant.

> What are we guarding?...
>
> ...We are guarding the very ideal of family, the bond, the web connecting us all, which gives rise to an energy, a lust for life lost to those for whom family has lost significance. We are guarding the buffer of protection that a family's love gives to each of its members, unconditionally, a magic covering that we can wear as armor but not notice its weight. (pp. 207–208)

In *Refuge*, there is a kinship and a love that cannot be cut off or gotten rid of, and which includes not only love within a particular family, but also the love of a member of the earth for all other members in the family of the earth. This love becomes a belief that will support modern society and human development, and is therefore worth guarding.

*Refuge* also conveys the importance of belief to the world through the author's grandmothers' words, and emphasizes that "Faith without works is dead" (p. 198). The author's mother practices what she believes in; as an advanced stage cancer patient she uses her own experience to help other cancer patients, even though she herself is in serious difficulty. In a letter to a young person with brain cancer, she writes,

> I want to live and think as actively and creatively as it is physically possible for me to do. This year, with two major surgeries, one year of chemotherapy, and six weeks of radiation, has been the most difficult year of my life, and also the most beautiful. It has enabled me to sense and see things I never did before. It brings life into focus one day at a time. You live each moment and when you see the sunset at the end of the day, you are so grateful to be part of that experience. (pp. 83–84)

When the author has sent the letter for her mother, her mother's kind action makes her think, "Our correspondences have wings—paper birds that fly from my house to yours—flocks of ideas crisscrossing the country. Once opened, a connection is made. We are not alone in the world." This is very different from the modern concept that "every man is an island"; it is human friendship and love beyond blood relationships that people are yearning for. However, right at this moment, she turns her attention in a new direction, "But how do we correspond with the land when paper and ink won't do? How do we empathize with the Earth when so much is ravaging here?" She thinks of the heartbeats she felt when she was in her mother's womb, her mother's and her own, heartbeats of the earth. She has found her channel of communication with Mother Earth,

> I can drum my heartbeat back into the Earth…
> …I drum back my return. (pp. 84–85)

At this moment, the human love and friendship extended to the land embodies everything of the natural world.

In *Refuge*, we see that Mother Nature will open her vast arms when humans are faced with tragedies in their lives. When her mother was terminally ill, it was the Colorado River and the ancient rocks of the Grand Canyon that gave her hope and strength to deal with the disease. When the author lost her family members, it was the waves, wind, sunshine and lake waters that gave her love and warmth like her mother's arms. We saw the Great Salt Lake, the Bear River Migratory Bird Refuge, the experiences of the author herself, and her family all became one. Therefore, this kinship went beyond the author's family to include the big family of the earth. After her mother's death, the author finds that the men in her family have changed. Her father and brothers no longer go hunting.

American poet Robert Hass (1941–) wrote, "You hear pain singing in the nerves of things; it is not a song." Williams summarizes, "For the men in my family, their grief has become their compassion." Out of great sorrow comes great love. As Williams says, "Grief dares us to love once more.... My refuge exists in my capacity to love" (pp. 251–252, 178).

This human love extends to care and love for the land, making us re-examine modern civilization. Egg specimens in museums whose insides have been extracted make the author think of the earth that has been left as an empty shell through overdevelopment by humans. As her grandmother who is suffering from cancer says, "The hollow eggs translated into hollow wombs. The Earth is not well and neither are we. I saw the health of the planet as our own" (pp. 262–263).

At the same time, we also need to reconsider how we value nature. When Williams turns her eyes to nature, it is a communication between souls. What she sees is not the economic value; she notices: "[t]he eye of the cormorant is emerald; the eye of the eagle is amber; the eye of the grebe is ruby; the eye of the ibis is sapphire," and exclaims, "Four gemstones mirror the minds of birds, birds who mediate between heaven and earth," but we humans "miss the eyes of the birds, focusing only on feathers" (p. 95).

Williams is also concerned about the wetland that the birds rely on for survival—a very productive but also very threatened ecosystem. From waterbird biologists she learns that after the Great Salt Lake's water retreats, the recovery of the wetland would take three to seven years and the entire recovery process 15 to 20—in fact, the ecosystem would never regain its original condition. No other ecosystem could replace or absorb this complex wetland. As experts say, "There is a certain threshold that once crossed, we can never recover" (p. 114). Williams, in her vivid story, tells us to treat nature kindly and not go over this threshold.

Williams sees warnings about the development of human civilization in the behavior pattern of nesting groups of pelicans. She uses Brigham Young's management of the Mormon Church from success to failure as an example and explains that people who exclusively promote one pattern forget a critical element: diversity. Williams's comment is to the point. "Brigham Young's United Order wanted to be independent from the outside world. The Infinite Order of Pelicans suggests there is no such thing" (p. 103).

Modern society not only needs to accommodate multiple cultures, but also needs people to stay calm and have the stability and strength to respond to changes. Williams feels this deeply. "My fears surface in my isolation. My serenity surfaces in my solitude." She considers herself part of the desert, the mountains and the Great Salt Lake and regards the language of the wind, water and birds as integrated with that of humans, believing that "[p]eace is the perspective found in patterns" (p. 29).

As a female writer, Williams's understanding of civilization is unique. She repositions modern civilization from the perspective of family harmony and social stability. Modern civilization should not be isolated, selfish and self-centered, but closely linked with the family, society and the natural world. A Kenyan friend's words give Williams an insight, "Because we have forgotten our kinship with the land, our kinship with each other has become pale. We shy away from accountability and involvement. We choose to be occupied, which is quite different from being engaged" (p. 137). This phenomenon of disengagement with the land and with kinship and avoidance of accountability is a mistake of modern civilization. *Refuge* tells a story of human kinship with the land, the warmth of family and the deep affection in the human world. It uses the experiences of one family to tell us that even in modern western society, where time is money and speed determines survival, the kinship between human and human and that between humanity and the land are still the strong bonds that support society. From *Refuge* we see the cycle of love; human kinship extends to love for the land, and peace and stability in nature in its turn will feed back to humans as solace for the heart. We can say that nature writing has integrated human love for nature with human kinship, and turns land ethics to social ethics, and responsibility for the land to responsibility for society. What the book promotes is great love and the chain of love.

## *The Resonance Between the Humans and the Land*

*Refuge* is a stylistically unique book with a vivid language, the success of which, to a large extent, comes from the author's personal experience. Williams says frankly,

> The Bear River Migratory Bird Refuge has remained a constant. It is a landscape so familiar to me that there have been times I have felt a species long before I saw it…The birds and I share a natural history. It is a matter of rootedness, of living inside a place for so long that the mind and imagination fuse. (p. 21)

This special experience made it possible for her to adopt the extraordinary method of using a different bird as the title of each chapter, while the birds, the lake and the humans together create a story of the land. In doing this she creates a new textual organization, and uses a new language, which is also one of the special features of nature writing.

Chapter 7, "The Californian Seagull," is about the author's experience of seeking mental support from the California seagulls when her mother's condition deteriorates. Historically the Californian seagulls at the Great Salt Lake once saved the Mormons. In the 1840s, the author's ancestors fled religious persecution and came to the Great Salt Lake area led by Brigham Young, because on this isolated, sandy and desolate land they could establish their Mormon community as they wished and enjoy religious freedom. In 1848, when the land the Mormons had tilled was about to produce a bumper harvest, crickets invaded the wheat fields, and there was no way they could defend against this natural disaster. Then unexpectedly, Californian seagulls came and cleared the fields of the crickets and saved the crops. From then on, the Californian seagull has been the subject of legends and has been named the state bird of Utah. When the author looks at the California seagull by the side of the lake, she discovers another characteristic of the bird: to avoid the threat of predatory animals and humans, the California seagull has chosen as its breeding place the remote and desolate islands of the Great Salt Lake where food is scarce. In order to find food, they have to fly a 50- to 100-mile return trip from their breeding place to the Bear River Migratory Bird Refuge and back every day. The motive behind this action resembles that of the Mormons. Her ancestors' story and that of the Californian seagull give Williams courage and strength to face her difficult situation. When disasters and pressure come, she wants to be like the California seagull, improve her adaptability and give wings to her heart.

Chapter 10, "Snow Buntings," fuses natural scenery with inner scenery. Williams and her mother come to the frozen Great Salt Lake, and the snow and ice remind the mother of Tolstoy's short story *God Sees the Truth, but Waits.* In this story, the protagonist is falsely accused of murder and spends the rest of his life in a concentration camp in Siberia. Her mother suggests Williams read the story and explains,

> Each of us must face our own Siberia…
> …We must come to peace within our own isolation. No one can rescue us. My cancer is my Siberia.

At this moment, two white birds fly into their vision, seeking food at the edge of the melted snow. When it is confirmed that these are the snow buntings which breed in the Antarctic, Williams's mother says thoughtfully, "Tell me, Terry, are these birds Tolstoy may have known?" (pp. 93–94) Her mother's thoughts are wide ranging and the author's imagination is also very lively. In Chapter 15,

"Whistling Swan," Williams spends a lot of time by the lakeside and carefully decorates a white swan that had died in the flood, maybe in preparation for the funeral of her dying mother. Then she lies down beside the swan and imagines how he extended his wings and flew.

> I imagined the stars seen and recognized on clear autumn nights as they navigated south. Imagined their silhouettes passing in front of the full face of the harvest moon. And I imagined the shimmering Great Salt Lake calling the swans down like a mother, the suddenness of the storm, the anguish of its separation.... At dusk, I left the swan like a crucifix on the sand. I did not look back. (pp. 121–122)

Chapter 30, "Birds-of-Paradise," describes what happens after the death of the author's mother. Williams finds images of her mother in nature. When sorting out the things left behind by her mother, she notices her hairbrush on the counter. She pulls the short dark hairs out of the hairbrush and remembers that her mother once mentioned that she would like to become a bird's nest. To fulfil her mother's wish,

> I quietly open the glass doors, walk across the snow and spread the mesh of my mother's hair over the tips of young cottonwood trees—
> For the birds—
> For their nests—
> In the spring. (p. 233)

*Refuge* has been called a "classic" of American nature writing not only because it is telling a moving story in a unique style, but also because it uses a fresh and moving language.

Williams's writing style is unique in that she describes not only the landscape, but also the soundscape. Some critics call this "the echo system of writing," that is, the soundscape in nature writing echoes that in the natural world (Chandler and Goldthwaite, p. 84). In noisy modern society, people's hearing is increasingly desensitized, just as the Italian philosopher Gemma Corradi Fiumara says, "modern language has lost its 'eco-logical' dimension." She then explains that modern people have failed to listen to trees, animals, birds, mountains, insects, rivers, oceans and land before their cries become too palpable to ignore (quoted in Chandler and Goldthwaite, p. 91). In *Refuge,* through intimate contact with the land and through the presentation of the soundscape, Williams recovers the listening ability that modern people have lost. After her mother dies, she drives to the Great Salt Lake. As she passes by the Mormon Church, she heard the hymns broadcast over the square; however, it is not the religious sound that consoles her, but the sound of the wilderness:

> Once out at the lake, I am free. Native. Wind and waves are like African drums driving the rhythm home.... My hair is tossed, curls are blown across my face and eyes, much like the whitecaps cresting over waves. Wind and waves. (p. 240)

The repetition of "wind and waves" adds movement to the scenery of the lake, making the reader feel the waves of emotion in the author's heart. The landscape echoes the emotions, and this is a multidimensional landscape, as it also has sound. In Chapter 9, Williams links the mother's womb with the land and human heartbeats with the heartbeats of the land, seeking communication with the land:

> I can drum my heartbeat back into the Earth, beating, hearts beating, my hands on the Earth—like a ruffed grouse on a log, beating, hearts beating—like a bittern in the marsh, beating, hearts beating. My hands on the Earth beating, hearts beating. I drum back my return. (p. 85)

The heartbeat and the drumbeat provide a strong rhythm which leads us to a landscape of primitive forest where ruffled grouses wander around, and urges us to follow the drumbeats to communicate with the land and make human language echo with the language of the land.

In *Refuge*, you can see the charm of Williams's "prose poetry" language. She describes the Great Salt Lake,

> The sun set behind Antelope Island. Great Salt Lake was a mirror on the valley floor. One had the sense of water being in this country now, as the quality of light was different lending a high gloss to the foothill. (p. 34)

The Great Salt Lake in winter presents another scene:

> The eastern shore of Great Salt Lake is frozen, and for as far as I can see it translates into isolation. Desolation. The fog hangs low, with little delineation between earth and sky. A few ravens. A few eagles. And the implacable wind. (p. 91)

She describes the migratory birds that have come to spend winter at the lake, "Twelve bald eagles stood on the ice of Great Salt Lake, looking like white-hooded monks. From November through March, they grace northern Utah" (pp. 201–202). She calls snowy plovers the scribes of the salt flats. "Their tracks are cursive writing, cabalistic messages for the bird-watcher who cares enough to follow their eccentric wanderings. On this day, their calls are the only dialogue in the desert" (p. 259).

She describes the desert by the side of the lake:

> It's strange how deserts turn us into believers. I believe in walking in a landscape of mirages, because you learn humility. I believe in living in a land of little water because life is drawn together.... There is no place to hide, and so we are found. (p. 148)

When climbing the canyon in southeastern Utah, she falls down a cliff. Though she doesn't lose her life, her forehead is scarred. She described this experience, "I have been marked by the desert. The scar meanders down the center of my forehand like a red, clay river. A natural feature on a map. I see the land and myself in context" (pp. 243–244).

She writes about seeking motherly love in nature,

> I am reminded that what I adore, admire, and draw from Mother is inherent in the Earth. My mother's spirit can be recalled simply by placing my hands on the black humus of mountains or the lean sands of desert. Her love, her warmth, and her breath, even her arms around me—are the waves, the wind, sunlight, and water. (p. 214)

Of course, Williams's rich imagination and her delicate fine feminine feelings make her words and phrases like clouds floating overhead, "A thin green line appears on the horizon. Bulrushes. The liquid, lambent stage for birds.... Tonight these mountains are lavender with blue creases that fall like chintz" (pp. 108–109).

Williams's language has a lingering and evocative beauty and it also gives us hope, "When Emily Dickinson writes, 'Hope is the thing with feathers that perches in the soul,' she reminds us, as the birds do, of the liberation and pragmatism of belief" (p. 90).

*Refuge* can be read from multiple perspectives. It is a nature writing classic, describing in vivid prose the close connection between humans and nature and urging us to reflect on modern civilization and to fuse with the natural world, which is far greater than humans; it contains eco-psychological content, describing how nature relieves pains and heals wounds; and it shows eco-feminist features. The shapes of nature under Williams's pen are all symbols of women, and she believes that human ravaging of nature is like the unfair treatment of women.

It is also a book that promotes traditional virtues and accommodates multiple cultures, reminding us that at a time of modern technology and development, family, kinship, values and a variety of cultures remain the supporting foundation of society.

Williams's unique experiences and her rich imagination make her writing non-linear and associative. In *Refuge* she mentions "mirage" and "illusions" many times, and as a result the sparks of her thoughts are sometimes difficult to capture. However, the story of land and family she describes is a story every one of us experiences. We can follow the author's emotional journey, we can experience it and we find it difficult to stop. Maybe, inspired by Williams, we will be able to find our own "refuge." As she explains,

> The landscapes we know and return to become places of solace. We are drawn to them because of the stories they tell, because of the memories they hold, or simply because of the sheer beauty that calls us back again and again. (p. 244)

## Lecture 28. Snyder: Wilderness is Nature

*The Gary Snyder Reader: Prose, Poetry and Translations*, published in 1999, collects the author's prose, poetry and translation works into one book and shows all aspects of his writing. However, when you open the book, what you see on the dedication page is a quotation from the *Analects of Confucius*, the ancient Chinese philosopher, "To learn and then put it in practice—isn't that a delight? To have friends come from afar—isn't that a joy?" Examining Snyder's life, his writing and his spiritual journey, this quotation reveals the author's life philosophy. Gary Snyder has been called "a rarity in the United States" (in Snyder 1999, p. 321). He is a simple person who loves and experiences nature, a western scholar who is interested in eastern cultures, a poet who was once part of the so-called beat generation, and a philosopher who looks at the relationships between humanity and nature, between a specific place and the whole earth from a perspective that cuts across time and space. He is like a multifaceted prism, reflecting varied images. This chapter seeks to approach Snyder from a literary perspective.

Snyder's unique contribution to nature writing lies in that, first of all, he is one of the few who observe nature and describe it from the perspectives of various cultures, due to his interest in eastern culture and also in the indigenous American Indian culture. His writing shows a communication that crosses both time and borders, between eastern and western culture, on the issue of humanity's relationships with nature. Secondly, he links the land on which people live with the whole world in his exploration, against the background of multiple cultures, of how to protect the environment, extending the understanding of the value of wilderness in the nature writing tradition. He practises what he says, and promotes a "re-inhabitation" lifestyle. He not only believes in bioregionalism, but also tries to describe the landscape of a particular region with a particular bioregional language.

Snyder is a prolific writer. His works include *Myths & Texts* (1960), *The Back Country* (1967), *Riprap and Cold Mountain Poems* (1969), *Turtle Island* (1974), *Mountains and Rivers Without End* (1996) and *Danger on Peaks* (2005). *Turtle Island* won the Pulitzer Prize for Poetry in 1974. His prose collections include *Earth House Hold* (1969), *The Practice of the Wild* (1990), *A Place in Space* (1995) and *Back on the Fire* (2007). He became a member of the American Academy of Arts and Letters in 1987, and was awarded the John Hay Award for Nature Writing in 1997.

It can be said that Snyder has devoted his whole life to describing the relationship between man and nature in poetry and prose essays. However, Patrick D. Murphy, a scholar on Snyder, says in his *A Place for Wayfaring: The Poetry and Prose of Gary Snyder* (2000, p. 12) that since the 1990s he has relied more and more on prose essays to call on people to reconnect with nature, with *The Practice of the Wild* and *A Place in Space* published in 1990 and 1995 respectively. Thomas J. Lyon, a well-known American nature writing scholar in the 2001 edition of his book *This Incomparable Land: A Guide to American Nature Writing*,[3] not only lists Snyder's *Earth House Hold*, *The Practice of the Wild* and *A Place in Space* as important American nature writing references in his book, but also lists the publication of *The Practice of the Wild* as an event in the chronology of the development of American nature writing. Another American scholar, James I. McClintock, in his monograph on modern American nature writing *Nature's Kindred Spirits: Aldo Leopold, Joseph Wood Krutch, Edward Abbey, Annie Dillard, and Gary Snyder* (1994), considers Snyder to be in the same order of importance as the other four writers included, and points out that of the five, Snyder's

3 It is worth pointing out that the 2001 edition of *This Incomparable Land: A Guide to American Nature Writing* is a revised edition, and is not only a collection of American nature writings as its first edition in 1989. As its subtitle indicates, Lyon discusses American nature writing's development, history, categories, its major writers and their representative works. It also has detailed necessary and supplementary references.

connections to Thoreau "are most obvious and far reaching" (p. 9). *Poets on the Peak* and *The Practice of the Wild* respectively describe Muir's and Leopold's influences on Snyder's life and his writing (in American writer John Suiter's *Poets on the Peaks: Gary Snyder, Philip Whalen & Jack Kerouac* 2002, p. 83; Snyder 1999, p. 215), which is also why I will approach Snyder from the perspective of nature writing in this chapter.

## Music from Far Away

Chinese readers are probably all familiar with the Tang Dynasty poet Zhang Ji's poem "Mooring by Maple Bridge at Night," and the Han Shan Temple mentioned in the poem, which may also remind people of a less well-known Tang poet named Han Shan. Surprisingly many contemporary American nature writers, including Snyder, are very much interested in Han Shan, a Chinese poet of more than 1,000 years ago, and try to communicate with this poet of long ago and far away. Snyder has translated 24 of Han Shan's poems, and they are collected in his *Riprap and Cold Mountain Poems* (1969). Snyder admires Han Shan (Cold Mountain) the person, his poems and his lifestyle. In Jack Kerouac's novel *Dharma Bums* (1958) the protagonist Japhy Ryer, modelled on Snyder, is called the "American Cold Mountain." In 1984, when Snyder visited China, he made a special trip to the temple in Suzhou, gave the temple a copy of his translation of Han Shan's poems, and also wrote right there and then a poem titled "At Maple Bridge." Literary critic Timothy Gray says that Snyder incorporated his own experience and imagination in the mountains of California while he was translating Han Shan's poems, and therefore these poems are not only semantically accurate translations but also a combination of Han Shan's Chinese Zen mentality and mountain soul and Snyder's own experience of the wilderness (quoted in Murphy 2000, p. 62). As an American writer and a poet Snyder could look at and describe nature from the perspectives of both eastern and western cultures, as a result of his life experience, and once said frankly, "When I was young, I had an immediate, intuitive, deep sympathy with the natural world, which was not taught me by anyone. In that sense, nature is my 'guru' and life is my Sadhana" (Snyder 1999, p. 91).

Snyder was born in San Francisco in 1930, and 18 months later his family moved to a King County north of Seattle, where his father managed a small dairy farm surrounded by woods. From his home he could clearly see the snow-covered mountain ranges all around. Snyder says that "[t]he childhood landscape is learned on foot" (p. 184). He wandered around the farm and the woods, mountains and wetlands, and when he was about 10, he would spend nights in the woods alone; to him, "The woods were more of a home than home." The surrounding mountains attracted him even more, as "[t]hose unearthly glowing floating snowy summits are a promise to the spirit"; and at the age of 15, he climbed to the top of Mount St Helens (9,000 feet above sea level) and experienced the ecstasy of a mountaineer. He exclaims,

> Being above all the clouds with only a few other high mountains also in the sunshine, the human world still asleep under its gray dawn cloud blanket, is one of the first small steps toward Aldo Leopold's "thinking like a mountain." (Snyder 1999, pp. 214–215)

In 1947 Snyder went to Reed College in Ohio to study anthropology and literature. While in college he showed an interest in American Indian culture, and his thesis on the folklores of the American Indians—which became "[t]he most copied Reed thesis of all time"—was later published as *He Who Hunted Birds in His Father's Village* in the 1970s (Suiter, p. 7). After graduating from college he went to Indiana University to study linguistics and anthropology at postgraduate level. He was a very good student and was given a position as a teaching assistant. From there Snyder could have followed the usual career path in academia; however, his heart was pulled in another direction. Suiter describes an enlightening experience that Snyder had by a roadside in Shoshone Mountains in Nevada in the autumn of 1952, on his way to Indiana University. There were very few cars on the road, and to while away his time he took out a book by a Japanese Zen master. In that clear September afternoon, a story about a Chinese Tang Zen master moved him, revealing to him the importance of direct experience and self-awareness, and how they differ from knowledge in books. This enlightenment in the mountains gave Snyder a new belief; several years later he recalled, "And although I didn't know it at the moment, that was the end of my career as an anthropologist" (quoted in Suiter, pp. 19–20)—his study at Indiana University only lasted one semester. Pioneer of nature writing Thoreau wrote in *Walden*,

> Why should we be in such desperate haste to succeed and in such desperate enterprises? If a man does not keep pace with his companion, perhaps it is because he hears a different drummer. Let him step to the music which he hears, however measured or far away. (quoted by James E. in Miller in his *Heritage of American Literature: Beginnings to the Civil War*, 1991, p. 1268)

Indifferent to fame and interested in mountains and rivers, Snyder looked to the far East. From primitive nature and ancient eastern civilization he saw the mistakes of western civilization, and found his spiritual home in the East. After he ceased his studies at Indiana University, he worked as a fire lookout over summer on the West Coast, and enrolled to study Chinese and Japanese in the University of California, Berkley in preparation for his trip to Asia. He was particularly interested in China, but could only go to Japan at the time due to the Cold War. From 1956 to 1969 he studied Zen meditation in Japan and wrote poems and essays describing the relationship between man and nature based on his own experiences in nature. In 1969, he returned to America with his Japanese wife and they settled down in northern California and lived a very simple life there. Or in other words, from then on he took the path of "thinking like a mountain" advocated by Leopold and experienced the carefree lifestyle and spiritual pursuit of Han Shan's "with a born love for mountains and rivers, and an innate will to admire the Tao" (Xiang 2000, p. 582). More than 1,000 years ago, Han Shan already had a sense of the "wilderness." By being born with "a love for mountains and rivers" he actually meant a sense of finding joy in the wilderness and of the integration of humanity and nature; it is the demonstration of human nature. This is why Han Shan, a Tang poet who is not very well-known in China, is so loved by Snyder: the sense and concept of wilderness are shared.

Through translating Han Shan's poems, Snyder tried to communicate with this ancient Chinese poet across time and space. His connection with Han Shan also had something to do with Chinese landscape paintings; in an interview in 1977, Snyder mentioned that he visited the Chinese room in the Seattle Art Galley when he was about 12 and was surprised to see that the scenery in the Chinese landscape paintings there was very similar to the mountain ranges near his hometown. He exclaims,

> The waterfalls, the pines, the clouds, the mist looked a lot like the northwest United States. The Chinese had an eye for the world that I saw as real. In the next room were the English and European landscapes, and they meant nothing. (Snyder 1999, p. 92)

Suiter linked Snyder's fascination with Chinese landscape paintings in his teens and his translation of Han Shan's poems in adulthood, because in the paintings Snyder also saw the Chinese hermits together with mountains and rivers (Suiter, p. 163). The impact of Chinese landscape paintings on Snyder greatly influenced his life, leading him to follow Han Shan, translate his poems, and try to live his carefree life and take on his austere style of writing.

Snyder's "Lookout's Journal," included in his first collection of essays *Earth House Hold*, is a record of his life as a fire lookout at Crater Mountain and Sourdough Mountain in the North Cascades in the summers of 1952 and 1953—an experience that was not only significant to his writing, but also attracted the interest of the literary world. Kerouac's novel *Dharma Bums* was based on it, and Snyder was described as the American Cold Mountain who "lives in a thatched hut, drinks Chinese tea and translates Cold Mountain's poems." Paul comments appreciatively on "Lookout's Journal" and compares it favourably with Thoreau's *Walden* (quoted in Murphy 2000, p. 90). However, it differs from Thoreau in that Snyder was at a lookout more than 8,000 feet above sea level in a desolate area in the American West, and that his life was influenced by eastern culture, reflecting the simplicity typified in Chinese landscape paintings. In *Poets on the Peaks* the following scene is described: outside the window, the bright moonlight, swift creek water, birds and buzzing insects, while in the room, Snyder reading Bai Juyi's poems under a kerosene lamp, as if the ancient poet were also in the room (Suiter, p. 13). The influence of Chinese culture also infiltrated Snyder's language. In "Lookout's Journal," we notice that he uses the Chinese word to spell the word "mountain," that is "Crater Shan" instead of "Crater Mountain." In *Poets on the Peak*s, there is a photograph of the mountain path up the soaring Crater Mountain. Underneath the photograph, there is a line by Snyder: "I asked the packer, 'Where does the trail go from here?' And he said, 'There isn't any trail from here. There'll be some marks painted on the rocks to show you which way to climb'" (Suiter, p. 27). In this scene Snyder is taking his provisions to be a fire lookout in the mountains, an experience that laid a solid foundation and gave him the mood and the language necessary for his translation of Han Shan's poems. This personal experience in the mountains enabled him to communicate in his heart and poems with the ancient Chinese poet who resided among the cold rocks in another time and place. When he translates "People ask Han Shan the way, Han Shan says there is no through trail," maybe he incorporates his own feelings about Crater Mountain as described in his "Lookout's Journal." No wonder people comment that through translating Han Shan's poems, Snyder "created a tough, knowing, lonesome persona for a modern-day Han Shan" (Suiter, p. 163).

This section of the essay is called "Music from Far Away" not only because of Snyder's attitude toward life, away from the madding crowd, but also because of his interest in the far-off ancient cultures of the East. This interest made it possible for him to look at the relationship between nature and humanity from a broader and more integrating perspective. For example, in *The Practice of the Wild*, he discusses the Chinese "mountains and rivers" and specifically points out that the "mountains and rivers" were the landscape. To the Chinese, mountains are hard and dry, strong and bright, and are a symbol of the masculine *yang*, while rivers are soft, wet and dark, and are a symbol of the feminine *yin*. He concludes,

> "Mountains and waters" is a way to refer to the totality of the process of nature. As such, it goes beyond dichotomies of purity and pollution, natural and artificial. The whole, with its rivers and valleys, obviously includes farms, fields, villages, cities and the (once comparatively small) dusty world of human affairs. (in Snyder 1999, pp. 203–204)

Maybe it is because of his understanding of Chinese "mountains and rivers" and his appreciation of this music from far away that Snyder can give the wilderness a new meaning and carry out his "practice of the wild."

## *Interpretations of the Wild*

Snyder finds the values he pursues in Han Shan, who he regards as his soulmate. He writes, "We need a civilization that can live fully and creatively together with wildness.... Sacred refers to that which helps take us (not only human beings) out of our little selves into the whole mountains-and-rivers mandala universe" (quoted by David M. Robinson in his "Wilderness and the Agrarian Principle: Gary Snyder, Wendell Berry, and the Ethical Definition of the 'Wild,'" 1999, pp. 20, 24). This broad vision is fully demonstrated in his extension of the value of wilderness and expansion of ecological ethics. In *The Practice of the Wild*, he explains in detail the definitions of "wild," "wildness" and "wilderness." He believes that being "wild and free" is "[a]n American dream-phrase loosing images: a long-maned stallion racing across the grasslands, a V of Canada Geese high and honking, a squirrel chattering and leaping limb a limb overhead in an oak" (in Snyder 1999, p. 168). In other words, "wild" is another word for "inner freedom." Snyder quotes Thoreau, "Give me a wildness no civilization can endure" (in Snyder 1999, p. 169). From this we can see that the "wildness" Thoreau and Snyder refer to is not a physical place, but a carefree, energetic and healthy mental state. Snyder specifies that "[w]ilderness is a place where the wild potential is fully expressed, a diversity of living and nonliving beings flourishing according to their own sorts of order." He explains further, "To speak of wilderness is to speak of wholeness. Human beings came out of that wholeness" (in Snyder 1999, p. 173).

In fact, as early as the mid-nineteenth century, Thoreau explains the close connection between "wildness" and "the wilderness." In Thoreau's essay "Walks," readers will notice the sentence "in wildness is the preservation of the world." After explaining in detail the importance of the wildness to the human heart and body, Thoreau points out that the wilderness is the foundation of nurturing and maintaining wildness, and is also the foundation of human culture, "I enter a swamp as a sacred place, a sanctum sanctorum. There is the strength, the marrow, of Nature" (Thoreau 2002, p. 78). Perhaps we can say that the English word "wilderness" is the physical body of the wildness and "wildness" is mental wilderness in the consciousness and the inner wildness of humans. Buell summarizes this. He thinks that in some literary works "wildness" and "wilderness" are synonyms, with the former emphasizing more the features and the latter more space. Wildness and wilderness can be used synonymously, but wilderness literally refers to a spatial area, whereas wildness is a term of quality rather than location. Traditionally people do not have a positive feeling toward the word "wildness." Snyder is unique in regarding "wildness" as a cultural value. He also refers to a similar view by Thoreau, who states that he is interested in the wilderness because he wants to absorb "wild nourishment" from it to fight against "excessive civilization." In other words, wildness is a feature shared by humans and non-human living things (quoted in Buell 2005, pp. 148–149). Snyder inherits Thoreau's wilderness theory and succeeds him as a spokesperson for the wilderness. He sees that wilderness plays an irreplaceable role in human culture and civilization. In a speech titled "The Wilderness" he says,

> The reason I am here is because I wish to bring a voice from the wilderness, my constituency. I wish to be a spokesman for a realm that is not usually represented either in intellectual chambers or in the chambers of government.

Then he points out that the problem of western culture is that it has inherited too many errors, and has alienated itself from both "the wilderness outside" and "the wilderness within." This is the root of the environmental crisis and is self-destructive (quoted in Begiebing & Grumbling 1990, p. 645). In the high-tech advanced society in which humanity and nature have been objectified and commodified, Snyder sees the mistakes of modern civilization. He links protection of the environment and nurturing humanity's wildness with the survival of human culture, in an attempt to prompt people into thinking from a cultural perspective and taking a strong ecological viewpoint that advocates harmonious coexistence between nature, society and spirit. He revises and updates Thoreau's saying "In wilderness is the preservation of the world" into "Wilderness is not just the 'preservation of the world', it *is* the world" (in Snyder 1999, p. 169), which has become the motto of the Sierra Club, originally established by John Muir. Snyder urges people to safeguard the realm of spiritual freedom through protecting the wilderness and advocating individualism. He explores American Indian cultures and studies the eastern Zen, all in the hope that humans will not become the slaves of machines, and will keep their diversity, making the human world as full of life as the wild natural world.

Snyder not only provides a new interpretation of the wilderness, but also puts forward the concepts of "a place in space" and "re-inhabitation," thus pointing out a practical way of protecting the wilderness. As mentioned above, one of the characteristics of nature writing is a "sense of place." Snyder's personal experience and his works not only show this characteristic of nature writing, but also expand the concept of "sense of place." In *A Place in Space*, he links the land he knows with the land and sky that are essential to the common survival of humans, thereby expanding Leopold's land ethics and showing a broader eco-vision. It should be pointed out that Snyder differs from Leopold in that he seeks a kinship with the land from the perspectives of ancient eastern and American Indian cultures.

American nature writer Wendell Berry (1934–) once used the phrase "a restless America" to describe the contemporary American penchant for moving around. As a matter of fact, historically speaking, Americans' feet have never stopped, moving from Europe to the new American continent, then from the East to the West Coast. In modern society, the nature writers are concerned that this America on wheels has made people lose their close connection with, and their sense of responsibility for, their place. Snyder believes that humans have to go native like the American Indians if they want to restore their connection with the land. He says, "As a poet, I hold the most archaic values on earth" (quoted in Begiebing and Grumbling, p. 641). In the foreword to his poetry collection *Turtle Island*, he explains that the names of the country, the states and the counties are not suitable. "'Turtle Island' is a better 'address' because it is the American Indian name for North America linked to 'creation myths of people living here for millennia'" (quoted in McClintock, p. 111). In an interview Snyder also says that people should live as cautiously as if they were going to live at their place for 10,000 years. Snyder's "a place in space," therefore, comes from "a sense of place." His advice to restless Americans is, "Find your place on the planet. Dig in, and take responsibility from there" (in Snyder 1999, p. 110). In *The Practice of the Wild* he points out, "I want to talk about the place of survival as a life experience and recommend a life style that is 'rooted in the land'" and says, "Our place is part of what we are" (in Snyder 1999, pp.183, 184).

Snyder proposes the concept of "re-inhabitation" in *A Place in Space.* He believes that many people who live on this earth are only passers-by and not real residents, as they have lost close connection with the land. "Re-inhabitation" does not mean "no traveling," it is not only limited to the concept of place. The "re-inhabitant" here is a person who escapes from industrial society and modern civilization and returns both physically and spiritually to the land and to a place of residence. Snyder puts forward three practical steps toward "re-inhabitation": feel gratitude for everything around you; start from yourself and take up the responsibility; and keep close connection with the sources of your physical and mental energy (such as soil, water, bodies of flesh) (in Snyder 1999, pp. 183–191).

In *The Practice of the Wild* Snyder writes, "There are tens of millions of people in North America who were physically born here but who are not actually living here intellectually, imaginatively, or morally" (p. 194). This is similar to Leopold's "land ethics," which advocates that humans be citizens of the land community and have a sense of duty and responsibility. Snyder declares,

> The treasure of life is the richness of stored information in the diverse genes of all living beings. If the human race, following on some set of catastrophes, were to survive at the expense of many plant and animal species, it would be no victory. (p. 254)

In the 1960s, Abbey put forward the concept of "opposition-compromise-balance" in regard to the relationship between humanity and nature. Snyder, however, seeks to find a solution to the conflict between development and environmental protection from American Indian cultures. He believes that there is nothing wrong with wishing to develop; the key is to seek a way of development that is not destructive. In this respect, American Indian culture is 'mature': protection as against production, stability as against growth, quality as against quantity (p. 255).

American scholar Ursula K. Heise in her book *Sense of Place and Sense of Planet: The Environmental Imagination of the Global* (2008), explains the complex and interrelated relationship between a local culture and ecological system on the one hand and the global culture and ecological system on the other. Heise emphasizes Snyder's contribution in the relationship between the place and the globe. She believes that he is in a privileged position to address issues of transnationalism, and that some of his works and concepts "highlight a planet-wide perspective" (Heise, p. 43). Heise's comments are appropriate; from the titles of Snyder's two prose collections *Earth Home Hold* and *A Place in Space* alone we can see the dialogical relationships between the place and the globe. Snyder himself points out, "We seek the balance between cosmopolitan pluralism and deep local consciousness" (Snyder 1999, p. 196), and his opinions are widely quoted by scholars of nature writing and eco-criticism.

## *The Practice of the Wild*

In the view of American critic Dan McLeod, Snyder attracts the attention of the American people not only because of the views he expresses in his books, but also because the example of his experience and values provide a constructive option for the American mainstream (in Murphy 2000, p. 10). In his childhood Snyder developed a love of mountains, and in his adulthood he regards the Chinese Tang poet Han Shan as his "soulmate" because Han Shan has become the symbol, or soul, of mountains. The mountain is a synonym of wildness, and wildness is in turn the synonym of sacredness and freedom. In the natural world it seems the mountain is the most definite contrast to the noisy city. In opposition to the modern metropolitan city, where sensual pleasure and impulsiveness rule, is the silent, long-existing and profound mountain.

In the North Cascades mountain range in Washington state there are still some forest fire lookouts at the tops of mountains in the clouds which make mountaineers look up and feel fascinated. Among them are the lookouts at Crater Mountain and Sourdough Mountain where Snyder lived and worked as a forest fire lookout. Suiter, the author of *Poets on the Peaks*, comments,

> Now, too, as solitude, self-reflection, and plain silence have become ever more fleeting and rare in our lives, the lookouts have emerged as symbols of a nearly lost and longed-for American simplicity and integrity—Thoreau's cabin transferred to the mountaintop. (Suiter, p. 268)

In June 1952, when Snyder applied to be a fire lookout on Crater Mountain, he asked to be sent to the "highest, most remote, and most difficult-of-access lookout" (quoted in Suiter, p. 3), and was sent to Marblemount Ranger Station which, at 8,129 feet above sea level, was then the highest lookout still in use in the state of Washington. Looking out from there, no roads or towns could be seen, only ranges of mountains, and in the middle of July it was still frozen and snowing. Unlike ordinary mountaineers, Snyder climbed the mountain not to conquer it, but to understand himself. He found the solitude he needed, and transformed the work of looking out for forest fires into a watch over the mountains. In an interview in 1997, Snyder recalled that

> Aldo Leopold uses the phrase "Think like a mountain." I didn't hear that until later, but mountain watching is like mountain being or mountain sitting. How do you watch a mountain? Nothing's going to happen in any time frame that you can consider—except the light changes on it. And so that was my mountain watching. The changing light on the mountain was like the changing thoughts in my mind, just these little shifting shadows, that's all that it is. (quoted in Suiter, p. 45)

Through his personal experience Snyder explains the concept of "think like a mountain," and describes the wisdom of the silence and movement of the mountains. To him, Crater Mountain was his teacher and the whole mountain range was the lesson. This cannot help but remind us of the Chinese poet Li Bai's poem "Sitting Alone at Mount Jingting": "The birds have all flown away, / and the lonely clouds have also gone. / Looking at each other without ever tired of each other, / only Mount Jingting." (Li Bai. "Sitting Alone at Mount Jingting" in (1705) 2008. *Quan Tangshi*. Vol. 182. No. 028; *Zhong niao gao fei jin, guyun du qu xian. Xiang kan liang bu yan, zhi you Jingting shan*). Li Bai, in ancient China, personified the Mount Jingting he saw; he and the mountain he saw were communicating emotionally, even though they were both silent. What is rare is that Snyder, who lived in a modern society, also communicates spiritually with the mountain in the same way as the ancient Chinese literati, yet with the characteristics of a modern western poet. In his "Lookout's Journal" Snyder writes, "Don't be a mountaineer, be a mountain" (Snyder 1999, p. 20), which vividly shows his lifelong attitude and spiritual vision.

Snyder's experience as a forest fire lookout in the Cascades Range is the kind of simple yet rich life advocated by Thoreau, a perfect combination of work and interest, a poetic residence. When looking around for any signs of fire, Snyder climbs mountain after mountain, admiring the mountains and the grandeur of nature; in his spare time, he boils snow to make tea, learns meditation, reads Tang poems and practises Chinese calligraphy. For him, this simple, solitary life which ordinary people would find difficult to endure is charming and beautiful.

Snyder has been called the "Mountain Saint." This relates to his place of residence after his marriage. Snyder's description of this place is unique.

> On the western slope of the northern Sierra Nevada, in the Yuba River watershed, north of the south fork at the 3000-foot elevation, in a community of black oak, incense cedar, madrone,[4] Douglas fir, and ponderosa pine. (p. 195)

When Snyder first went there, there was no electricity or telephone; the nearest town was 25 miles away across valleys, and the life he lived was really a mountain life. He named his mountain home Kitkitdizze—the American Indian name for a local shrub whose small flowers give off a fresh fragrance. At Kitkitdizze Snyder practised the "re-inhabitation" he advocated, seeking the meaning of living wisely and carefully in a local area. He says frankly, "we were cash poor and land rich." Like Leopold, Snyder and his family treated the land under their feet very kindly, although the land under Snyder's feet had been damaged by gold mining in the nineteenth century, not farming. He built his house to suit the local climate and environment, combining the simplicity of Japanese farm cottages, the primitiveness of the American Indian homes and the practicality of modern eco-houses, and says, "Living in a place like this is absolutely delicious. Coyote-howl fugues, owl exchanges in the treetops, the almost daily sighting of deer." Snyder lived in Kitkitdizze for more than 20 years, and summarizes his understanding of life in his communication with an oak tree.

> After twenty years of walking right past it on my way to chores in the meadow, I actually paid attention to a certain gnarly canyon live oak one day. Or maybe it was ready to show itself to me. I felt its oldness, suchness, inwardness, oakness, as if it were my own. Such intimacy makes you totally at home in life and in yourself. (Snyder 1999, pp. 277–283)

Maybe Snyder's words will open a new window for the hearts of people who live in modern society, where obsession with greed and quick profits prevails. Snyder's life and art are integrated, and even today, the years he spent in the wilderness have not disappeared but live on in his poems and prose essays.

Snyder is known as an eco-poet, but his writing style—based on his own personal experience—is the same as that of nature writing, which generally employs a simple, direct and free style. Murphy, commenting on Snyder's writing style, says that Snyder "wants his reader to think about 'wild writing,' with the idea that the natural world should instruct writers about their craft" and believes that "[g]ood writing is wild language" (Murphy, p. 171).

4 Madrone is an evergreen in North America that has white flowers and red berries.

However, wild language comes from a wild soul and from the practice of the wild. Like Thoreau and Abbey, Snyder explores a language close to the land, a language that is in step with the all-inclusive wilderness teeming life. This language also comes from John Muir's "university of the wilderness" and John Burroughs's Riverby study. For Snyder, the wild language he explores is a fusion of eastern and western cultures. Although located in the mountains of the American West and describing north American nature, Snyder's writing has the charm of Chinese landscape paintings, in which the ink has five shades and colors and the flavor of the Chinese Tang poet Han Shan. His writing has the wild imagination of the West, but also the calm simplicity of the East's Zen.

Snyder is a writer and poet who comes from "nature's workshop," and his works are products of nature, having the colors of the seasons, the dews and fragrance of flowers and the songs of the wilderness. Opening his "Lookout's Journal," the language and descriptions are unique and refreshing:

> Sourdough Mountain Lookout, Elevation: 5977 feet 17 July 53.
> This is the place to observe clouds and the gradual dissolution of snow....
> ..."because there are no calendars in the mountains"—shifting of light & cloud, perfection of chaos, magnificent *jiji mu-gel* interlacing interaction. (Snyder 1999, pp. 16–17)

Snyder takes colors from nature:

> Right now looking down on the Skagit—pink clouds—pale rose-water pink, with soft shadings of gray and lavender, other combinations of pastel reds and blues, hanging over Pyramid Peak. (p. 18)

"Nature's workshop" has rid Snyder's writing of any embellishment and renders it simple and clear:

> What happens all winter; the wind driving snow; clouds—wind, and mountain—repeating.
> This is what always happens here. (p. 10)

Snyder believes that using nature as the background is beauty, and this beauty is a vivid and lively beauty shown in his poetic sketch:

> Two butterflies
> A chilly clump of mountain
> Flower

He also puts himself into the natural scenery:

> Leaning in the doorway whistling
> A chipmunk popped out
> Listening (p. 10)

A storm scene in the book could only have been written by someone who experienced the storm:

> Last night: thunderstorm. A soft piling of cumulus over the Little Beaver in late afternoon—a gradual thickening and darkening. A brief shower of hail that passes over & went up Thunder Creek valley: long gray shreds of it slowly falling and bent in the wind—while directly above Ruby Creek sunlight is streaming through. Velvety navy blue over Hozomeen, with the sun going down behind Mt. Terror and brilliant reds and pinks on the under-clouds, another red streak behind black Hozomeen framed in dark clouds. Lighting moving from Hozomeen slowly west into red clouds turning gray, then black; rising wind. Sheet lightning pacing over Little Beaver, fork lightning striking Beaver Pass. (Snyder 1999, p. 18)

American scholar Tom Lynch, in his book *Xerophilia* (2008), advocates an aesthetic tradition that is based on eco-regions, including "bioregional literature." He explains,

> Bioregional literature, by telling the stories of place, by promoting the imagery of place, by helping to dispense information about place in memorable and vivid ways, can help to cultivate a bioregional imagination that will help us live in our life-space with the kind of intellectual, imaginative, and moral dimension that we, and our places, deserve. (pp. 19, 22)

The bio-regional literature that Lynch explores uses literature as a medium for humans to communicate with the place where they live. Snyder's "Lookout's Journal" has all the characteristics of this kind of bio-regional literature, and brings our hearts closer to nature. Reading Snyder, as Jim Dodge's foreword to *The Gary Snyder Reader* promises, "You'll find that wild glittering throughout this book" (in Snyder 1999, p. xx).

However, Snyder shows us not only the wild glittering; he is deeply influenced by western thoughts, but also eastern philosophy. His interpretation of the wilderness shows communication across time and space, between eastern and western cultures, on the issue of the relationship between humanity and nature. He is a poet and writer who describes the wilderness, using poetry and prose to show the intimate connection between humanity and nature and the spiritual value of the wilderness. He has put forward the concept of "re-inhabitation" of the earth, and personally practices the wild and cares for the land. In Snyder and his works, we see the exploration of nature by both eastern and western cultures, showing us a greater eco-vision.

## References

Begiebing, Robert J., and Owen Grumbling, eds. 1990. *The Literature of Nature: The British and American Traditions*. Medford, OR: Plexus Publishing.

Beston, Henry. (1928) 1992. *The Outermost House: A Year of Life on the Great Beach of Cape Cod*. A reprint of the 1988 edition with an introduction by Robert Finch. New York: Henry Holt and Company.

Beston, Henry. (1928) (1992) 2007. *The Outermost House: A Year of Life on the Great Beach of Cape Cod*. Translated by Hong Cheng. Hong Kong: Joint Publishing.

Beston, Henry. 1970. *Especially Maine: The Natural World of Henry Beston from Cape Cod to the St. Lawrence*. Selected and with introductions by Elizabeth Coatsworth. Brattleboro, VT: Stephen Greene Press.

Bradford, William. (1651) 2002. *Of Plymouth Plantation* (1620–1647). New York: Alfred A. Knopf.

Chandler, Katherine R., and Melissa A. Goldthwaite, eds. 2003. *Surveying the Literary Landscapes of Terry Tempest Williams: New Critical Essays*. Salt Lake City, UT: University of Utah Press.

Heise, Ursula K. 2008. *Sense of Place and Sense of Planet: The Environmental Imagination of the Global*. New York: Oxford University Press.

Li Bai. "Sitting Alone at Mount Jingting" in (1705) 2008. *Quan Tangshi* (《全唐诗》) Vol. 182. No. 028. Beijing: Zhonghua Book Company.

Li Shangyin, "Untitled" in (1705) 2008 *Quan Tangshi* Vol. 539. No. 130. Beijing: Zhonghua Book. Company.

Lyon, Thomas J. 2001. *This Incomparable Land: A Guide to American Nature Writing*. Minneapolis, MN: Milkweed Editions. Revised edition.

McClintock, James I. 1994. *Nature's Kindred Spirits: Aldo Leopold, Joseph Wood Krutch, Edward Abbey, Annie Dillard, and Gary Snyder*. Madison, WI: University of Wisconsin Press.

Miller, James E., Jr., ed. 1991. *Heritage of American Literature: Beginnings to the Civil War*. Vol. 1. San Diego, CA: Harcourt Brace Jovanovich.

Murphy, Patrick D. 2000. *A Place for Wayfaring: The Poetry and Prose of Gary Snyder*. Corvallis, OR: Oregon State University Press.

Robinson, David M. 1999. "Wilderness and the Agrarian Principle: Gary Snyder, Wendell Berry, and the Ethical Definition of the 'Wild.'" *ISLE* 6 (1): pp. 15–27.

Paul, Sherman. *For Love of the World: Essays on Nature Writers*. Iowa City, IA: University of Iowa Press, 1992.

"Ronald Perera's Biography." n.d. Ronald Perera. Accessed May 16, 2013. http://www.ronaldperera.com/biography.html.

Slovic, Scott. 2008. *Going Away to Think: Engagement, Retreat, and Ecocritical Responsibility*. Reno, NV: University of Nevada Press.

Snyder, Gary. 1990. *A Place in Space*. Washington, D.C.: Counter Point.

Snyder, Gary. 1995. *A Place in Space*. Washington, D.C.: Counter Point.

Snyder, Gary. 1999. *The Gary Snyder Reader: Prose, Poetry and Translations*. Washington, D.C.: Counter Point.

Snyder, Gary. 2004. "Ecology, Literature, and the New World Disorder." *ISLE* 11 (1): pp. 1–13.

Spence, Jonathan D. "What Confucius Said." *The New York Review of Books*. 10 April 1997. pp. 8–13.

Suiter, John. 2002. *Poets on the Peaks:* Gary Snyder, Philip Whalen & Jack Kerouac. Washington, D.C.: Counter Point.

Thoreau, Henry David. (1865) 2008. *Cape Cod*. Illustrated edition with a foreword from Thoreau biographer Robert Richardson. Houghton Mifflin Harcourt.

Thoreau, Henry David. (1862) 2002. *Wild Apples and Other Natural History Essays*. Ed. William Rossi. Athens, GA: University of Georgia Press.

Wilding, Don. 2001. *On its Solitary Dune: Dedicated to Henry Beston and his Literary Classic*. The Outermost House.

Wilding Don. 30 May 2008. "Finch Introduces 'A Sense of Place' Exhibit." Accessed December 30, 2020. http://blogs.capecodonline.com/cape-cod-henry-beston/2008/05/30/finch-introduces-a-sense-of-place-exhibit/

Wilding, Don. 27 September 2008. "The Music of 'The Outermost House' in Dennis." Accessed May 16, 2013. http://blogs.capecodonline.com/cape-cod-henry-beston/2008/09/27/the-music-of-theoutermost-house-in-dennis/.

Williams, Terry Tempest. 1991. *Refuge: An Unnatural History of Family and Place*. New York: Vintage Books.

Williams, Terry Tempest. 2008. *Finding Beauty in a Broken World*. New York: Pantheon Books.

Xiang, Chu. 2000. *Han Shan Poems and Notes* (《寒山诗注》). Beijing: Zhonghua Book Company.

# 9

# Ecocriticism—An Extension of Nature Writing

## Lecture 29. The Origin and Current Situation of Ecocriticism

### *The Concept and Origin of Ecocriticism*

Ecocriticism is a genre of literary criticism that studies the relationship between literature and the natural environment. It started in America and has since spread to the rest of the world. It has inherited the tradition of nature writing, but differs from nature writing in that it aims to comment on and study nature writing, environmental writing and other literary writings that explore the relationship between humanity and nature. At the same time, it promotes a reading of literary works old and new from the perspective of ecology, to encourage a strong eco-awareness and a sense of crisis. Naturally, there are other interpretations of ecocriticism. Some people regard it as synonymous with environment literature, eco-literature and green literature—which indicates the extensive scope and depth of literature about the relationship between humanity and nature—while Lawrence Buell in his *The Future of Environmental Criticism: Environmental Crisis and Literary Imagination* (2005, p. viii) prefers to call it "environmental criticism." However, the themes and aims are similar and they belong to the same field.

The term "ecocriticism" first appeared in 1978, in William Rueckert's published thesis *Literature and Ecology: An Experiment in Ecocriticism*. Cheryll Glotfelty, in her introduction to *The Ecocriticism Reader: Landmarks in Literary Ecology* (Glotfelty & Fromm 1996, pp. xix–xx), quotes Rueckert's definition of ecocriticism: "the application of ecology and ecological concepts to the study of literature." Rueckert's view did not attract much attention when he published his thesis in 1978, but at the 1989 conference of the American Western Literature Association conference ecocriticism became a focus. Glotfelty, a postgraduate student at Cornell at the time and now a professor of literature and environment at the University of Nevada, reintroduced Rueckert's term and insisted on using it to replace the term "the study of nature writing," to expand the scope of this research area—a suggestion that was met with an immediate positive response. At the conference Glen A. Love, then president of the Western Literature Association and a professor of English at the University of Oregon, spoke on "Revaluing Nature: Toward an Ecological Literary Criticism" (published in Glotfelty and Fromm, pp. 225–240), and since then the term ecocriticism has appeared frequently in dissertations, essays, monographs, literary collections and university teaching programs. The first collection of ecocriticism, Glotfelty and Fromm's *The Ecocriticism Reader*, was published in 1996. In 2002, Peter Barry added a chapter on ecocriticism to the reprinted edition of his 1995 book on literature and cultural studies *Beginning Theory: An Introduction to Literary and Cultural Theory*, explaining that this was due to changes and development in literary theory. Barry's addition marks the first registration of the term in a literary theory book of ecocriticism which aims at interpreting literature from a green perspective. (Barry [1995] 2002, p. xii). From then on various kinds of scholarly works appeared, including Glen Love's book *Practical Ecocriticism: Literature, Biology and the Environment* (2003) and Greg Garrard's *Ecocriticism* (2004).

Ecocriticism has the same origin as nature writing; both have been greatly influenced by pastoral literature, romanticism and transcendentalism, but ecocriticism focuses on analysis while nature writing focuses on description. In this lecture, the focus will be the relationship between ecocriticism and nature writing. Lawrence Buell says that ecocriticism has focused on nature writing from the very beginning. He vividly describes the development of ecocriticism in waves, separating it into a "first wave" and a "second wave." In his *The Future of Environmental Criticism: Environmental Crisis and Literary Imagination* (2005), he cites his own early book *The Environmental Imagination: Thoreau, Nature*

*Writing, and the Formation of American Culture* (1995) and his later *Writing for an Endangered World: Literature, Culture, and Environment in the U.S. and Beyond* (2001) as representative works of the first and second waves respectively. The former, through the study of Thoreau and American nature writing, shows that literature can help people to form eco-centered values. The latter examines broader visions, more diverse critical methods and a greater variety of genres, including nonfiction nature writing, nature poems, wilderness novels and other complex forms (Buell 2005, pp. 22, 138). From both the titles and the content of two books edited and published by Michael P. Branch, an American scholar on ecocriticism, we can again see the close connection between ecocriticism and nature writing. In 1998 Branch published *Reading the Earth: New Directions in the Study of Literature and the Environment*, a collection of essays that presents an overview of the theory, categories and stages of ecocriticism. In 2004 Branch published *Reading the Root: American Nature Writing before Walden*, another collection of essays which explains the value of studying early American nature writing and its significance in current ecocriticism. From the titles of these two collections alone we can see the connection between ecocriticism and nature writing. Therefore, we can say that ecocriticism and nature writing, with the former containing the other, cannot be separated. For example, Leopold—who put forward the theories of "land ethics" and "eco-conscience"—was also a nature writer himself. His nonfiction work *The Sand County Almanac* (1949) is considered a representative work of American nature writing, and he himself was also a pioneer of ecocriticism. This shows that "nature writing" and "ecocriticism" overlap. We can also say that nature writing is, in a certain sense, more specifically literary, while ecocriticism has a broader, more inclusive range. On the basis of Buell's "two waves," some scholars divided the development of ecocriticism into three waves: the first wave, from 1980 onwards, is characterized by American nature writing; the second wave, from 1995 onwards, is characterized by the diversity of nature writing, the internationalization of ecocriticism and environmental justice;[1] and the third wave, from 2001 onwards, is characterized by a global view of place, with an emphasis on multicultural, multi-country and multiethnic connections in their view of and care for the environment.

1 Sometimes this is also called "environmental ethnicism." Its purpose is to challenge the unjust distribution of environmental risks in the world, particularly concerns about the effect of pollutants on poor people, disadvantaged groups and people of color.

From the above, we can see that ecocriticism has expanded from the basis of nature writing and has broadened its scope. Although ecocriticism has not changed the theme of reflecting on the relationship between humans and nature, the focus of its interest has expanded from nonfiction nature writing to novels, poems, plays, English teaching and research, and even to movies, television, news and other fields. The research perspective is also a little different: it has linked life science with literature. For example, it advocates a study of literature from the understanding of Darwin's evolutionary theory and ecology, and an ecocritical or eco-centered reading of literary texts. In addition, with the changes and development over time, the meanings of the words "nature" and "environment" have also expanded. There is no doubt that literary criticism has also been changing and operating in a much broader context.

Ecocriticism is a product of the development of the times. The damage to the ecosystem and the greenhouse effect caused by modernization are threatening human lives and quality of life. Studying literature from an ecological point of view, and through literature's influence on the formation and implementation of policies that are beneficial to ecological balance and sustainable development, has become a moral duty for literary researchers and literature professors in universities.

## *The Current Situation of Ecocriticism*

Since the end of the 1980s, ecocriticism has positioned itself as a study of the relationship between literature and the natural environment. Over time the word "ecocriticism" has been used more and more frequently, its usage scope has expanded, and it has been included as a term in specialized works of western literary theory.

At present, ecocriticism has expanded the definition of "nature" in terms of its scope and place, thereby extending its subject matter from nature to ecology. Barry, in the chapter that he added to *Beginning Theory* (1995) in 2002, "Ecocriticism," divides the "natural environment" into four types: the wilderness (desert, oceans, uninhabited continents); the scenic sublime (forests, lakes, mountains, cliffs, waterfalls); the countryside (hills, fields, woods); and the domestic picturesque (parks, gardens, lanes) ([1995] 2002, p. 255). Buell, in his 2001 book on the environment and literature *Writing for an Endangered World*, expands the "environment" to include both green and brown landscapes, that is, Barry's countryside plus wilderness landscapes of remote rural areas and those of industrialized and urban areas. Buell's environment includes both the "natural" and the "human-built" spheres, the grand wilderness and the cityscape ravaged by environmental problems, as both have been affected by the greenhouse effect. He says that the "environmental imagination" (as advocated in his first book) should not stop at the edge of the woods but should extend to all natural and urban environments under threat in the world (Buell 2001, pp. 1–29). Jonathan Bate bases his book *The Song of the Earth* ([2000] 2002) on William Wordsworth's view that "human thoughts can be part of nature," and shows the relationship between the poet and nature. Bate writes, "The dream of deep ecology will never be realized upon the earth, but our survival as a species may be dependent on our capacity to dream it in the work of our imagination" (pp. 37–38), thus locating ecocriticism in a much wider intellectual space.

The research perspective of ecocriticism links literature with life sciences. Based on Darwin's evolutionary theory and the theory of ecology, ecocritics believe that the external environment's effect on humans is closely related to everything. They advocate, therefore, that participants in and students of ecocriticism should cross the borders between literature and life sciences and look at literature from the angles of human evolution and the changes in the environment, thus making ecocriticism both interdisciplinary and multidisciplinary. Among books of this kind worth mentioning, in addition to Barry's *Beginning Theory*, are the works of Joseph Carroll and Joseph Meeker. Carroll wrote *Evolution and Literary Theory* (1995) and *Literary Darwinism: Evolution, Human Nature, and Literature* (2004). He is regarded as a pioneer in combining Darwinian evolutionary theory with literary theory, and also modern biology with the study of humanities, in his research. Meeker is a scholar of comparative literature with a background in evolutionary biology. His *The Comedy of Survival: Studies in Literary Ecology* (1974) and *The Comedy of Survival: Literary Ecology and a Play Ethic* (1997) reread literary works from the ecological perspective, focusing on the relationship between ecology and literature. In Meeker's view, if literary creation is an important characteristic of humans, then we should carefully study the influence that literature has on human behavior and on the natural environment; we should explore the role literature plays in the connection between humans, other living forms and the natural world; and we should examine the insights these offer. Literature cannot be separated from nature or from the ecological environment. In Meeker's opinion, the "comedy of (human) survival" is dependent upon humans changing themselves instead of the natural environment and acknowledging their own limitations instead of cursing them. Literature should play a role in this comedy of human survival. Therefore, we can say that ecocriticism has not only inherited the tradition of pastoral literature, romanticism and transcendentalism, but has also been influenced by Darwinian evolutionary theory and the theory of ecology.

As the scope of ecocriticism and its analysis expands, its content has become richer. It advocates an "ecocritical reading of a literary text" or an "eco-centered reading of a literary text" (Barry, [1995] 2002, pp. 257, 260), and includes comments on contemporary works that predict and imagine future ecological disasters, as well as rereadings of historical classics. Perhaps the most representative contemporary work is Don Delillo's *White Noise* (1985), which captures a strange phenomenon in contemporary life, namely, the fact that humankind lives in a time when it can destroy itself with the knowledge it has acquired. In it the Gladneys, a college art professor couple with their four postmodern children, live in a society where consumerism is the most salient feature. Toxic gases leak from the chemical factory near their home, while "white noise" inside it (from radio, microwave, sound system, TV and other ultrasonic household appliances) infuse their life and fill it with danger and fear. It has been commented that this novel not only exposes the consumerism of capitalism, but also shows the changing role of humans in post-industrialized societies. According to Cynthia Deitering, a professor of English at Indiana University, humans have become producers of waste, while the so-called "nature" is scarred and exhausted and has become the "empire of waste" (Deitering, pp. 198–199). Love, on the other hand, uses Jack Gladney's initial reaction to the toxic gas to show that even college professors cannot avoid disaster in a foul environment. At first Gladney does not believe that the toxic gas will reach the elegant residential area that the family live in; he thinks that only poor, uneducated people are affected by natural and man-made disasters. Using typhoons as an example of how only people in poor areas are affected by flood, he asks, "In all those TV shots of flood, have you ever seen a college professor on a boat escaping in a flooded street?" However, Gladney is proved wrong, and Love uses this fictional example to illustrate his point that literature and researchers cannot live in a fantasy world and avoid the impacts of the natural world (Love 2003, p. 17). He urges humans to develop a strong ecological sense and be aware of potential danger. Many other contemporary works have attracted the attention of ecocriticism, including American novelist Coiraghessan T. Boyle's *A Friend of Earth* (2000), British novelist Doris Lessing's *Mara and Dann* (1999), Russian writer Tatuyanan Tolstaya's *The Slynx: A Novel* (2000) and Canadian writer Margaret Atwood's *Oryx and Crake* (2003).

Rereading and commenting on literary classics from the perspective of contemporary ecological culture is an interesting characteristic of ecocriticism. W. J. Keith has published three books in this field: *The Rural Tradition: A Study of the Non-fiction Prose Writers of the English Countryside* (1974), *The Poetry of Nature: Rural Perspectives in Poetry from Wordsworth to the Present* (1980) and *Regions of the Imagination: The Development of British Rural Fiction* (1988). In these books he re-examines classics of English literature in prose essay, poetry and novel respectively, and explores their deep connection to land. Love uses three chapters in *Practical Ecocriticism* to explain the practical use of ecocriticism, using the works of Willa Cather, Ernest Hemingway and W. D. Howells as examples. From Cather's narrative method in her 1925 novel *The Professor's House* we can see that the literary language that represents cultural aesthetics is interconnected with the ecological environment, demonstrating the close connection between humanity and their geographical location. Hemingway's *The Old Man and the Sea* (1951) is based on another kind of contact of humanity with nature, an ecological conflict that has a unique tragic significance. Howells' *The Landlord at Lion's Head* (1897) and *A Traveler from Altruria* (1894) show from the respective perspectives of realism and idealism that evolutionary theory and practice lead to a puzzling problem: the conflict between human nature and self-interest on one hand, and the altruism on the other. From these examples we can see that the eco-oriented reading does not simply consider literary works at the level of ecology; rather, based on traditional reading, it gives the works new significance from an ecological perspective, thereby enriching them.

Ecocriticism is at an emerging stage, with a focus on America and Britain, where it originated. In America it came into being toward the end of the 1980s. Its recognized founder there was Glotfelty, the first to advocate ecocriticism as a type of literary criticism that studies the relationship between literature and the natural environment. She was one of the founders of the Association for the Study of Literature and Environment in 1992, and compiled *The Ecocriticism Reader* (1996). With the publication of the *Reader*—which introduced the definition, the background, the current situation and the prospect of ecocriticism—ecocriticism was no longer the call of the wild or an isolated effort by an individual, but an awareness and collective cause shared by a group. In Britain, ecocriticism began at the beginning of the 1990s, founded by Jonathan Bate at Liverpool University. In his works *Romantic Ecology: Wordsworth and the Environmental Tradition* (1991) and *The Song of the Earth* ([2000] 2002), Bate find close connections between humanity and nature in the English literary tradition, showing the characteristics of English and other European ecocriticism. In his introduction to *Romantic Ecology*, Bate calls the book a "preliminary sketch toward ecocriticism" (1991, p. 11). In Britain, unlike in America, ecocriticism is referred to as "green studies." In terms of its characteristics, English "green studies" focuses on the threat that the environmental crisis poses to humanity, and warnings about this threat (Barry [1995] 2002, p. 251). For example, Bate, in *The Song of the Earth*, points out, "At the beginning of the third millennium of the Christian era, the state of nature is parlous.... We are bound to ask the old question, but in changed circumstances: where did we begin to go wrong?" ([1995] 2002, p. 24). Bate's question has been quoted widely by ecocritics. As we can see, the characteristics of ecocriticism vary from country to country, the field is still evolving, and the direction in which it is trending can be seen in the three waves it has already experienced.

## Lecture 30. The Concept and Literary Value of Ecocriticism

### *The Concept of Ecocriticism*

The most fundamental concept of ecocriticism is that human culture is closely related to the natural world. It influences nature, and is also influenced by it. Ecocriticism, therefore, takes the interconnection between nature and culture—especially language, literature and nature—as its main theme (Glotfelty, p. xix). Buell considers Darwinian theory (originally proposed by Englishman Charles Darwin), American Aldo Leopold's "eco-ethics," and European "deep ecology" as the three (western) theories that have influenced ecocriticism (Buell 2005, p. 100), although from the works of American nature writer Gary Snyder and others we can also see influences of eastern philosophies.

Ecocriticism is a reflection on the relationship between humanity and nature, between culture and nature. Influenced by postmodernism, both humanity and nature have tended to be commodified and commercialized in a high-tech society. According to Harold Fromm, both have taken on new forms and become objects of current supply and demand (Fromm, p. 34). The side effects of high tech not only include physical and mental risks to humans, but also directly threaten the earth we rely on for our survival, causing us to fall into the contradiction of belonging to and depending on the earth while at the same time destroying it. Whether from Buell's book title *Writing for an Endangered World* or Bate's question "where did we begin to go wrong," ([1995] 2002) we can feel the coming of eco-disaster and are aware of the mistakes of modern culture; therefore, exploring a healthy cultural model is inevitable.

Snyder, in one prose "Plain Talk" titled "The Wilderness" in his *Turtle Island* (1974), points out that western culture has inherited too many mistaken ideas, has been separated from both the inner and the external wilderness, is self-destructive, and is the cause of the crisis (p. 106). Kent C. Ryden, in his book *Landscape with Figures: Nature & Culture in New England* (2001), explores a healthy cultural model from the perspective of a close relationship between culture and nature. He believes that both nature and culture are in a perpetual state of change and are intricately interrelated. In modern society culture still reflects images of nature, thus breaking down the barriers between nature and culture, showing that humanity and nature, culture and nature are inseparable (Ryden 2001, pp. 10–11). It is precisely because ecocritics see the mistakes of modern civilization and realize that there is still a logical foundation in nature that they advocate an ecological vision and a healthy culture where nature, society and spiritual life coexist harmoniously. They believe that humans are a unique species in having culture, that the stories we tell will, in the end, affect our survival or death, and that changes in literature and language will inevitably influence the language of law. Words can move people, can enter their minds, and can lead to right actions.

Ecocriticism advocates that humans change from ego-consciousness to eco-consciousness, maintaining that humans are not the masters of nature but members of the land community, and live or die together with other members of the natural world. This can be seen in the position that ecocritics assume in their ecocriticism. The combination of eco and criticism shows that the study considers the relationship between humanity and nature an equal relationship. In other words, when we study language and landscape, text and place, words and woods, we are not studying two separate entities, but two interdependent entities and what we call life. They are interconnected and rely on natural elements like sunshine, air and water to survive. Therefore, language and literature are inseparable from the ecological environment.

Ecocriticism also attempts to reshape humanity's relationship with nature and elucidate the trend of culture and thought from the perspectives of ethics and social progress. Roderick F. Nash, who published his *Wilderness and the American Mind* in 1967, published his second book *The Rights of Nature: History of Environmental Ethics* more than 20 years later, in 1989. In the latter book, which is part of the History of American thought and culture series, Nash discusses the ethics of ecology with the eye of a cultural historian and from the perspective of changing ethics and the expansion of legal rights, showing trends in American culture and thinking. He uses images and writing to explain the progression of ethics from self to family, then country, animals and plants, and to the whole ecosystem. In the case of America, from the War of Independence and the abolishment of slavery to the feminist movement, human rights law and the *Endangered Species Act* of 1973, is shown the gradual expansion of the scope of rights. Some theorists link ecofeminism with ecocriticism, as ecofeminism considers women under oppression to be the same as nature under control, and therefore seek equality for both women and nature. Looking back over the history of humankind, various groups and their supporters around the world have fought hard on issues of rights, including equal rights for men and women and the abolition of racial segregation, and have succeeded. In turn, humanity should also give nature rights. From Muir's advocacy that we should respect "the rights of all other living things" in the late nineteenth century and Leopold's "land ethics" in the 1940s to today's calls for the "liberation of nature" and the "rights of the earth," we can see the progress of human civilization. Nash explains that nature itself cannot fight for its rights; humans are its moral agents and have an obligation to ensure and protect the rights of other residents of the earth. Such a concept of rights implies that humans have responsibility for and duty toward nature (Nash 1989, pp. 4–10). British scholar and advocator of ecoliterature Terry Gifford, in his book *Pastoral* (1999, pp. 160–61), believes that literature must convey an awareness that nature and culture are complementary. He believes that this awareness gives us conscience and ability, and makes us responsible for human behavior toward other species on the earth and toward the earth itself.

Ecocriticism also incorporates some of the concepts of life sciences and emphasizes interdisciplinary studies. E. O. Wilson and Joseph Carroll's opinions lay the foundation for studies in this area. Wilson, in his book *Consilience: The Unity of Knowledge* (1998), considers the world as a united entity in which everything that it is made up of is interrelated. Knowledge itself is made up of natural sciences, social sciences and human sciences; and according to Wilson future knowledge is converging, and in the process human sciences will gradually come nearer to natural sciences until the two eventually merge (in Love, p. 59). The pioneers of nature writing, including White in eighteenth-century Britain and Thoreau in nineteenth-century America, were both natural scientists and writers, concerned with what current ecocriticism calls interdisciplinary study in the green fields. Ecocriticism has, therefore, inherited from nature writing the traditions of Romanticism and transcendentalism, and also absorbed Darwinism and ecological theory to become a bold experiment in the field of literature. Carroll, in a speech titled Adaptation, Environment and Imagination that he gave at the annual British environment and literature study conference 2004, links evolutionary biology, ecology and literature to explain the relationships between human nature and environment, culture and nature. He points out that human mentality is formed in the process of adapting to the natural environment, and human thoughts and cultural activities are therefore controlled by the natural environment and by natural rules. Our bodies are imprinted by the natural environment, and this imprint will be reflected in the literature which expresses human thoughts. In his book *Literary Darwinism* he uses Bronte's *Wuthering Heights* to explain the influence of the natural environment on literary works. The plot of the novel progresses between two sharply contrasting geographical locations: the pleasant, tree-sheltered farm of the gentle but weak Linton family and the desolate and wind-swept Wuthering Heights, home of the rough but strong and passionate Shawns. The temperaments of the characters in the novel are very much in keeping with the eco-environments of their residences. The heroine's description of love in the novel is also imprinted by nature: she compares her love for Linton to "foliage in the woods," and her love for Heathcliff to "the eternal rocks beneath." In her eyes, a person's personality and feelings are determined by the features of their place of residence (quoted in Carroll 2004, p. 98). Ecocritics also consider language to be a product of evolution. Like the earth, it is not isolated from nature and it has evolved in step with the earth. Even the generation of the simplest sentence, therefore, is linked with our history, with the lives around us, and with the earth we rely on for survival. Carroll's work is regarded as a breakthrough in literary theory because he challenges the traditional literary theories and links literary criticism with evolutionary theory, ecology, psychology and other life sciences, thereby providing a foundation for ecocriticism's claim that there is interdependence between culture and nature and between humans and ecology, and also expanding the horizon of literary study and giving it a new vitality.

## *The Literary Value of Ecocriticism*

Ecocriticism is different from other literary criticism, which traditionally explores the relationship between the author, the text and the world (in most literary criticism, "world" being a synonym for "society"). Ecocriticism expands the "world" to the entire eco level, refusing to accept that everything is linked only socially and linguistically. According to Glotfelty, if we believe in the primary principle of ecology, that "everything is connected to everything else," then "literature does not float above the material world in some aesthetic ether, but, rather, plays a part in an immensely complex global system, in which energy, matter, and ideas interact" (Glotfelty, p. xix).

In a book titled *A Century of Early Ecocriticism*, published in 2001, editor David Mazel declares that even though ecocriticism is the product of the recent 20 years, it actually has a longer history. After giving a brief overview of the history of American literary criticism, Mazel points out that the problem of previous criticism is that "they [literary critics] paid so little attention to natural environment, the "world" they believed that literature should closely connect to is the social but not natural world." He goes on to explain that even though ecocriticism is still marginalized in many academic circles, it is in fact central and plays a decisive role, and without it our concept of "American literature" might not even exist. He cites Norman Forster's *Nature in American Literature* (1923) and F. O. Matthiessen's *American Renaissance* (1941), Henry Nash Smith's *Virgin Land* (1950) and Leo Marx's *The Machine in the Garden* (1964), to show that even before the 1960s and 1970s, in the process of the development of American literary criticism, there had always been a nascent impetus towards ecocriticism (Mazel, pp. 1–8).

Some ecocritics believe that ecocriticism helps us to lay a foundation for literary study from the perspective of the essence of life itself, from the perspective of the earth that surrounds us and supports us. It reconnects literary study with the problems of survival on this earth, which can no longer bear its heavy burden and is ridden with crises, making us aware of the earth under our feet, its relationship with us, and the consequences of our behavior toward it. It liberates literature from the realm of a game of words, from various literary self-obsessions, and requires us to live sensibly, wisely and beneficially. Other criticisms do not have this significance (in Ryden, "What is Ecocriticism?"). Examination of the history of academic studies shows a tendency to treat human life and non-human life separately and differently, and very rarely study the two together. To reconnect the relationship between humanity and its place of survival has become a focus of ecocriticism. Carroll comments that biology provides a consistent and persuasive explanation for the close relationship between humans and their place of survival in nature. Certain literary figures had realized earlier that human character is shaped by a particular natural environment; their intuitive understanding of human experiences and the experiences of the author, characters and the reader has always been the focus of the literary experience (Love, pp. 89–90). The relationship between humanity and its place of survival, therefore, has become one of the points of interest to literary scholars. American prose writer Rockwell Gray once wrote, "All experience is placed experience; all human experience, that is, literally takes place" (quoted in Ryden, "What is Ecocriticism?"). American nature writer Mary Austin believed that geographical environment shapes human consciousness and literature, as geographical environment has a greater cultural influence than common language and political relationships. In her view, the best writers are those who show the color of their own regions. This does not mean their writing has to be rustic and peasant-like, but that it is the inevitable product of the natural environment of the author (in Mazel, p. 11). Ecocritics believe that to fully understand human experience, close reading of the words in the text is not enough: we need to listen to the stories of the earth and appreciate how they shape and change the earth, to go to nature and become readers of scenery. We can say that ecocritics reestablish literary theory and its background. As poet Walt Whitman says in his "A Song of the Rolling Earth," "there can be no theory of any account unless it corroborates the theory of the earth."

Ecocritics therefore advocate for the above-mentioned theories to be incorporated into the teaching of English literature. In research, they advocate a reading of the text from the perspective of ecology. Barry interprets this way of reading as shifting attention from the inner to the outer, as a result of which, the setting has now moved from the margin to the center of criticism. He uses Edgar Allan Poe's *The Fall of the House of Usher* (1839) as an example: usually when people read this story, their attention is on the inner, that is, the owner of the house of Usher and his pathological mentality. A reading from the perspective of ecology places the focus on the outer, that is, the house and its outer environment. The house of Usher exists as a declining and isolated system that is completely cut off from the larger biological world beyond. The still water of the lake reflects the image of the rigid body of the house, which is breathing its last breath. What is more terrifying is that Usher has everything cultural but nothing natural, and cannot tolerate any contact with the natural world, his only contact with the outside world being through art. He is allergic to natural light, and can only endure the light shown in paintings. He cannot endure natural sounds, only the "processed" sounds of music. This story describes an ecosystem that is damaged beyond repair; from this reading, this story does not focus on a restless person at night, but an eco-disastrous long night (Barry, pp. 258–261). Bate comments on the English romantic poets' works from the perspective of ecology; he says that "Byron's 'Darkness' proposes that when ecosystems collapse, human bonds do so too. Keats's 'To Autumn' and Coleridge's 'Frost at Midnight' are thinkings of our bonds with each other and the earth, thinkings of fragile, beautiful, necessary ecological wholeness" ([2000] 2002, p. 112). It can be seen that this eco-oriented reading enhances the vitality of literary study instead of weakening it.

In the mid-1980s, with the emergence of nature writing, topics of literature and environment were introduced to the university curriculum. Ecocriticism made the teaching and learning deeper and more extensive. In the last 10 years or so, classes on the environment and literature are on the rise. It is not only top universities like Harvard, Yale, Princeton and Brown, but also others that have introduced this field of study. Buell's *The Environmental Imagination* was written based on his lectures on "American Literature and American Environment" for his Harvard students. It should be pointed out that the originating place of ecocriticism, Nevada University in America's West, is the center of such teaching and learning practice. Not only have ecocritic professors, scholars and visiting scholars who teach and research ecoliterature gathered there; the university also houses the literary journal *Interdisciplinary Studies in Literature and Environment*, or *ISLE* for short, which publishes research results and progress in the field. In their teaching, ecocritics believe that scholars' and students' activities in the wilderness are beneficial to both the teaching and the learning of ecocriticism. They advocate that students experience nature, read the land closely and observe its rhythms, patterns, and complex phenomena. The rationale behind this is that this personal contact with nature or experience of the wild will add vitality and imagination to students' reading of texts, and even make their thinking sharper and their writing more distinct. Ecocritics believe that the provider of information is the earth itself. This is, perhaps, due to the distinctive characteristic of ecocriticism. Where other literary forms usually keep their researchers and students indoors and lead them to the inner world, this kind of literary criticism leads people often and joyfully to the outside world, to nature, to absorb spiritual and artistic inspirations from the source.

In short, ecocriticism is a study of the relationship between literature and the natural environment. It holds an earth-centered attitude toward the study of literature, raising the issues of whether literature makes humans better adapted to or further alienated from the natural world, and whether the role of literature is more beneficial to eco-balance and harmony. The fundamental basis of ecocriticism is that human culture is closely connected with the natural world, influences the natural world, and is influenced by it. As a form of literary criticism it is rooted in literature, but also in nature. Ecocriticism differs from all other kinds of literary criticism in that it extends traditional literary criticism's mainly social sense of the world to everything in the entire natural world. Using literature to awaken people's sense of the need for environmental protection is one of the main tasks of ecocriticism; it therefore urges people to think carefully about the relationship between humans and nature. Most works of ecocriticism have a common understanding that we have already entered an era of extreme environment; human activities are destroying the earth, which can no longer bear the burden;

and dealing with the global crisis that we are facing requires the intervention of human ethics rather than expecting the ecosystem itself to find a solution.

Even though ecocriticism as a unique and new literary trend has attracted the attention of the literary circles and people in all walks of life, there are a variety of views of this literary trend. It is suspected of: being sensational in its use of fashionable terms and phrases; having obviously political inclinations and having the purpose of setting up a new "ism"; in essence an anti-theory with an inadequate theoretical basis; too ambitious, attempting to incorporate the complexities of evolutionary theory and ecotheory into literary criticism. Dana Phillips's 300-page *The Truth of Ecology: Nature, Culture and Literature in America* (2003) is a representative work that holds different views to that of ecocriticism, although this book itself is also controversial in the literary world. In her preface to the book Phillips says,

> We assume that *the truth* about *nature is straightforward.* Many of us still believe that ecologists can meet our need for a better understanding of natural processes simply by thinking "like a mountain".… And so round and round the mountain we go, trying to sense the presence of wolves and read the mountain's thoughts, yet secretly afraid we won't be able to do either. (Phillips, pp. vii–viii)

However, ecocritics seem to hold an accommodating attitude toward different views; they believe literary criticism is a house with many rooms and therefore should be inclusive, not exclusive. Of course, they also see that ecocriticism faces two main challenges in the future: how to deal with the relationship between economic globalization and ecocriticism, and how to establish a constructive relationship between green human sciences and environmental science. In a word, to literary researchers and teachers, ecocriticism is an interdisciplinary study and a new frontier of knowledge in the twenty-first century. It is of great significance in the reconsideration of the role of humans in the natural and the social environment, and of how humans can healthily and wisely survive on the earth.

## References

Barry, Peter. (1995) 2002. *Beginning Theory: An Introduction to Literary and Cultural Theory.* A new edition with a new chapter titled "Ecocriticism" added, Manchester: Manchester University Press.

Bate, Jonathan. 1991. *Romantic Ecology: Wordsworth and the Environmental Tradition.* New York: Routledge.

Bate, Jonathan. (2000) 2002. *The Song of the Earth.* Cambridge, MA: Harvard University Press.

Begiebing, Robert J., and Owen Grumbling, eds. 1990. *The Literature of Nature: The British and American Traditions.* Medford, OR: Plexus Publishing.

Branch, Michael. 1998. *Reading the Earth: New Directions in the Study of Literature and the Environment.* Moscow: University of Idaho Press.

Branch, Michael, ed. 2004. *Reading the Roots: American Nature Writing before Walden.* Athens, GA: University of Georgia Press.

Buell, Lawrence. 2001. *Writing for an Endangered World: Literature, Culture, and Environment in the U.S. and Beyond.* Cambridge, MA: Belknap Press of Harvard University Press.

Buell, Lawrence. 2005. *The Future of Environmental Criticism: Environmental Crisis and Literary Imagination.* Oxford: Blackwell.

Carroll, Joseph. 1995. *Evolution and Literary Theory.* Columbia, MO: University of Missouri Press.

Carroll, Joseph. 2004. *Literary Darwinism: Evolution, Human Nature, and Literature.* New York: Routledge.

Deitering, Cynthia. 1996. "The Postnatural Novel: Toxic Consciousness in Fiction of the 1980s." In Glotfelty and Fromm, *The Ecocriticism Reader*, pp. 196–203.

Delillo, Don. 1985. *White Noise.* New York: Viking Press.

Fromm, Harold. 1996. "From Transcendence to Obsolescence: A Route Map." In Glotfelty and Fromm, *The Ecocriticism Reader*, pp. 30–39.

Garrard, Greg. 2004. *Ecocriticism.* London: Routledge.

Gifford, Terry. 1999. *Pastoral.* London: Routledge.

Glotfelty, Cheryll. 1996. Introduction to Glotfelty and Fromm. *The Ecocriticism Reader*, Athens: University of Georgia Press; pp. xv–xxxvii.

Glotfelty, Cheryll, and Harold Fromm. eds. 1996. *The Ecocriticism Reader: Landmarks in Literary Ecology.* Athens, GA: University of Georgia Press.

Keith, W. J. 1974. *The Rural Tradition: A Study of the Non-fiction Prose Writers of the English Countryside.* Toronto: University of Toronto Press.

Keith, W. J. 1980. *The Poetry of Nature: Rural Perspectives in Poetry from Wordsworth to the Present.* Toronto: University of Toronto Press.

Keith, W. J. 1988. *Regions of the Imagination: The Development of British Rural Fiction.* Toronto: University of Toronto Press.

Love, Glen A. 1996. "Revaluing Nature: Toward an Ecological Criticism." Published in Glotfelty and Fromm, *The Ecocriticism Reader*, pp. 225–240.

Love, Glen A. 2003. *Practical Ecocriticism: Literature, Biology, and the Environment.* Charlottesville, VA: University of Virginia Press.

Mazel, David, ed. 2001. *A Century of Early Ecocriticism.* Athens, GA: University of Georgia Press.

Meeker, W. Joseph. 1974. *The Comedy of Survival: Studies in Literary Ecology.* New York: Charles Scribner.

Meeker, W. Joseph. 1997. *The Comedy of Survival: Literary Ecology and a Play Ethic.* Tucson, Arizona: University of Arizona Press.

Nash, Roderick F. (1967) 1968. *Wilderness and the American Mind.* Second print. New Haven, CT: Yale University Press.

Nash, Roderick F. 1989. *The Rights of Nature: A History of Environmental Ethics.* Madison, WI: University of Wisconsin Press.

Phillips, Dana. 2003. *The Truth of Ecology: Nature, Culture and Literature in America.* Oxford: Oxford University Press.

Rueckert, William. (1978). "Literature and Ecology: An Experiment in Ecocriticism." Republished in Glotfelty and Fromm, *The Ecocriticism Reader*, pp. 105–123.

Ryden, C. Kent. 2001. *Landscape with Figures: Nature & Culture in New England.* Iowa City, IA: University of Iowa Press.

Ryden, C. Kent. n.d. "What is Ecocriticism." *The Association for the Study of Literature & Environment (ASLE).* Accessed May 16, 2013. http://www.asle.org/site/resources/ecocritical-library/intro/defining/ryden/.

Snyder, Gary. 1974. *Turtle Island.* New York: New Directions Publishing.

Wilson, Edward O. 1999. *Consilience: The Unity of Knowledge.* New York: Vintage Books.

Whitman, Walt. ca. 1881. "A Song of the Rolling Earth." The Walt Whitman Archive. Accessed April 16 2016. org/published/LG/1891/poems/95.

# Conclusion

In his book *Landscape with Figures: Nature & Culture in New England* (2001, p. 19), contemporary American nature writer Ryden once wrote, "I'm trying to think about nature on the small scale in which people confront it every day, a scale limited by the frame of a window or windshield, the direction of a path and the spring in their legs, the nearness of the horizon." This collection of lectures, apart from attempting to sort out the origins, development and current situation of American nature writing, also aims to share my pleasure in nature and nature writing and to express the philosophical ideas derived from the experience of nature's microscopic details.

Raymond Williams once called nature "perhaps the most complex word in the [English] language" (quoted in Coates 1998, p. 1). If we look at early nature writing, we see that most of it focuses purely on the description of mountains and waters, animals and plants. However, in contemporary nature writing, nature has taken a deeper and broader meaning: it can be seen as a universe that includes us humans; as wilderness untrampled by humanity and standing in contrast to human culture; and as either an energetic and inspiring spiritual symbol for humanity or a cold and indifferent entity that operates mechanically, rejecting humanity. It can even be seen as a completely abstract concept: an eco-awareness, the existence of which even people who don't live in the wilderness can still feel, and willingly work hard to protect. In short, nature can be understood in both a narrow sense and a broad sense.

In a narrow sense, nature is everything in the natural environment that is not created by humans. In a broad sense, it is a kaleidoscope that reflects in many directions: nature on a gigantic scale interacts with the complicated human survival environment to create a vivid, changing picture. In this sense nature is complex and everlasting, while natural phenomena can undergo myriad changes in an instant. It is this complexity and changeability of nature that make it charming and attractive to nature writers.

As the subject matter of nature writing is the rich and mysterious nature, it provides writers with not only endless scope for imagination and an inexhaustible fount of artistic inspiration, but also a fresh, healthy and wonderful field for reading. Therefore, as nature writing continues to grow it has a vast capacity for further development.

The theme of nature writing has evolved over the years, from the relationship between nature and human thought and behaviour at the beginning, to the relationship between nature and the whole human civilization and culture now. This current theme is an unavoidable issue that humanity must face and address, not only within a single country but across the whole world. During the development of nature writing we have also seen both the growth of Western nature writers' interest in ancient Eastern cultures, and the meeting of many different cultures. Whether in Emerson's *Nature* or Thoreau's *Walden*, we can see the influence of Eastern thoughts. Contemporary nature writer Susan Zwinger expresses her admiration for Chinese philosopher and author Laozi in her works; she considers Laozi's *Tao Te Ching* to be full of wisdom about life, teaching people to live a quiet, simple and contented life close to the land (Zwinger 1991, p. 56). American nature writers have been interested not only in Confucianism and Taoism, but also in various Chinese literary works. They regarded the *Book of Songs*, a collection of ancient Chinese songs translated into English, as the "mother of poems." They saw the works of Xie Lingyun, Wang Wei, Su Shi and other Chinese poets, especially of the Tang Dynasty, as "Chinese wilderness poems," and writer and translator David Hinton has translated them into English and published them in a collection titled *Mountain Home* (first published in 2002). The works of these ancient Chinese poets continue to attract the attention of contemporary American nature writers because they incorporate nature and spirit into the inner wilderness (Snyder 2004, pp. 6–7). This cultural communication across space and time has helped us further understand wilderness: wilderness is spiritual and soulful, even human. Humans have wildness and wilderness. This wilderness is not chaotically wild, but wilful and confined within rules. Humans in the wilderness are in a natural state: humans and the heavens have fused into one. It is just as Emerson said: "In the wilderness, I find something more dear and connate than in streets or villages. In the tranquil landscape, and *especially* in the distant line of the *horizon*, man beholds somewhat as *beautiful* as his own *nature*" (Emerson, "*Nature*," in [1836] [1903] 1979, p. 10).

Some say that the theme of the twenty-first century will be harmonious coexistence between nature and humanity. The theme of nature writing research, therefore, is not only in the present but also in the future. Furthermore, nature itself has a continuous and continuing history; it is never going to end. As a result, study of the relationship between nature and humanity is study of an everlasting theme. As long as humans and nature exist, literary description of their relationship will persist, like love, war and death, with language continuing to merge the natural scenery with the scenery in the heart. Or in other words, familiarity with natural scenery and obtaining its spiritual benefits are rewards of human evolution. The inner human "scenery" is continually being nourished by natural scenery.

The relationship between nature and humanity can be described from the perspective of philosophy, but also from that of science. However, as humans are animals with thoughts and feelings, we cannot help but use literary language. To date, the most typical nature writing is still nonfiction prose written in the first person, describing the author's own experience of nature. This experience is at two levels: it is an external experience through the five senses, but also an inner, spiritual experience; it is a communication between individuals and nature, but also reflects the various aspects of nature and of human civilization. While modern culture indicates a collective consciousness, going to nature means going to the individual, to the self and to a borderless spiritual realm, which provides a vast imaginative space for literary writers. Nature writing is collective if taken on by a large number of writers working on a common theme, but as far as their writing styles are concerned, it is also individual: different writers have different interpretations of nature.

Moreover, nature writing is usually created from simple, easily accessible literary forms such as diaries, journals, notes and sketches, and this has laid the artistic foundation for its development. This simple, easy style makes it possible for many people to attempt nature writing, and although not everyone will succeed, this simplicity allows nature writers to obtain materials for their writing both directly and indirectly. As far as nature is concerned, every river, rock, tree and meadow has its own story and its own memory. When everyone can record and tell the story of the land, when people have sung the songs of all things, the natural landscape will merge with the spiritual landscape to become everlasting. This permanency in nature literature does not mean that every writer or work is necessarily everlasting, but it is a grand and symbolic permanency—it is the story and the myth of the land passed from generation to generation and supported by writers and the communities they live in that is everlasting.

As nature writing is still developing, it inevitably has its own contradictions and problems. For example, although the theme of nature writing is exploration of the relationship between humanity and nature, its exponents have sometimes been confused by the issue of what role humans play in this relationship. Many nature writers have emphasized nature centeredness. However, this nature is, after all, the nature in their eyes and their hearts. It is inseparable from the humans who create nature writing. Besides, the nature observed by nature writers cannot exclude human achievements, such as "natural" landscapes that have been transformed by humans. For this reason, some American literary critics believe that nature writing should not avoid society. From what we have discussed, we can see that love of nature and yearning for spiritual peace have been common human traits across time and space. Vertically there is the communication between modern and ancient people, while horizontally there is the contrast between eastern and western approaches to the same theme, as well as the expansion of nature writing itself; and "environmental literature" and "ecocriticism" have developed out of nature writing. In other words, nature cannot be separated from human society, and the relationship between nature and humanity can be expressed and explored in many ways.

Nature writing is a narrative form or concept in nonfiction prose essays. The relationship between nature and humanity inevitably has many aspects, and is shown through subjectivity and in diverse ways; as the Chinese poet of Song Dynasty Su Shi (1037–1101) once described Mount Lu, "It is a mountain range if viewed from the front but mountain peaks from the side" (Su Shi, *Written on the Wall of Xilin Temple* [1084], in *Selected Poems of Su Shi*, selected and annotated by Chen Erdong 1984, p. 199; "*Heng kan cheng ling ce cheng feng*"), there is infinite variety depending on the perspectives. Research in this area, therefore, should be open and nonexclusive. For example, one of the chief editors of *The Norton Book of Nature Writing*, John Elder, not only published many nature books and selections of nature writing, but also wrote a book on the imagination of poetry and nature titled *Imagining the Earth: Poetry and the Vision of Nature* (1985). In its second edition, published in 1996, the author added a chapter on the contemporary nature poet Mary Oliver, pointing out that her poems inherit the tradition of Emerson and Thoreau and share similarities with modern and contemporary nature writers such as Edward Abbey, Barry Lopez and Annie Dillard. He also points out that Terry Williams, in her representative work *Refuge* (1991), quotes Oliver's poem "Wild Geese" as the introduction to her book (Elder 1996, pp. 216–228, 217). It can be seen that nature writing is compatible with poetry, and is also related to environmental literature and ecocriticism, showing its inclusiveness and room for development.

It is true that nature writing has its own contradictions and confusions and that people have differing views of the genre, given that its theme is the complex and varied relationship between humanity and nature. However, it cannot be denied that nature writing not only exists, but has an increasing influence in the literary world. Perhaps we can see the vitality and the life force of nature writing in Susan Zwinger's *Ode to Old Growth*. Zwinger is awed by the vigorous life force present in an ancient forest. She sees all kinds of plants crisscrossed and knitted together, with sunlight shining through the gaps in the leaves onto the layers of trees and bushes. "'Living' here is not a stop-and-go proposition, as we humans perceive it, but huge trees constantly reconnecting themselves to Earth" (Zwinger 1991, pp. 206–208). In fact, American nature writing itself is like an evergreen tree, relying on itself to be rid of old roots and bark and to add new branches and leaves and develop freely in new and unprecedented directions. Some American scholars, therefore, prefer literature that studies humanity and nature to always remain "large and loose and suggestive and open" and every text to be "a dialogue open for further comments from other points of view. There is no conclusion" (in Glotfelty & Fromm 1996, pp. xxii, 387). This evergreen-like vitality, this free development without the constraints of rules and regulations, and this inclusiveness that does not seek uniformity but promotes diversity distinguish American nature writing from other types of literature and give it much charm. It is precisely due to its inclusiveness that American nature writers are no longer an isolated group, and their influence is no longer limited within America. Their description and praise of nature, their yearning for it, and their persistence in keeping the land under our feet from being harmed have already gone beyond national borders.

## References

Coates, Peter. 1998. *Nature: Western Attitudes Since Ancient Times.* Berkeley: University of California Press.

Elder, John. (1985) 1996. *Imagining the Earth: Poetry and the Vision of Nature.* Athens, GA: University of Georgia Press.

Emerson, Ralph Waldo. "Nature" (first published in 1836), in Ralph Waldo Emerson. (1903) 1979. Centenary Edition, *The Complete Works of Ralph Waldo Emerson.* Vol. I: *Nature Addresses and Lectures.* 2nd edition. Ogden, UT: AMS Press.

Glotfelty, Cheryll, and Harold Fromm, eds. 1996. *The Ecocriticism Reader: Landmarks in Literary Ecology.* Athens, GA: University of Georgia Press.

Miller, James E., Jr., ed. 1991. *Heritage of American Literature: Beginnings to the Civil War.* Vol. 1. San Diego, CA: Harcourt Brace Jovanovich.

Ryden, C. Kent. 2001. *Landscape with Figures: Nature & Culture in New England.* Iowa City, IA: University of Iowa Press.

Zwinger, Susan. 1991. *Stalking the Ice Dragon: An Alaskan Journey.* Tucson, AZ: University of Arizona Press.

# Index

## A



## B

## C



## D



## E



## F



## G



## H

## J



## L



## M



## N



## O



## P



## R



## S

www.ingramcontent.com/pod-product-compliance
Ingram Content Group UK Ltd.
Pitfield, Milton Keynes, MK11 3LW, UK
UKHW021827100926
8556IPUK00003B/195